Online Collaboration and Communication in Contemporary Organizations

Ditte Kolbaek
Aalborg University, Denmark

A volume in the Advances in Human Resources Management and Organizational Development (AHRMOD) Book Series

IGI Global
DISSEMINATOR OF KNOWLEDGE

Published in the United States of America by
IGI Global
Business Science Reference (an imprint of IGI Global)
701 E. Chocolate Avenue
Hershey PA, USA 17033
Tel: 717-533-8845
Fax: 717-533-8661
E-mail: cust@igi-global.com
Web site: http://www.igi-global.com

Library of Congress Cataloging-in-Publication Data

Names: Kolbaek, Ditte, 1955- editor.
Title: Online collaboration and communication in contemporary organizations / Ditte Kolbaek, editor.
Description: Hershey, PA : Business Science Reference, [2018]
Identifiers: LCCN 2017028177| ISBN 9781522540946 (hardcover) | ISBN 9781522540953 (ebook)
Subjects: LCSH: Organizational learning. | Organizational behavior. | Communication in organizations. | Telematics. | Virtual work teams.
Classification: LCC HD58.82 .O55 2018 | DDC 651.7/9--dc23 LC record available at https://lccn.loc.gov/2017028177

This book is published in the IGI Global book series Advances in Human Resources Management and Organizational Development (AHRMOD) (ISSN: 2327-3372; eISSN: 2327-3380)

British Cataloguing in Publication Data
A Cataloguing in Publication record for this book is available from the British Library.

For electronic access to this publication, please contact: eresources@igi-global.com.

Advances in Human Resources Management and Organizational Development (AHRMOD) Book Series

Patricia Ordóñez de Pablos
Universidad de Oviedo, Spain

ISSN:2327-3372
EISSN:2327-3380

Mission

A solid foundation is essential to the development and success of any organization and can be accomplished through the effective and careful management of an organization's human capital. Research in human resources management and organizational development is necessary in providing business leaders with the tools and methodologies which will assist in the development and maintenance of their organizational structure.

The **Advances in Human Resources Management and Organizational Development (AHRMOD) Book Series** aims to publish the latest research on all aspects of human resources as well as the latest methodologies, tools, and theories regarding organizational development and sustainability. The **AHRMOD Book Series** intends to provide business professionals, managers, researchers, and students with the necessary resources to effectively develop and implement organizational strategies.

Coverage

- Human Relations Movement
- Organizational Learning
- Personnel Policies
- Succession Planning
- Executive Education
- Performance Improvement
- Outsourcing HR
- Human Resources Development
- Process Improvement
- Collaborative Method

IGI Global is currently accepting manuscripts for publication within this series. To submit a proposal for a volume in this series, please contact our Acquisition Editors at Acquisitions@igi-global.com or visit: http://www.igi-global.com/publish/.

The Advances in Human Resources Management and Organizational Development (AHRMOD) Book Series (ISSN 2327-3372) is published by IGI Global, 701 E. Chocolate Avenue, Hershey, PA 17033-1240, USA, www.igi-global.com. This series is composed of titles available for purchase individually; each title is edited to be contextually exclusive from any other title within the series. For pricing and ordering information please visit http://www.igi-global.com/book-series/advances-human-resources-management-organizational/73670. Postmaster: Send all address changes to above address.

Titles in this Series

For a list of additional titles in this series, please visit: www.igi-global.com/book-series

Management Strategies and Technology Fluidity in the Asian Business Sector
Patricia Ordóñez de Pablos (University of Oviedo, Spain)
Business Science Reference • copyright 2018 • 338pp • H/C (ISBN: 9781522540564) • US $225.00 (our price)

Arab Women and Their Evolving Roles in the Global Business Landscape
Ebtihaj Al-A'ali (University of Bahrain, Bahrain) Minwir M. Al-Shammari (University of Bahrain, Bahrain) and Hatem Masri (University of Bahrain, Bahrain)
Business Science Reference • copyright 2018 • 321pp • H/C (ISBN: 9781522537106) • US $200.00 (our price)

Creativity in Workforce Development and Innovation Emerging Research and Opportunities
Sally Blake (Flagler College, USA) and Candice M. Burkett (University of Illinois at Chicago, USA)
Business Science Reference • copyright 2018 • 152pp • H/C (ISBN: 9781522549529) • US $145.00 (our price)

Evaluating Media Richness in Organizational Learning
Albert Gyamfi (Aalborg University Copenhagen, Denmark) and Idongesit Williams (Aalborg University Copenhagen, Denmark)
Business Science Reference • copyright 2018 • 328pp • H/C (ISBN: 9781522529569) • US $185.00 (our price)

Teaching Human Resources and Organizational Behavior at the College Level
John Mendy (University of Lincoln, UK)
Business Science Reference • copyright 2018 • 305pp • H/C (ISBN: 9781522528203) • US $195.00 (our price)

Handbook of Research on Organizational Culture and Diversity in the Modern Workforce
Bryan Christiansen (PryMarke LLC, USA) and Harish C. Chandan (Argosy University, USA)
Business Science Reference • copyright 2017 • 506pp • H/C (ISBN: 9781522522508) • US $255.00 (our price)

Handbook of Research on Human Factors in Contemporary Workforce Development
Bryan Christiansen (PryMarke LLC, USA) and Harish C. Chandan (Argosy University, USA)
Business Science Reference • copyright 2017 • 563pp • H/C (ISBN: 9781522525684) • US $255.00 (our price)

Driving Multinational Enterprises Through Effective Global Talent Management
Khaled Tamzini (University of Sousse, Tunisia) Tahar Lazhar Ayed (Umm Al-Qura University, Saudi Arabia) Aisha Wood Boulanouar (Sultan Qaboos University, Muscat, Oman) and Zakaria Boulanouar (Umm Al-Qura University, Saudi Arabia)
Business Science Reference • copyright 2017 • 191pp • H/C (ISBN: 9781522525578) • US $170.00 (our price)

IGI Global
DISSEMINATOR OF KNOWLEDGE

701 East Chocolate Avenue, Hershey, PA 17033, USA
Tel: 717-533-8845 x100 • Fax: 717-533-8661
E-Mail: cust@igi-global.com • www.igi-global.com

Editorial Advisory Board

List of Reviewers

Table of Contents

Detailed Table of Contents

Section 1
Definitions and Methodology

To understand practices of online collaboration and communication, a theoretical foundation is needed. The theoretical basis for working online is explored through a literature review and a relational perspective. This is followed by methodological considerations on conducting research on online work practices with linkages to learning.

Contemporary organizations operate in highly connected structures, with work extending beyond physical borders. As such, organisations experience the emergence of virtual teams: teams that collaborate virtually and work together, despite differences in time-zones, physical proximity or cultural backgrounds. This chapter aims to shed light on this contemporary work arrangement from a relational perspective, leading to an understanding of the dimensions of virtuality underpinning work. For doing so, the chapter begins by describing the emergence of dispersed teams as a contemporary business practice. The discussion, then, turns to ways of relating to one another on a virtual landscape, and focuses on the virtuality dimensions of time-space, leadership, trust, and the struggle between cohesion and flexibility. The chapter concludes with a projection on the future around dispersed collaboration, and with an outline of future research for advancing the understanding and practice of virtual teams.

This chapter presents a suggestion for design-based research (DBR) as a methodology for organizational studies. Although DBR was developed for investigating classroom training, this chapter discusses the methodological issues that are involved when DBR is employed for investigating learning in the context of work. DBR seems to be suitable in this complex context as it is an authentic learning environment. The purpose of this chapter is to provide new perspectives on DBR, including suggestions for guidelines regarding how to conduct DBR for studying learning from experience in the context of work. The research question is: What is needed to utilize DBR to explore learning from experience in the context of work? The theoretical framework is based on Yrjö Engeström and John Dewey's work on learning, and the exploration of DBR is based on a literature review and findings in a seven-year DBR study on learning from experience in the context of work.

Section 2
Challenges in Online Leadership and Online Collaboration

Distance matters when people work together. The lack of bodily clues along with high diversity are challenges that must be overcome to create a successful online team. This second section explores challenges in online leadership and online collaboration by linking theoretical considerations of online collaboration and online leadership with descriptions and cases that illustrate practitioners' ways of handling the online work environment. The researches apply different methodologies and methods for studying online collaboration and leadership in different settings, and Section 2 provides insights into different approaches to understand and to handle online collaborarion and online leadership including cultural understanding, perceptions of trust, perceptions of contributions to the organization, perceptions of being present, as well as recommendations for overcoming the challenges in online collaboration.

Web 2.0 brought a participatory potential with it, prompting organisations to ask themselves how to design social media platforms that can engage external stakeholders in the organisation's process of knowledge creation and innovation. This chapter presents an analysis of the data from a case study of such a platform, the World Bank's online game Urgent Evoke, which has been designed with the purpose of engaging citizens in developing innovative solutions for sociopolitical problems like poverty. The analysis is based on Nonaka's concept of Ba, which means "place" and is described as a platform for advancing the creation of knowledge. The analysis suggests that, in order to create a digital Ba, the design should not only facilitate the four characteristics of Ba that Nonaka has described –socialization, externalization, combination, and internalization. The design should also allow multiple forms of participation, as well as recognise external stakeholders' contributions to the process of knowledge creation and innovation.

Cross-cultural management is challenging. However, online leadership across cultures demands even more intercultural competence, as the actors involved neither share physical space nor everyday experiences. This chapter offers a new interdisciplinary perspective on global collaboration, which addresses the complexity in online cross-cultural leadership. The perspective is formed as a post-cultural prism comprising practice theory, positioning theory, and postcolonial theory. It is argued that this interdisciplinary approach is capable of filling the gap in the field of intercultural competence, as it offers tools for reflection on online interactions in everyday situations. The aim of this chapter is to contribute to a conceptualization in cross-cultural leadership and to introduce basic intercultural practices for practitioners.

In this chapter, the authors explore how power processes can promote or hinder effectiveness in online leadership. The chapter offers a theoretical and empirical analysis of the power processes in online leadership, using the triangle of structural, personal, and discursive power as an analytical lens. The theoretical outset for understanding online leadership is the literature on boundaries and boundary-spanning leadership. The authors define online leadership as social influence processes mediated by technology to produce individual and collective efforts to accomplish shared objectives. The chapter is empirically based on material from one of the authors' PhD work on online leadership in a global Danish manufacturing company. The chapter illustrates how power can be used both constructively and destructively in online leadership.

An interview study focusing on online collaboration in geographically distributed IT development teams in Danske Bank revealed seven problem areas. To cope with the problems the authors applied a design science research approach to construct a conceptual framework for improving online collaboration. The conceptual framework combines a six-phase teambuilding model with six elements of social capital. Thus, in each phase of teambuilding, the online collaborators aim at building up all six elements of social capital. The complete six-by-six framework was successfully tried and diffused throughout Danske Bank. This chapter gives an account of the framework content and the results from the evaluation.

Mobile workers are professionals who frequently work on the move, far from a fixed workplace, often performing knowledge-intensive activities. Mobility challenges creation and sharing of knowledge among these professionals, and the existing literature lacks powerful theoretical frameworks conducive to creating supportive learning pathways for them to meet these challenges. This chapter is a theoretical and practical contribution to fill this gap by analyzing a case of expansive learning initiated by mobile workers themselves. Based on longitudinal participant observation, the study traces the steps undertaken by these professionals to create a new artifact that helped them to know what, when, where, and how they needed to perform their work activities.

Online working environments develop and change continuously, meaning that online leaders and online team members must learn to adapt to change and should utilize emerging possibilities for doing their jobs. The purpose of this chapter is to explore how online leaders learn from experiences developed by and with online teams; the chapter also provides a metaphor by which online leaders' learning is likened to a spiral. The theoretical foundation rests in activity theory and proactive review. The methodological approach involves a case study and netnography, the latter of which is utilized for researching phenomena that occur in the interaction between humans and information and communication technology. Data were gathered online from an online team in a global IT company classified as big business. The following research question was addressed: How do online leaders learn from experiences generated by their online teams by utilizing proactive review as a practice of learning?

This chapter presents and discusses the concept of presence in online dialogues and interactions. The point of departure in this discussion is that of multitasking and social presence. Then, the chapter frames different e-learning scenarios that give participants the impression of being—or not being—present in the same room although geographically dispersed. On the basis of theoretical inspiration from symbolic interactionism and ethnomethodology, educators' new roles in e-learning are discussed in the context of empirical material collected from an e-learning program in a physiotherapy school. In addition, the concept of multi-presence is presented. This chapter argues that multi-presence occurs in both online and offline interactions and that, as a result of technology, is utilized in both professional and learning situations.

Section 3 provides rich cases and recommendations for overcoming challenges in online collaboration and leadership. Practitioners share their experience and insights in online leadership, and they provide hands-on recommendations for overcoming challenges in the online work environment. In addition, an interview with a digital nomad is presented for further research and inspiration. Practitioners may benefit from Section 3 as a guidebook in handling challenges in online collaboration and communication in contemporary organizations. These practitioners could be global managers, people who work online, or HR staff that want to improve the organization's online working practices.

Online leadership largely develops through practice: taking the first steps into what seems a relatively uncontrollable environment is scary. Learning to adapt one's leadership style to suit conditions in the online world is an experimental process that benefits from coaching and guidance from experienced online leaders. This chapter distils 10 years of recent experience the first author gained helping senior managers develop an influential online presence at the same time as implementing digital strategies in three organizations. It starts with a brief review of face-to-face leadership theory, which most leaders encounter in traditional development programs, identifies how online practice differs, explores the phases of a typical leader's journey from conscious incompetence to conscious competence online, and highlights practical interventions that both develop capability and diffuse change organization wide. It concludes with a summary of useful characteristics for change agents supporting the enterprise transformation, which is usually the aim of a move to online working.

A large number of studies show that trust is crucial to success as a virtual leader. The purpose of this chapter is to provide a theoretical and practical perspective on the challenges of creating trust and strong relationships to the virtual leader. The chapter presents the consequences of virtual collaborations if trust is low or high. Next, it points out how the virtual leader can work to create trust. The chapter presents "The Circle of Trust," which consists of several important behavioural attitudes for the virtual leader. The meaning of metacommunication in the form of clear agreements on the use of communication is outlined. Several communication forms—e-mail, videoconferencing, and online informal communication—are examined, focusing on how these forms of communication can consciously promote trust in virtual collaborations. The chapter provides a collection of the most important solutions and recommendations for the virtual leader.

This chapter contributes to the understanding of how online collaboration and cooperation can take place between geographically dispersed people by providing insights into how people work from the Hubud co-working space on Bali in Indonesia. The chapter revolves around the presentation of an interview with a person working from this co-working space. The case represents an example of how people, who, in a radical way, embrace the new possibilities of information and communication technology (ICT) work and live their professional lives. The aims are to provide insight into how people in the co-working space collaborate and work in a primarily online world and to present material that other researchers, students, and people who are interested in online collaboration can use in their work. The chapter introduces a framework that is intended as a tool to explore the material by introducing some of the key concepts, a possible theoretical approach, and a context for the case.

Preface

ONLINE COLLABORATION AND LEADERSHIP

The objective of this book is to provide, in one publication, a comprehensive resource that would enable readers to apply theories and practices of online communication and collaboration in a practical sense and/or in further research. The aim of this book is to link theoretical considerations of online collaboration and leadership with descriptions and cases presenting best practices. The intention is to inspire practitioners, researchers and students in the field by presenting theories and cases that are fairly easy to read.

Online leadership means leading geographically dispersed teams that rarely or never meet face-to-face, but cooperate and collaborate online to accomplish common objectives. Distance matters, as the communication is mediated by available information technology (IT) stemming from video-conferences, intranet, phone calls, etc. (Kirkman & Mathieu, 2005).

The book was initiated in recognition of relationships between people and organizations challenged by globalization, in which work often crosses the physical boundaries of an organization in order to save costs (Boh, Ren, Kiesler, & Bussjaeger, 2007; Delbridge & Sallaz, 2015). Globalization entails that the organizations evolve from local entities to a business structure that crosses geographical borders (Flyverbom, 2012). Some of the implications are that employees at all levels communicate, cooperate, and collaborate online instead of face-to-face, and they may communicate in a foreign language or with peers of different cultural backgrounds (Cramton & Hinds, 2014). These changes require a new balance between various dimensions of work, including online leadership of teams (Martins et al., 2004). This book explores online leadership and collaboration from the perspectives of leadership and organizational learning, as huge change in leadership practices necessitates learning at all levels in the organization—from leaders to employees.

The book looks into theories and methodology for understanding and present challenges and best practices for online leadership and collaboration. This book pays less attention to specific technologies because they change so rapidly, whereas challenges and best practices are more stable. However, working online is still a fairly new work condition, so this book explores facets such as lack of shared physical space, synchronous chronological time, and the crossing of multiple boundaries (Brannen & Salk, 2000).

The editor has brought together a collection of chapters from researchers and practitioners to support understanding of online collaboration and communication, as well as to help promote the implementation of comprehensive best practices. The ambition is to identify ways of working that leaders and employees will find beneficial when collaborating online, and also to develop new theories for understanding the online working environment.

POINT OF DEPARTURE

The first step in the voyage of developing this book was a literature review in JSTOR of books and articles published from 2013 to 2016. The first attempt included the search topics "online leadership," or "e-leadership," or "virtual leadership," that revealed no texts about research in organizations, except for two articles about a school and a university.

A literature review based on the search topic "global virtual teams" returned 112 results and, after skimming the abstracts, the search revealed 20 texts about online collaboration and leadership in private companies or governmental organizations.

This research claims that distance matters when people work together (Cramton & Hinds, 2014; Richardson, Beck, LaFrance & McLeod, 2016; Wildman & Griffith, 2015). When an organization grows global, it heavily depends on IT, because that technology becomes the basis of communication and collaboration. Consequently, the organization is vulnerable to technology break-down or varied work habits in different geographical places (Pinjani & Palvia, 2013). In global organizations, national borders may appear as discontinuities in space and time (Beugelsdijk & Mudambi, 2013), and these discontinuities require specific leadership skills and employee abilities to be able to successfully collaborate across multiple types of boundaries (Brannen & Salk, 2000). Employees need a common language, common IT, and the ability to utilize the technology. They must develop cross-cultural competencies and strengthen their communication skills (Wildman & Griffith, 2015), and they must also be flexible with regard to working hours and styles of communication and collaboration. Technology mediates the relationship between employees who may never meet face-to-face. A lack of body language clues, along with a high diversity influences the ability to build strong relationship and the ability to trust in and share knowledge with far-away colleagues (Pinjani & Palvia, 2013; Richardson et al., 2016; Wildman & Griffith, 2015).

These challenges for collaboration and communication require special attention from online leaders, who must possess the same abilities as traditional leaders plus additional competencies related to leading, communicating, and collaborating online (Wildman & Griffith, 2015). Online leaders must be flexible and adaptable to changes in technology and unforeseen events (Richardson, 2016). They also need solid knowledge about their organization's objectives in order to develop a clear strategy and create a common understanding of the organization's vision among the online team members. The online leader structures the team by setting simple, measurable, and achievable goals that enable employees to work in the same direction at a pace they agree to (Cramton & Hinds, 2014; Korzynski, 2013; Richardson et al., 2016; Wildman & Griffith, 2015). The online leader and the online team develop processes and best practices for collaboration and communication. The lack of physical proximity makes it important to develop practices for both formal and informal communication, i.e. formal meeting structures that utilize available IT and informal gatherings, like virtual coffee breaks, to foster social relations and trust among the online team members (Beugelsdijk & Mudambi, 2013; Korzynski, 2013; Richardson et al., 2016; Wildman & Griffith, 2015).

The online leader communicates team member roles and expectations for collaboration. Informal communication is important because it supports the feeling of belonging and common understanding that allows online team members to depend on colleagues they've never met face-to-face to achieve a common objective (Wildman & Griffith, 2015). This leads to considerations on how to give and perceive feedback and coaching online, and to understand the impact of feedback in the online working environment. In traditional working environments, feedback between peers is often given spontaneously during a break. This is not an option in online teams, so the online leader should make this issue explicit and, together with the online team, develop beneficial practices for online feedback and coaching (Korzynski, 2013; Wildman & Griffith, 2015).

If the online team members are spread over the globe, they may represent culturally diverse ideas of living and perceiving work and colleagues. The online leader should possess cultural awareness and be able to enhance cross-cultural competencies in the team (Wildman & Griffith, 2015).

SCOPE

The lack of literature indicates that organizations and researchers, in general, tend not to pay specific attention to online leadership and collaboration. It seems as if best practices are lacking, and that both academics and practitioners could benefit from more theoretical development in this field.

The second step in the voyage of developing this book was to invite authors with expertise to contribute their newest research and experiences. Luckily, 20 authors from 11 countries wrote the 16 chapters for this book, and they provided peer reviews of the other chapters to ensure high quality of all contributions. The result is a book that focuses on theories, challenges and best practices for online leadership and collaboration. The book explores ways in which online collaboration, communication, and leadership can be understood in order to improve everyday practices, based on theory and illustrated through case studies.

The chapters investigate applicable methodology for exploring online leadership and collaboration, as well as the application of practical and meaningful theory. The authors examine practices that help or hinder successful online collaboration and leadership. These chapters collectively contribute to multiple disciplines and fields of academia and practice, including psychology, communication, learning, rhetoric and methodology. Thus, the authors bring forward academic and practical expertise that provides a holistic approach to the topic. The authors' general approach is to present a model founded in theory and tested in practice, which leads to recommendations for improved online leadership and collaboration. The result is, therefore, a faceted collection of case studies as the basis of in-depth trend analysis. The authors' contributions fill the gap discovered in the literature review by providing new theories, models, and methodological approaches for studying online leadership and collaboration, which may result in more knowledgeable practitioners, enlightened and insightful online managers, fertile future research fields, and new and improved resources for trainers, educators, and students.

The book should be used as a guidebook or a textbook. As a guidebook, it serves for implementing good online work practices. These practitioners could be project managers of global projects, employees, or organizations that want to improve current online working practices. As a textbook it may serve for students in leadership and collaboration in contemporary organizations, or as inspiration and a source of new insight for researchers who investigate IT-mediated leadership and teamwork, organizational development, and learning in contemporary organizations.

The book can be read from beginning to end or, after an orientation with the introduction, the reader may find a non-linear approach works better. The literature (including this book) labels geographically dispersed leadership and collaboration in different ways. Here are some examples: virtual teams (Gibson, Gibbs, Stanko, Tesluk, & Cohen, 2011), global teams (Cramton & Hinds, 2014), e-leadership (Lee, 2009), and online leadership or online team work (Huffaker, 2010). The authors in this book use the terms "virtual" and "online" interchangeably.

OVERVIEW OF CHAPTERS

The chapters explore online leadership and online collaboration, and the book is divided into three sections:

- Definitions and Methodology
- Challenges in Online Leadership and Online Collaboration
- Cases and Good Practices

The first section of the book begins with a comprehensive literature review in Chapter 1 by Dr. Chrysavgi Sklaveniti, and it summarizes studies that contribute to our understanding of online leadership and collaboration. The chapter aims to shed light on this contemporary work arrangement from a relational perspective, leading to an understanding of the dimensions of work underpinned by virtuality. The chapter concludes that technology-aware generations are coming into the workplace, and they carry with them new sets of practices about how to use technology for managing collaboration and for organizing everyday life. Consequently, this book should be followed up by continuous research in this rapidly-evolving field.

The radical changes in leadership required to adjust to online practices initiates learning at all organizational levels. Chapter 2, by Dr. Ditte Kolbaek, provides a methodology for studying online leadership and collaboration. Although design-based research (DBR) was developed for investigating classroom training, this chapter discusses methodological issues involved when DBR is employed for investigating work learning. The chapter offers new perspectives on DBR and suggests guidelines for DBR when studying learning in the context of work.

The second section explores challenges in online leadership and collaboration and presents models and recommendations for working and leading online. Chapter 3, by Dr. Mette Wichmand, presents a case study of World Bank's online game Urgent Evoke, which has been designed with the purpose of engaging citizens in developing innovative solutions to sociopolitical problems like poverty. The participants are situated in various countries, and they collaborate online. The chapter digs into the challenges of knowledge creation between "strangers," and it utilizes Nonaka's concept of "Ba." The analysis suggests that, in order to create a digital Ba, the collaborative design should facilitate the four characteristics of Ba being socialization, externalization, combination, and internalization and allow multiple forms of participation, including recognition of external stakeholder contributions to the collaboration. Chapter 4, by Professor Iben Jensen, contributes to a conceptualization in cross-cultural collaboration and introduces basic intercultural practices. The chapter provides a postcultural prism comprising practice theory, positioning theory, and postcolonial theory. This interdisciplinary approach fills the gap in the field of intercultural competence, as it offers tools for reflection on online interactions in everyday situations and addresses the complexity of online cross-cultural collaboration. In Chapter 5, Dr. Benjamin J. Cline provides insights into the classical rhetorical concept of ethos that can solve difficulties of presence and trust. Dr. Cline exemplifies and elaborates on ethos by analyzing the online collaborative tool Slack, and he provides suggestions for establishing e-ethics.

The following three chapters include longitudinal studies and focus on online leaders and how they contribute to successful online collaborations. Chapter 6, by Dr. Ditte Kolbaek, explores ten years (2002–2012) of online leadership development in a global IT company. The chapter looks into social, cultural, and economic capital. The study employs methods of memory work and document analysis and provides insights into four themes, namely establishment of common ground for cooperation, working practices, critical competencies, and stock valuation trends. The findings indicate that information and communications technology (ICT) supports the enhancement of all three types of capital. In Chapter 7, Dr. Kimiz Dalkir describes a multi-year study on how online collaboration could be effective in the design of a research equipment sharing lab with university and industry partners. The chapter focuses on challenges such as establishing a shared vision and trust among diverse partners. The analysis uncovers differences between academic and industrial organizational cultures, as well as differences in leadership style. Chapter 8, by Dr. Morten Fogsgaard, Dr. Claus Elmholdt and Rikke Lindekilde, defines online leadership as social influence processes mediated by technology to produce individual and collective efforts to accomplish shared objectives. The chapter explores how power processes can promote or hinder effectiveness in online leadership and offers a theoretical and empirical analysis of the power processes in online leadership in a global Danish manufacturing company. The chapter presents a model for understanding power relations, which are a triangle of structural, personal, and discursive power. Chapter 9, by Professors Jan and Lene Peies-Heje, presents a framework for improving online collaboration in geographically-distributed IT development teams. The model combines a six-phase teambuilding model with six elements of social capital. The implementation of the model at Danske Bank in Denmark and India showed that each phase of teambuilding enhanced all six elements of social capital. The interview method revealed seven problem areas. To cope with these problems, the authors applied a design science research approach to construct the conceptual framework for improving online collaboration. The result is a framework of six times six elements that seems to fit the cultures of Denmark and India and the technology available to cooperate and collaborate online across both countries.

The second section continues with three chapters that study different perspectives of learning in online environments. Chapter 10, by Dr. Rosemary Francisco, Professor Amarolinda Zanela Klein, Professor Yrjö Engeström, and Professor Annalisa Sannino presents a case study of mobile workers—professionals who frequently work on the move, far from a fixed workplace, often performing knowledge-intensive activities. The chapter analyzes a case of expansive learning initiated by mobile workers themselves. Based on longitudinal participant observation, the study traces the steps undertaken by these professionals to create a new artifact that helped them to know what, when, where, and how they needed to perform their work activities. The entire learning process may be seen as an "intravention" rather than a traditional intervention, but the downside is that mobile workers often remain isolated or are interrupted due to lack of awareness and support among the management. In Chapter 11, Dr. Ditte Kolbaek studies how online leaders learn from experiences developed by and with online teams. The theoretical foundation rests in activity theory and proactive review, and the methodological approach involves a case study and netnography. Data were gathered from an online team in a global IT company. The chapter includes a model of the online leader's learning as a spiral. The online leader and teams learn from experience when participating in proactive reviews that enable organizations to learn from the bottom up or the top down. Chapter 12, by Dr. Anne-Mette Nortvig, presents the new "multi-presence" concept. To understand presence in digital settings, presence across space must be distinguished from presence across time, as both potentially multiply with the use of technology. The chapter presents and discusses the concept of presence in online dialogs and interactions, and then presents the concept of multi-presence. Different e-learning scenarios are framed, which give participants the impression of being—or not being—present in the same room, although geographically dispersed. On the basis of theoretical inspiration from symbolic interactionism and ethnomethodology, teachers' new roles in e-learning are discussed in the context of empirical material collected from an e-learning program in a physiotherapy school. This chapter argues that multi-presence occurs in both online and offline interactions and, as a result of technology, multi-presence is utilized in both professional and learning situations.

Section three provides cases and good practices for online leadership and collaboration. In Chapter 13, Dr. Bonnie Cheuk and Professor Jane McKenzie distill ten years of recent experience Dr. Cheuk gained in helping senior managers develop an influential online presence while implementing digital strategies in three organizations. This chapter outlines some principles for developing leadership judgment to intervene effectively within a complex ecosystem. Online leadership largely develops through practice. Taking the first steps into what seems a relatively uncontrollable environment is scary. Learning to adapt one's leadership style to suit conditions in the online world is an experimental process that benefits from coaching and guidance by experienced online leaders. This chapter explores the phases of a typical leader's journey from conscious incompetence to conscious competence online, and it highlights practical interventions that both develop capability and diffuse organization-wide change. Chapter 14, by Psychologist Mads Schramm, digs into the challenge of creating trust online. The chapter presents consequences for online collaborations when trust is low or high, and the author provides a model showing important behavioral attitudes for creating trust online. This model is called "The circle of trust" and includes a collection of the most important solutions and recommendations for online leaders. Employees at any hierarchical level must train themselves to use technology to create trust and good relationships, and they must also be aware of the possibilities and limitations of working online. The chapter examines several communication forms, such as e-mail, videoconferencing, and online informal communication, and describes how these forms of communication can consciously promote trust in online collaborations. The chapter recommends improved metacommunication about the use of technology. Chapter 15, by Anders Sandborg Nielsen, presents the case of a co-working space in Bali. This chapter is a plain case study based on interviews describing online collaboration and cooperation between geographically dispersed people situated in the Hubud co-working space in Bali, Indonesia. The chapter introduces a framework intended for use as a tool to explore the material by introducing some of the key concepts, a possible theoretical approach, and a context for the case. The interviewee calls himself a digital nomad, and he is an example of the increasing fluidity of what it means to be an organization and how people in an organization can work together and create value. The case shows that creating a product or service must not necessarily be the result of traditional work in a traditional organization. This fluidity and complexity will hardly decrease in the near future, which means that scientific methods, academic genres, and understandings of online collaboration may undergo radical changes in the years to come.

The Conclusion summarizes the key recommendations from these studies and provides suggestions for future studies in online leadership and online collaboration.

REFERENCES

Beugelsdijk, S., & Mudambi, R. (2013). MNEs as border-crossing multi-location enterprises: The role of discontinuities in geographic space. *Journal of International Business Studies*, *44*(5), 413–426. doi:10.1057/jibs.2013.23

Boh, W. F., Ren, Y., Kiesler, S., & Bussjaeger, R. (2007). Expertise and collaboration in the geographically dispersed organization. *Organization Science*, *18*(4), 595–612. doi:10.1287/orsc.1070.0263

Brannen, M. Y., & Salk, J. E. (2000). Partnering across borders: Negotiating organizational culture in a German-Japanese joint venture. *Human Relations*, *53*(4), 451–487. doi:10.1177/0018726700534001

Cramton, C. D., & Hinds, P. J. (2014). An embedded model of cultural adaptation in global teams. *Organization Science*, *25*(4), 1056–1081. doi:10.1287/orsc.2013.0885

Delbridge, R., & Sallaz, J. J. (2015). Work: Four worlds and ways of seeing. *Organization Studies*, *36*(11), 1449–1462. doi:10.1177/0170840615612021

Flyverbom, M. (2012). Globalization as it happens: On globalizing assemblages in tax planning, *Distinktion. Scandinavian Journal of Social Theory*, *13*(3), 295–309.

Gibson, C. B., Gibbs, J. L., Stanko, T. L., Tesluk, P., & Cohen, S. G. (2011). Including the "I" in virtuality and modern job design: Extending the job characteristics model to include the moderating effect of individual experiences of electronic dependence and copresence. *Organization Science*, *22*(6), 1481–1499. doi:10.1287/orsc.1100.0586

Huffaker, D. (2010). Dimensions of leadership and social influence in online communities. *Human Communication Research*, *36*(4), 593–617. doi:10.1111/j.1468-2958.2010.01390.x

Kirkman, B. L., & Mathieu, J. E. (2005). The dimensions and antecedents of team virtuality. *Journal of Management*, *31*(5), 700–718. doi:10.1177/0149206305279113

Korzynski, P. (2013). Online social networks and leadership: Implications of a new online working environment for leadership. *International Journal of Manpower*, *34*(8), 975–994. doi:10.1108/IJM-07-2013-0173

Lee, M. R. (2009). E-ethical leadership for virtual project teams. *International Journal of Project Management*, *27*(5), 456–463. doi:10.1016/j.ijproman.2008.05.012

Martins, L. L., Gilson, L. L., & Maynard, M. T. (2004). Virtual teams: What do we know and where do we go from here? *Journal of Management*, *30*(6), 805–835. doi:10.1016/j.jm.2004.05.002

Pinjani, P., & Palvia, P. (2013). Trust and knowledge sharing in diverse global virtual teams. *Information & Management*, *50*(4), 144–153. doi:10.1016/j.im.2012.10.002

Richardson, J. W., Beck, D., LaFrance, J., & McLeod, S. (2016). Job attainment and perceived role differences of cyberschool leaders. *Journal of Educational Technology & Society*, *19*(1), 211–222.

Wildman, J. L., & Griffith, R. L. (Eds.). (2015). *Leading Global Teams: Translating Multidisciplinary Science to Practice.* SpringerLink. Retrieved from http://dl.pgu.ac.ir/handle/Hannan/8470

Acknowledgment

It is important to me, the editor, to thank the authors for their great work, which has resulted in an interesting and insightful book that has been thoroughly reviewed by fellow authors and other competent researchers, who contributed tremendously to the high quality, coherence, and content presentation of each chapter. Their efforts are greatly appreciated.

It is overwhelming that this idea is now a reality, thanks to the authors' original contributions, the reviewers' enhancing comments, the skillful support from IGI Global, and everybody's remarkable online cooperation.

Ditte Kolbaek
Aalborg University, Denmark

Section 1
Definitions and Methodology

To understand practices of online collaboration and communication, a theoretical foundation is needed. The theoretical basis for working online is explored through a literature review and a relational perspective. This is followed by methodological considerations on conducting research on online work practices with linkages to learning.

Chapter 1
Theorizing Virtual Teams:
Relationality in Dispersed Collaboration

Chrysavgi Sklaveniti
University of St. Gallen, Switzerland

ABSTRACT

Contemporary organizations operate in highly connected structures, with work extending beyond physical borders. As such, organisations experience the emergence of virtual teams: teams that collaborate virtually and work together, despite differences in time-zones, physical proximity or cultural backgrounds. This chapter aims to shed light on this contemporary work arrangement from a relational perspective, leading to an understanding of the dimensions of virtuality underpinning work. For doing so, the chapter begins by describing the emergence of dispersed teams as a contemporary business practice. The discussion, then, turns to ways of relating to one another on a virtual landscape, and focuses on the virtuality dimensions of time-space, leadership, trust, and the struggle between cohesion and flexibility. The chapter concludes with a projection on the future around dispersed collaboration, and with an outline of future research for advancing the understanding and practice of virtual teams.

INTRODUCTION

There is much interest today in transformations at the interface between technology and the workplace (Griffith, 2012; Orlikowski, 2000; Orlikowski & Scott, 2008; Shapira, Lewin, Schoonhoven, Argote, & Levinthal, 2014), and this chapter outlines a theoretical agenda for one such transformation; that of virtual teams. The term 'virtual teams' has been widely taken up to indicate work teams that collaborate online across boundaries, whether organisational, functional, geographical, temporal, national or cultural (Al-Ani, Horspool, & Bligh, 2011; Gilson, Maynard, Jones Young, Vartiainen, & Hakonen, 2015; Kelley, 2001; Wilson, Crisp, & Mortensen, 2013). To be fair, this wide uptake reflects the significant changes in infrastructure that have occurred across many organisations; but it also reflects the aspiration that the virtual makes possible new ways of conducting and knowing work life. While positioning the understanding of virtual teams in terms of their structural characteristics undoubtedly helps in familiarising with their context, it may not be equally helpful in other respects. To frame virtual teams with their physical

DOI: 10.4018/978-1-5225-4094-6.ch001

implications for work arrangements allows a narrow appreciation of the relation of technology and the workplace; one that does not convey how relationality itself is undergoing transformation in dispersed collaboration (Puranam, Alexy, & Reitzig, 2014; Wilson et al., 2013).

As has been pointed out over years of research (Delbridge & Sallaz, 2015; Dewett & Jones, 2001; Griffith, 2012), technological integration in the workplace does not only facilitate dispersed collaboration, or renders it commonplace, it also makes virtual teams amenable to new virtual practices. However, only acknowledging the networking possibilities of virtual teams or the mobile workplaces they make available does not enable to investigate the wider contexts used by virtual teams. As put forward by Beyes and Steyaert (2012), there is a need to instil a relational perspective to advance the understanding of organisational phenomena beyond physical distance. Following on from here, this chapter goes on to discuss virtuality from a relational perspective (Gergen, 2009), arguing that the technology, for which the rise of the virtual serves as occurrence, is not collaborative in and of itself. It is only in the 'doing' of relational practices virtually that technology becomes collaborative. Thereupon, the chapter puts forward an understanding of how virtual teams change the relation between technology and relationality.

The recognition that it is critical to acknowledge and account for the virtual context in which dispersed collaboration takes place is, obviously, anything but breakthrough. However, despite the simplicity in this observation, the field of organisation studies has yet to express sensitivity towards the virtual context (Griffith, 2012; Martins, Gilson, & Maynard, 2004). More explicit attention to the subtle ways relationality developed in virtual dimensions is arguably more relevant to practice, and closer to the changes in the workplace. For doing so, the chapter is concerned with letting go off the anchor of physicality that pervades the literature, and suggests instead centring on relationality to grasp the context *in use* of virtual teams, through a discussion of time-space, leadership, trust, and team cohesion versus flexibility. Thereupon, the main concern in the chapter is to provide a relational theorising of collaboration in virtual teams; one that addresses their context, together with the ways such a context shapes up relationality. This in itself is an important endeavour, in view of the lack of theoretical approaches in organisation studies that take the work arrangement of virtual teams beyond traditional framings (Bakken, Holt, & Zundel, 2013; Golden & Geisler, 2007; Hatch, 2002; Hernes, Simpson, & Söderlund, 2013; Vesa & Franck, 2013). From this vantage point, the chapter approaches the virtual turn in contemporary organisations by centring important virtuality dimensions in dispersed collaboration; namely, those of time-space, leadership, trust, and team cohesion versus flexibility. In doing so, it sheds light on the ways work life in virtual teams is organised, conducted and controlled with technology. The chapter develops as follow. To begin with, there is a discussion about how virtual teams have come to signify an important organisational phenomenon. Then, the theoretical approach of the chapter is presented, giving way to addressing the dimensions of virtuality: time-space, leadership, trust and the struggle between cohesion and flexibility. Finally, the implications for future theoretical and empirical developments in research on dispersed collaboration are discussed.

BACKGROUND

In the modern world of technology and internationalisation, work often crosses the physical boundaries of an organisation. (Boh, Ren, Kiesler, & Bussjaeger, 2007; Delbridge & Sallaz, 2015). As such, the work arrangement of virtual teams emerges and points to rethinking traditional concepts of theorising team-work and collaboration in organisation studies (Kirkman & Mathieu, 2005). Although separated in physical distance, virtual teams are called to collaborate and accomplish common objectives. Indeed, understanding the contours of contemporary workplaces demands repositioning traditional theorising of co-located teams (Martins et al., 2004). This includes delineating the specific context of virtual teams, and encountering how relationality develops across physical distance.

The majority of research on dispersed teams highlights their growing significance, and expanding use in organisations (Sewell & Taskin, 2015; Wageman, Gardner, & Mortensen, 2012), investigates conditions of efficiency (Wai Fong, Yuqing, Kiesler, & Bussjaeger, 2007), addresses knowledge management (Cramton, 2001; Faraj, Jarvenpaa, & Majchrzak, 2011; Hwang, Singh, & Argote, 2015; Sole & Edmondson, 2002), and overall refers to the structural particularities of this contemporary work arrangement (Bailey, Leonardi, & Barley, 2012; Cummings, 2004; Gibson & Gibbs, 2006; Henfridsson & Bygstad, 2013). While an analysis of the structural characteristics concerning virtual teams focuses on their technological capabilities or mobile functionalities, it may not be equally helpful for understanding the everyday life of dispersed collaboration. This is because such a focus puts forward the comparison with co-located teams, but critically misses to address the unconventional relationality, which makes up dispersed teams.

By definition, dispersed teams work in conditions of novelty: the lack of same physical space and synchronous chronological time, the crossing of multiple boundaries, and the struggle between autonomy and control (Brannen & Salk, 2000). Conceptualising work teams beyond the physical encounter entails a move from traditional theorising of workplaces to experiential approaches of the connections shaping up dispersed work (Beyes & Steyaert, 2012). To the extent that physicality is the key denominator of collaboration, we arguably neglect the relationality that underpins work in dispersed teams, thus overlooking the complexities and messiness of modern work life (Delbridge & Sallaz, 2015; Denyer, Parry, & Flowers, 2011).

Many labels are used in the literature to refer to dispersed collaboration, such as virtual (Gibson, Gibbs, Stanko, Tesluk, & Cohen, 2011; Kirkman, Rosen, Gibson, Tesluk, & McPherson, 2002), global (Cramton & Hinds, 2014; Neeley, Hinds, & Cramton, 2012), international (Jonsen et al., 2013), or geographically dispersed teams (O'Leary & Mortensen, 2010; Wai Fong et al., 2007), and with different purposes, such as structural characteristics (Raghuram, Garud, Wiesenfeld, & Gupta, 2001), innovation (Malhotra, Majchrzak, Carman, & Lott, 2001; Yoo, Boland, Lyytinen, & Majchrzak, 2012) or efficiency (Maynard, Mathieu, Rapp, & Gilson, 2012; Montoya-Weiss, Massey, & Song, 2001). For the purposes of this chapter, it is worth putting forward a definition of virtual teams as comprising a dynamic work arrangement characterised by people who collaborate across a set of boundaries (organisational, functional, geographical, temporal, national and cultural) in pursuit of common goals, using technology to communicate and coordinate work. The proposed definition aims to open up the focus on physical distance, by adding the use of asynchronous means of communication (Saunders, Van Slyke, & Vogel, 2004; Wiesenfeld, Raghuram, & Garud, 1999), as well as the reliance on technology to provide a platform for collaboration (Zammuto, Griffith, Majchrzak, Dougherty, & Faraj, 2007). This is to indicate the context in use underpinning the nature of virtual teams.

STATE OF RESEARCH

Virtual collaboration is not a new phenomenon; however, as the workplace becomes increasingly connected via IT technologies, the growth and popularity of virtual teams have intensified in the past two decades

(Hinds, Liu, & Lyon, 2011). Responding to the trends and pressures of our modern era, a growing number of organisations now establish virtual teams to produce more efficient and skilled work units (Baden-Fuller & Haefliger, 2013). Economic pressures to minimise operational costs, a greater development in the area of knowledge work, people's preferences for flexibility in work arrangements together with impressing technological evolution have given way to virtual dimensions in worlds of work (Dell & Intel, 2016). Dell and Intel (2016) report in their Future Workforce Study that 51% of global employees work virtually at least a few times on a weekly basis. Remarkably, although organisation studies have shown interest in the phenomenon of virtual teams, the literature stands far apart from re-orienting traditional notions of work teams (Leung & Wang, 2015). Now that the virtual workplace dominates organisations, understanding the nature of dispersed collaboration becomes imperative.

There is a broad range of possibilities of how virtual teams are set up in terms of physical diffusion and technological reliance, indicating various orientations to virtuality. Kirkman and Mathieu (2005) explain that there are teams operating in different cities within the same country, teams operating in different locations within the same city, or teams that operate in different countries and in different continents. Likewise, there are teams that operate solely by means of technology and those that accompany technological mediation with face-to-face meetings. While earlier research regarded virtuality as a contrast between co-located and distributed collaboration, more recent work has treated it on a continuum (Kirkman & Mathieu, 2005; Kolb, Caza, & Collins, 2012). The one extreme of this continuum assembles teams with low technological reliance and a preference for face-to-face collaboration. The other extreme assembles teams with high technological reliance and a preference for asynchronous collaboration. As technology takes centre stage in the organisational arena, work teams increasingly fall into hybrid positions in the continuum with task parameters indicating the use of technology.

All of these arrangements implicate the need to re-interpret many taken-for-granted characteristics of co-located teams. Firstly, the proximity afforded by co-location is lost, as well as the rapport brought about by synchrony in time; both of which are seen as characteristics bringing team members together (Maynard et al., 2012; Wageman et al., 2012). The complexity of collaboration is eased in co-located teams, considering that communication may be initiated instantaneously, without, for example, needing to add to the equation considerations of time zones or cultures. In addition, with the possibility of face-to-face meetings, co-located colleagues can express and understand relationality with embodied cues, such as facial expressions. Secondly, to understand dispersed collaboration it is critical to acknowledge the multiple contexts needing to be crossed to achieve synergy: time, space, and culture are no longer given denominators of everyday life (Gibson & Gibbs, 2006; Leonardi, 2011; O'Leary & Cummings, 2007). As such, technology in virtual teams becomes the principal medium of collaboration, such that technological mediation becomes a recurrent and expected part of everyday work. More importantly, it is present not only in regulating tasks, but also in building relational processes, such as trust or collegiality (Jarvenpaa & Leidner, 1999; Sieben, 2007).

A further review of the literature indicates fragmentation and a plethora of topics explored in parallel, such as the adaptation of team members to virtual work arrangements (Majchrzak, Rice, Malhotra, King, & Ba, 2000; Malhotra & Majchrzak, 2014), leadership (Al-Ani et al., 2011; Joshi, Lazarova, & Liao, 2009; Zimmermann, Wit & Gill, 2008), or the distribution of knowledge (Baba, Gluesing, Ratner, & Wagner, 2004; Boh et al., 2007; Cramton, 2001; Haas, 2006; Sole & Edmondson, 2002). The primary debate on these topics concerns the communication and coordination of virtual teams (Wiesenfeld et al., 1999), specifically the questions about the use of IT (Malhotra & Majchrzak, 2014), the separation in time and space (Gibbs, 2009), and the synchronisation of work to achieve the team's objectives (Faraj,

Kudaravalli, & Wasko, 2015). These kinds of questions probe spatial and temporal implications for virtual teams. On the one hand, the nature of virtual teams is tied with increased fluidity (Maznevski & Chudoba, 2000), allowing dispersed collaboration to navigating effortlessly in space and time and overcoming obstacles of distance, related cost-dependencies, and information asymmetry in the job market. On the other hand, relational dynamics, otherwise taken-for granted in co-located teams, are extremely fragile (Hsiao, Tsai, & Lee, 2012). In virtual teams, notions such as trust, engagement or cohesion need to be negotiated in the organisation of everyday life.

Informed by such understandings about virtual teams, the emphasis of the literature is on technological mediation and independency (Golden & Fromen, 2011). While existing studies have emphasised the growing use of virtual teams, and have contributed significantly to understanding the multiple boundaries being crossed, it has not been their aim to understand how relationality is transformed in dispersed collaboration (Gibson & Gibbs, 2006; Kirkman & Mathieu, 2005; Martins et al., 2004; Raghuram et al., 2001). Importantly, approaching virtual teams in terms of their separation from the physicality of traditional workspaces is beneficial in recognising organisational, functional, geographical, temporal national and cultural boundaries. However, solely such a focus obscures certain important relational elements, and shifts attention away from apprehending how people connect across these boundaries. Examining virtual teams from a relational perspective helps illuminate and articulate relationality in dispersed collaboration, in light of its implications for organising work (Golden & Fromen, 2011; Kirkman, Cordery, Mathieu, Rosen, & Kukenberger, 2013; Malhotra & Majchrzak, 2014). Centring relationality in dispersed collaboration entails assumptions that are significantly different from traditional theorising of co-located teams (Puranam et al., 2014) which nonetheless continue to underpin the literature on dispersed teams (Josefy, Kuban, Ireland, & Hitt, 2015). Rather than assuming physicality as denominator of dispersed collaboration, a relational approach provides an alternative lens for studying distributed teams (Beyes & Steyaert, 2012; Gergen, 2009); one, which addresses and explains the connections across delineated boundaries, and the ways such connections play out in everyday life.

THEORETICAL APPROACH

Besides the rational choice to form a team of experts, virtual teams are characterised by other, equally interesting, aspects of collaboration. Next to their objectives, members of virtual teams share the outcomes of their work (Cummings, 2004; Sewell & Taskin, 2015), as well as engagement and commitment (Fineman, Maitlis, & Panteli, 2007; Kasper-Fuehrer & Ashkanasy, 2001). Including the social aspects in the conceptualisation of virtual teams highlights that they are not a sum of individuals, but rather a social unit described by Weick and Roberts (1993) as a collective mind. More importantly, the conception of virtual teams as a social unit draws attention to the transformation of relationality within a virtual context, which calls for the study of team emergent states and team processes. As far as the social aspects of virtual teams are concerned, research has primarily focused on issues of performance and effectiveness, highlighting that positive relationships among team members are important for success (Ahuja & Galvin, 2003). Social network theory has also been used to explore the links between people in virtual teams (Faraj & Johnson, 2011; Mohrman, 1999). What is missing to date is an exploration of the ways in which people in virtual teams relate to each other and are interconnected beyond physical boundaries. The social aspects of virtual teams mark an interesting theme of research, critical to the performance of organisations, which has not yet been addressed in the literature (Kirkman & Mathieu, 2005; Martins et al., 2004; Puranam et al., 2014).

Until recently, virtual teams have been met with criticism about their ability to achieve team processes and outcomes to the same extent as co-located counterparts (Wilson, O'Leary, Metiu, & Jett, 2008). The criticism stems from the challenges associated with the dimension of virtuality underpinning dispersed collaboration. The mediation of technology is seen to challenge the flow of dispersed collaboration (Wageman et al., 2012), when asynchronous connections dominate synchronous ones. Asynchronous communication (for example, via email) poses a number of hurdles that do not exist in co-located teams. When communicating asynchronously, people occupy different physical locations (boundary of space), they may potentially occupy different time zone (boundary of time), they can respond to one another at their own pace (boundary of simultaneity), they cannot see each other (boundary of visibility), nor can they hear each other (boundary of audibility). By putting forward these boundaries, the greater assumption of claims that regard co-located teams as superior to dispersed ones is that collaboration takes on a sequential logic, where turn-taking in communication leads to more effective team processes.

According to the theory of communication bandwidth, brought forward from the field of engineering, the quality of communication is dependent upon the number of channels used and the amount of information communicated at any given time (Levi, 2015). Arguably, communication in co-located teams encompasses most communication channels (linguistic plus embodied cues) and, hence, enables the sharing of multiple information points. In a similar vein, social presence theory holds that communicative interaction is achieved when participants perceive others as taking part in the exchange (Latour, 2005). Extrapolating this to the context of virtual teams, the inference drawn by critics of virtuality is that the lack of co-location restrains communication. Despite these unfavourable outlooks, virtual teams not only are on the rise, but they also accomplish remarkable goals. From as early as the beginning of the millennium, when technological integration had not yet fully taken over organisational life, virtual teams are seen to innovate and collaborate effectively. For example, Boeing's SLICE rocket was a project carried out by a virtual team, who only met face-to-face once, before the launch of the project (Malhotra et al., 2001). This is not to say that successful outcomes in virtual teams cannot co-exist with complications, obstacles or tensions.

Undoubtedly, communication consists of rich embodied cues, such as facial expressions, or tone of the voice; the so-called body language compliments face-to-face communication (Chidambaram & Jones, 1993; Faraj et al., 2015). Nevertheless, when resorting to asynchronous communication, there is time and space for reflection, giving people the opportunity to analyse the situation at hand in more detail (Fayard & Metiu, 2014). Therefore, if the absence of embodied cues can be seen to hinder dispersed collaboration, the affordance of reflection may make amendments for the loss of co-presence. Furthermore, asynchronous communication can be argued to facilitate collaboration between different cultures (Leung & Wang, 2015; Michailova & Hutchings, 2006; Pines & Zaidman, 2014; Ravlin, Ward, & Thomas, 2014). Given that one of the boundaries to be crossed in virtual teams is that of culture, the distance in communication may actually be more beneficial for team members. Because embodied cues do not have the same significant or meaning in different cultures, asynchronous communication may actually enable, rather than hinder dispersed collaboration. In addition, the many boundaries present in dispersed collaboration facilitate the elimination of traditional workplace cases of discrimination (Greckhamer, 2011; Leonardi, 2011). The gender, ethnic origin, age or physical outlook become irrelevant in the virtual dimension, thus leading to a revolution in the ways collaboration develops beyond embodied physical cues. Besides, the expansion and development of technology now allows the diffusion of embodied cues in a synchronous manner. For example, contrary to the asynchronous means of communication via email, audio conferencing eliminates the boundary of audibility, and video conferencing further eliminates the boundary of visibility.

Therefore, instead of holding that co-located teams can achieve greater processes and outcomes compared to virtual teams, what about giving way to parallel possibilities? Collaboration is an expression of relational engagement, and as such, it cannot be confined to turn-taking sequences; it rests on the relationality of the collaborating team. By exploring the ways relationality is transformed within a virtual context to achieve dispersed collaboration, it possible to appreciate how virtual teams overcome, or perhaps override such challenges to create team processes and outcomes. As such, the chapter sheds light on virtual teams in contemporary organisations by emphasising the distinctive relationality in dispersed collaboration in the ways people use technological integration to make connections with each other and organise the connections in use.

THEORETICAL ADVANCEMENTS

For moving into a discussion about relationality in virtual teams a relational ontology of becoming (Gergen, 2009; Hosking, 2011) is adopted, following an interest in process thinking (Helin, Hernes, Hjorth, & Holt, 2014). Such a relational focus means that no action has meaning in itself; it is the responsiveness to one another that gives meaning to communication. Therefore, attention turns to the responsiveness among members of virtual teams. Rather than zooming-in on individual action, conceptual focus is on the relatedness to one another: "with a relational intelligibility in place, we can shift our attention to what transpires between people, not what is contained within them" (McNamee, 1998, p. 102). From this vantage point, relationality foregrounds the virtuality dimension of dispersed collaboration, entailing a shift not only in the ways of knowing relationality, but also in the ways relationality is tied to technological integration. This conceptual characterisation of dispersed collaboration places a specific requirement to understanding virtual teams: the virtuality dimension should be rendered more specific with the view to appreciating relationality not just *by means of*, but *with* technology.

The relational perspective put forward here significantly unsettles technology-centred theorising of virtual teams (Ding, Levin, Stephan, & Winkler, 2010; Malhotra & Majchrzak, 2014; Nyberg, 2009), which focuses on describing how technology itself takes centre stage in dispersed collaboration, and only accounts for what people do with technology in later instances. From another perspective, a relation approach focuses on the *doing* of technology (Baralou & Tsoukas, 2015) and puts to the fore an analysis of the making of connections between dispersed collaborators, together with the making of content to be shared. What is remarkable, from this vantage point, is the extent to which it recognises the participation of technology in the doing of virtual work, and the intensity with which it exposes the virtuality of dispersed collaboration with un-bracketing the creation, sustainment and/or erosion of relational ties among members of virtual teams. Therefore, the chapter takes the discussion to the ways work life in virtual teams is organised, conducted and controlled with technology, calling into question a relational un-bracketing of important virtuality dimensions in dispersed collaboration because of the relational ties among team members; namely, those of time-space, leadership, trust, and team cohesion versus flexibility.

With technology enabling work across geographic locations and making the physical encounter irrelevant, conventional parameters underpinning work are challenged; time and space take on relational conceptualisations (Maznevski & Chudoba, 2000). In dispersed collaboration, discussions, plans, negotiations, agendas, together with the creativity and outcomes of action that they are later connected to, may retrospectively seem readily obvious (Bailey & Madden, 2015). However, at the moment of their occurrence, this is hardly the case. Firstly, dispersed collaboration rests in constant everyday tensions, as team members attempt to control work flows (Gibbs, 2009). Secondly, when people become engaged in the doings of everyday organisational life, chronological time disconnects from their work mode (Czarniawska, 2004; Lee & Liebenau, 1999). Immersion into work entails the extension of the past into the future (Gergen, 2010): acting in the present never occurs in isolation; being part of a team implicates that certain occurrences, events, behaviours and so on become more meaningful in time than others.

When focusing on virtual teams, the understanding of space is also put to question (O'Leary & Cummings, 2007). The ability opened by technology to work across physical distance suggests the need to look beyond the idea of space as given coordinates, or the idea of space as the static backdrop of organisational life. This conceptual re-orientation invites to focus instead on how space is experienced as people in dispersed teams are immersed in everyday life, and as they encounter technology, which through its use, mediates collaboration and connects one with the other. Space does not contain work (Czarniawska, 2004); it encompasses organisational life, as relationality travels across organisational, functional, geographical, temporal, national and cultural boundaries. This relational view suggests that space for virtual teams holds significance as the sphere in and through which collaboration is achieved. Therefore, instead of talking about the spaces and locations that occupy organisational life, the proposition put forward here is to talk about relational closeness or remoteness in dispersed collaboration. What is important is to understand that relationality does not occur in space; it is relationality itself that creates 'space' in a configuration of human and technological interrelations (Gergen & Thatchenkery, 2006; Hosking, 2006).

This relational 'space' has implications on the way leadership is understood and expressed in virtual dimensions. Commonly referred to as e-leadership (Avolio, Kahai, & Dodge, 2000; Avolio, Sosik, Kahai, & Baker, 2014), there is a striking revolution introduced into the leadership agenda. Whereas, typically, the strong presence of the individual leader is praised in co-located teams, virtual teams decentre the visibility of the individual. From leadership to e-leadership, relationality holds similar premises. Self and other are in a process of mutual and continuous (re)construction, conceptualising (e)leadership as connecting venture participants, rather than separating them (Hosking, 2006; Hosking, Dachler, & Gergen, 1995). Gergen (2009, p. 397) explained that "in all that we say and do, we manifest our relational existence. From this standpoint, we must abandon the view that those around us cause our actions. Others are not the causes, nor we their effects. Rather, in whatever we think, remember, create, and feel, we participate in relationship". Therefore, the notion of the 'individual' becomes an intersection of multiple relations (Bradbury & Lichtenstein, 2000).

E-leadership brings a fresh outlook into the leadership agenda, requiring the achievement of direction in a dispersed horizon and a virtual dimension (Joshi et al., 2009). The lack of face-to-face contact together with the asynchronous nature of communication, call for renewed practices of connecting with one another. Besides the usual expression of relationality, new expressions come into place in e-leadership: greater sensitivity towards the other, a multi-cultural collaboration mindset, together with a-temporal and a-spatial orientations (Mailhot, Gagnon, Langley, & Binette, 2016; Raelin, 2016). These give a unique rhythm to e-leadership. Members of dispersed teams are 'invisible' to one another. At the same time, each team member is 'blind' of the others' action. From here, collaboration needs to be made explicit: everyone is called to share with one another work progress. In a similar vein, collaboration also needs to be specifically planned. Most often, a spontaneous meeting to deal with an unexpected turn of events is almost unattainable. E-leadership, then, is oriented towards an already established direction (the work output bringing the dispersed team together) and moves in subtle rhythm, passing through boundaries and resorting to technological mediation.

Described as such, the conception of e-leadership presents itself as both a paradox and an opportunity to open up the usual leadership agenda. The multiple boundaries cutting across team members in virtual teams implicate a flat hierarchical structure (Zimmermann et al., 2008). This flat structure does not necessarily mean a more democratic, or a 'better' leadership. What it draws attention to is the sharing of leadership in all facets of organisational life. Given the nature of dispersed collaboration, all members of the team are a priori involved in the mundane aspects of leadership, structuring routine everyday work. Concurrently, all are involved in the emotional aspects of leadership, such as motivating participation, or easing feelings of solitude. Therefore, e-leadership is not to be framed around the traditional reference points of leader-followers. Instead, it is more helpful to conceive e-leadership in the responsiveness to one another as team members collaborate dispersedly towards direction, alignment and commitment (Drath et al., 2008).

Re-orienting the study of virtual teams towards a relational ontology, leads to a broader understanding of trust not as an a priori individual attribution, but as a relational characteristic (Gergen, 2009). In traditional organisational research, trust involves individuals discovering each other's trustworthiness (Jarvenpaa & Leidner, 1999). However, the conceptual decentring of the role of the individual, together with the multiple boundaries among team members in virtual teams raises questions about the attribution of trust. From a relational perspective (Gergen, 2009), trust-building is a responsive process that plays out asymmetrically. On the one hand, trust develops gradually and incrementally, drawing on trusting experiences. On the other hand, there is no outright certainty that trust is to be bestowed. In understanding trust, participation in virtual teams forms a continuous flow, which requires drawing on relational resources, such as openness, in the practices of everyday life.

Given the shortage of time to familiarise with one another, team members in virtual teams develop swift trust (Crisp & Jarvenpaa, 2013; Gilson et al., 2015; McKnight, Cummings, & Chervany, 1998), which is an a priori bestowment of trustworthiness. The virtual dimension of dispersed collaboration is quite intensive of time, thus promoting the construction and maintenance of trusting connections among members of dispersed teams (Kasper-Fuehrer & Ashkanasy, 2001). This is because trust building brings the physically distant team members together, thus enduring relational ties, encouraging shared understanding, and establishing rapport. These are critical issues in virtual teams; the lack of past shared history together with the great diversity in them call for a connecting centre to create lasting collaboration. In this vein, the bestowment of swift trust accelerates connectivity, and removes the perception of boundaries.

While teamwork is traditionally expected to be viewed as a field of convergence and stability, dispersed collaboration thrives on tensions and contradictions (Gibbs, 2009). Rather than treating these ambiguities as danger in people's connections which must be dealt with or distinguished, a relational approach regards conflicts as an ordinary underpinning of the dilemmatic nature of virtual teams. It is not the animation of ambiguities per se that can erode relational connections, but the way these are experienced in dispersed collaboration (Hinds & Bailey, 2003). The boundaries present in the work life of virtual teams (organisational, functional, geographical, temporal, national and cultural) do not represent an option between two opposing poles, nor are they mutually exclusive possibilities. For example, people in dispersed teams are distant in physical space, but relationally close in work commitments. Therefore, a relational perspective invites the re-conceptualisation of boundaries as sites of creative medleys. Boundaries do not stand for either/or choices, but rather encompass a both/and perspective, where relationality points to generating fresh routes or alternatives to rise above the dilemmatic nature of dispersed collaboration.

Examining closely the boundaries upon which rests dispersed collaboration, there emerges a tension between the cohesion needed to work and move forward, versus the flexibility to allow polyphonic perspectives within the dispersed team (Bailey et al., 2012; Montoya-Weiss et al., 2001). This tension co-exists with the paradox of looking for team interdependence, versus the autonomy each member asks in one's work (Charlier, Stewart, Greco, & Reeves, 2016). Virtual teams are often set up specifically with the aim of bringing together diverse expertise, hence team members are united with the sense of sharing knowledge; each from their own stance (Cramton, 2001; Cummings, 2004; Faraj et al., 2011; Haas, 2006). While achieving dispersed collaboration requires synergy and integration, creative team outcomes benefit from diversity. From the perspective of team outcomes, diversity manifests itself as a larger struggle to build shared meanings at distance (Hwang et al., 2015), together with increased awkwardness in coordinating productive work (Malhotra & Majchrzak, 2014). Therefore, overall, examining the dilemmatic nature of dispersed collaboration from a relational take calls for recognising a fluctuating and vivacious balance between integration within a greater social structure, and diversity of differentiated dispersed positions.

DISCUSSION

Today's organisations have embraced technological integration, promoting a virtual workplace, where virtual teams comprise a dynamic and ever-growing business practice. Extensive explanations of the rise and future of dispersed teams have been produced using various angles (cost benefits, global exposure, client pressures and employees' preferences), different levels of analysis (organisational, team, individual) and themes of interest (team selections and composition, effectiveness, use of technology and boundaries to be crossed). This has resulted in a fragmented picture of the process that calls for dispersed collaboration. As has been established over the years of research (Gilson et al., 2015; Griffith, 2012), dispersed collaboration is approached from representational positions which examine, in differing depths and different ways, team processes and outcomes in line with, or perhaps despite, the lack of physical co-presence. However, acknowledging representational forms of physicality does little to enhance our understanding of the unconventional conditions of relationality (Gal, Jensen, & Lyytinen, 2014), which make up dispersed teams, or our understanding of the virtuality underpinning the nature of virtual teams (Gibson & Gibbs, 2006). Indeed, considering the abundance in the agenda of assumptions pertinent to co-located teams, there is a need to advance our thinking on the nature of dispersed collaboration. Following on from these starting points, the chapter has provided a relational theorising of collaboration in virtual teams by addressing their virtual context, together with the ways such a context shapes up and is shaped by relationality. Inspired by the idea of relationality, the principal issue the chapter has dealt with is the re-orientation of traditional notions of work teams, thus opening up the field of dispersed collaboration beyond representational forms.

Moving forward, the first and most basic thing is to revisit the assumptions discussed above about the mediating role of technology in dispersed collaboration (Malhotra & Majchrzak, 2014; Orlikowski, 1992; Zammuto et al., 2007). We are well beyond the point of debating the extent to which technology makes up virtual teams; we are now at a point where technology is integral to contemporary organisations. Still, we lag in research that addresses how dispersed collaboration takes shape through the dynamic interaction of technology and relationality (Gilson et al., 2015; Griffith, 2012). To fully grasp how technology is mediating the nature of collaboration in dispersed teams, it is desirable to consider technology as constitutive of this mediation; rather than a mere enabler of mediation. Such a perspective regards the duality of technology and relationality as entwined, within the context of virtual teams. Orlikowski and Scott (2008, p. 455) highlight that "entities (whether humans or technologies) have no inherent properties, but acquire form, attributes, and capabilities through their interpenetration". The manifestation of technology may be material: most people perform their work though computers, laptops, desktop monitors, Internet modems or hot-spots, mobile phones, tablets, and so on. However, these are not used simply to transform the material infrastructure of work; they are used as platforms of virtual worlds.

Understanding the experience of undertaking work in and through virtual worlds, it is also important to realise that there is still a lot to learn about the new generation of tech aware users (Mäntymäki & Riemer, 2014). People who experienced the virtual revolution only later in their lives may regard virtuality as an artefact of organisational reality. However, people who grew up in the virtual revolution regard virtuality as an inherent part of everyday life. While for the former, technology is a means to performing specific tasks, such as sending emails, for the latter technology is a facet of their lives, for example coining their presence in the world through social media. These differing levels of familiarity with technology impact significantly on the ways people understand relationships (Fineman et al., 2007). For the former, physical presence is the basis for establishing trusting relationships. For the latter, digital presence is an equally valid basis. Therefore, for people native to virtuality being part of a dispersed team is nothing extravagant or novel; it is a continuation to the usual and habitual ways of communicating with one another. Such a realisation not only calls for re-orienting traditional theorising, but also cuts through questions about the possibility to collaborate effectively beyond the physical encounter.

FUTURE RESEARCH DIRECTIONS

How can studies of dispersed collaboration move forward into a relational direction? Irrespective of their methodologies, future studies can benefit from meaningfully addressing the virtual dimension of dispersed collaboration, illustrating the transformation of relationality. To do so, researchers may use the themes of time-space, leadership, trust and cohesion versus flexibility as guiding questions in designing their research projects. In particular, researchers may explicitly specify what the time space of the virtual team(s) in their case is, what the particularities of leadership are when these teams connect, what kind of trust is being shared, and what the balance between convergence and divergence in the examined dispersed team(s) is. Rather than simply acknowledging virtuality by mentioning it as a distinctiveness of the context in dispersed collaboration, researchers may also reflect on how their findings are constitutive of this particular context. From their part, editors and reviewers may actively encourage researchers to reflect upon the particular context of their study, and demonstrate this reflective process.

Furthermore, it is important for future research to become sensitive to the interconnections between the dimensions of virtuality (Gilson et al., 2015; Griffith, 2012). Even though, each dimension is discussed separately to offer a deeper understanding of the context encountered in virtual teams, they should rather be thought of as part of the holistic experience of dispersed collaboration. Indeed, trying to disentangle the dimensions of virtuality, as if each stands on its own, only attests to the relational dynamics of dispersed teams: virtuality is mutually constitutive and inseparable of dispersed collaboration (Wilson et al., 2013). That is, the dimensions of virtuality do not exist in a vacuum; there are recurrent meeting points between them. The questions about the time-space of a dispersed team overlaps with the struggle of cohesion over flexibility. Similarly, an exploration on leadership is closely related to an exploration on trust. In other words, these four dimensions of virtuality represent a seamless whole, making up the relational 'reality' of dispersed teams.

There has seldom been research attention on these overlaps of virtuality (Kirkman & Mathieu, 2005), perhaps because representational images of context offer a neat contextualisation of virtual teams. One way to move forward on this is to embrace the overlaps of the virtuality dimensions not only by acknowledging them, but also by adding a layer of complexity and analysing the relationships between them. One possibility may be to explore the patterns of virtuality from which specific practices come to be significant in dispersed collaboration. Comparative analysis may also enable to gain insight into the ways certain patterns of virtuality produce variations of relationality.

Illustrative examples that elucidate the dimensions of virtuality in dispersed teams come from rich, qualitative studies; ethnographic ones, in particular (see, for example Baralou & Tsoukas, 2015; Jonsen et al., 2013). This hardly comes as a surprise given that ethnography by its nature encourages an immersion in the field, accompanied by fine grained analysis (Van Maanen, 1979). However, it is time to move beyond prescriptive mantras that rich analysis is limited to ethnographic orientations. A relational ontology may be conveyed with a plethora of methodologies, aiming to centre virtuality. For example, narrative inquiries (Hoyer & Steyaert, 2015) can be utilised to examine the experiences of dispersed collaboration, targeting explicitly the four dimensions of virtuality. Another possible direction at the methodological level is to include virtuality in the research design. While virtual teams use asynchronous means of communication to achieve collaboration, it surprising that research methodologies use synchronous methods to follow their collaboration (Dooley, Tsoukas, Lorino, Tricard, & Clot, 2011). The lag observed here between the dispersed teams' and the researchers' rhythms may be an important reason why there is such fragmentation in the literature on dispersed teams, and why relationality has not been a key theme in the research agenda.

CONCLUSION

Collaboration in virtual teams is far from a static endeavour; it is rather an energetic field of study in the organisational domain. Its evolution is in fact very rapid; unsurprisingly, people and technology paint a changeable picture. Technology aware generations are coming into the workplace, carrying with them new sets of practices about how to use technology for managing collaboration, and for organising everyday life. Simultaneously, technology keeps reinventing itself: communication tools are evolving and are coming to impact both personal and professional lives. Understanding the effect these changes have in the ways work, and work relationships are constructed is the path this chapter follows, hoping to motivate further travelling along its lines.

REFERENCES

Ahuja, M. K., & Galvin, J. E. (2003). Socialization in virtual groups. *Journal of Management*, *29*(2), 161–185. doi:10.1177/014920630302900203

Al-Ani, B., Horspool, A., & Bligh, M. C. (2011). Collaborating with 'virtual strangers': Towards developing a framework for leadership in distributed teams. *Leadership*, *7*(3), 219–249. doi:10.1177/1742715011407382

Avolio, B. J., Kahai, S., & Dodge, G. E. (2000). E-leadership: Implications for theory, research, and practice. *The Leadership Quarterly*, *11*(4), 615–668. doi:10.1016/S1048-9843(00)00062-X

Avolio, B. J., Sosik, J. J., Kahai, S. S., & Baker, B. (2014). E-leadership: Re-examining transformations in leadership source and transmission. *The Leadership Quarterly*, *25*(1), 105–131. doi:10.1016/j.leaqua.2013.11.003

Baba, M. L., Gluesing, J., Ratner, H., & Wagner, K. H. (2004). The contexts of knowing: Natural history of a globally distributed team. *Journal of Organizational Behavior*, *25*(5), 547–587. doi:10.1002/job.259

Baden-Fuller, C., & Haefliger, S. (2013). Business models and technological innovation. *Long Range Planning*, *46*(6), 419–426. doi:10.1016/j.lrp.2013.08.023

Bailey, C., & Madden, A. (2015). Time reclaimed: Temporality and the experience of meaningful work. *Work, Employment and Society*, *31*(1), 3–18. doi:10.1177/0950017015604100

Bailey, D. E., Leonardi, P. M., & Barley, S. R. (2012). The lure of the virtual. *Organization Science*, *23*(5), 1485–1504. doi:10.1287/orsc.1110.0703

Bakken, T., Holt, R., & Zundel, M. (2013). Time and play in management practice: An investigation through the philosophies of McTaggart and Heidegger. *Scandinavian Journal of Management*, *29*(1), 13–22. doi:10.1016/j.scaman.2012.09.003

Baralou, E., & Tsoukas, H. (2015). How is new organizational knowledge created in a virtual context? An ethnographic study. *Organization Studies*, *36*(5), 593–620. doi:10.1177/0170840614556918

Beyes, T., & Steyaert, C. (2012). Spacing organization: Non-representational theory and performing organizational space. *Organization*, *19*(1), 45–61. doi:10.1177/1350508411401946

Boh, W. F., Ren, Y., Kiesler, S., & Bussjaeger, R. (2007). Expertise and collaboration in the geographically dispersed organization. *Organization Science*, *18*(4), 595–612. doi:10.1287/orsc.1070.0263

Bradbury, H., & Lichtenstein, B. M. B. (2000). Relationality in organizational research: Exploring the space between. *Organization Science*, *11*(5), 551–564. doi:10.1287/orsc.11.5.551.15203

Brannen, M. Y., & Salk, J. E. (2000). Partnering across borders: Negotiating organizational culture in a German-Japanese joint venture. *Human Relations*, *53*(4), 451–487. doi:10.1177/0018726700534001

Charlier, S. D., Stewart, G. L., Greco, L. M., & Reeves, C. J. (2016). Emergent leadership in virtual teams: A multilevel investigation of individual communication and team dispersion antecedents. *The Leadership Quarterly*, *27*(5), 745–764. doi:10.1016/j.leaqua.2016.05.002

Chidambaram, L., & Jones, B. (1993). Impact of communication medium and computer support on group perceptions and performance: A comparison of face-to-face and dispersed meetings. *Management Information Systems Quarterly*, *17*(4), 465–491. doi:10.2307/249588

Cramton, C. D. (2001). The mutual knowledge problem and its consequences for dispersed collaboration. *Organization Science*, *12*(3), 346–371. doi:10.1287/orsc.12.3.346.10098

Cramton, C. D., & Hinds, P. J. (2014). An embedded model of cultural adaptation in global teams. *Organization Science*, *25*(4), 1056–1081. doi:10.1287/orsc.2013.0885

Crisp, B. C., & Jarvenpaa, S. L. (2013). Swift trust in global virtual teams. *Journal of Personnel Psychology*, *12*(1), 45–56. doi:10.1027/1866-5888/a000075

Cummings, J. N. (2004). Work groups, structural diversity, and knowledge sharing in a global organization. *Management Science*, *50*(3), 352–364. doi:10.1287/mnsc.1030.0134

Czarniawska, B. (2004). On time, space, and action nets. *Organization*, *11*(6), 773–791. doi:10.1177/1350508404047251

Delbridge, R., & Sallaz, J. J. (2015). Work: Four worlds and ways of seeing. *Organization Studies*, *36*(11), 1449–1462. doi:10.1177/0170840615612021

Dell & Intel. (2016). *Future workforce study*. Austin, TX: Authors.

Denyer, D., Parry, E., & Flowers, P. (2011). "Social", "open" and "participative"? Exploring personal experiences and organisational effects of enterprise 2.0 use. *Long Range Planning*, *44*(5-6), 375–396. doi:10.1016/j.lrp.2011.09.007

Dewett, T., & Jones, G. R. (2001). The role of information technology in the organization: A review, model, and assessment. *Journal of Management*, *27*(3), 313–346. doi:10.1177/014920630102700306

Ding, W. W., Levin, S. G., Stephan, P. E., & Winkler, A. E. (2010). The impact of information technology on academic scientists' productivity and collaboration patterns. *Management Science*, *56*(9), 1439–1461. doi:10.1287/mnsc.1100.1195

Dooley, K. J., Tsoukas, H., Lorino, P., Tricard, B., & Clot, Y. (2011). Research methods for non-representational approaches to organizational complexity: The dialogical mediated inquiry. *Organization Studies*, *32*(6), 769–801. doi:10.1177/0170840611410807

Drath, W. H., McCauley, C. D., Palus, C. J., Van Velsor, E., O'Connor, P. M. G., & McGuire, J. B. (2008). Direction, alignment, commitment: Toward a more integrative ontology of leadership. *The Leadership Quarterly*, *19*(6), 635–653. doi:10.1016/j.leaqua.2008.09.003

Faraj, S., Jarvenpaa, S. L., & Majchrzak, A. (2011). Knowledge collaboration in online communities. *Organization Science*, *22*(5), 1224–1239. doi:10.1287/orsc.1100.0614

Faraj, S., & Johnson, S. L. (2011). Network exchange patterns in online communities. *Organization Science*, *22*(6), 1464–1480. doi:10.1287/orsc.1100.0600

Faraj, S., Kudaravalli, S., & Wasko, M. (2015). Leading collaboration in online communities. *Management Information Systems Quarterly*, *39*(2), 393–412. doi:10.25300/MISQ/2015/39.2.06

Fayard, A.-L., & Metiu, A. (2014). The role of writing in distributed collaboration. *Organization Science*, *25*(5), 1391–1413. doi:10.1287/orsc.2013.0893

Fineman, S., Maitlis, S., & Panteli, N. (2007). Virtuality and emotion. *Human Relations*, *60*(4), 555–560. doi:10.1177/0018726707078346

Gal, U., Jensen, T. B., & Lyytinen, K. (2014). Identity orientation, social exchange, and information technology use in interorganizational collaborations. *Organization Science*, *25*(5), 1372–1390. doi:10.1287/orsc.2014.0924

Gergen, K. J. (2009). *Relational being: Beyond self and community*. Oxford, UK: Oxford University Press.

Gergen, K. J. (2010). Co-constitution, causality, and confluence: Organizing in a world without entities. In T. Hernes & S. Maitlis (Eds.), *Process, Sensemaking and Organization* (pp. 55–69). Oxford, UK: Oxford University Press. doi:10.1093/acprof:oso/9780199594566.003.0004

Gergen, K. J., & Thatchenkery, T. J. (2006). Organizational science and the promise of postmodernism. In D. M. Hosking & S. McNamee (Eds.), *The Social Construction of Organization* (pp. 34–53). Malmo, Sweden: Liber & Copenhagen Business School Press.

Gibbs, J. (2009). Dialectics in a global software team: Negotiating tensions across time, space, and culture. *Human Relations*, *62*(6), 905–935. doi:10.1177/0018726709104547

Gibson, C. B., & Gibbs, J. L. (2006). Unpacking the concept of virtuality: The effects of geographic dispersion, electronic dependence, dynamic structure, and national diversity on team innovation. *Administrative Science Quarterly*, *51*(3), 451–495. doi:10.2189/asqu.51.3.451

Gibson, C. B., Gibbs, J. L., Stanko, T. L., Tesluk, P., & Cohen, S. G. (2011). Including the "i" in virtuality and modern job design: Extending the job characteristics model to include the moderating effect of individual experiences of electronic dependence and copresence. *Organization Science*, *22*(6), 1481–1499. doi:10.1287/orsc.1100.0586

Gilson, L. L., Maynard, M. T., Jones Young, N. C., Vartiainen, M., & Hakonen, M. (2015). Virtual teams research: 10 years, 10 themes, and 10 opportunities. *Journal of Management*, *41*(5), 1313–1337. doi:10.1177/0149206314559946

Golden, A. G., & Geisler, C. (2007). Work-life boundary management and the personal digital assistant. *Human Relations*, *60*(3), 519–551. doi:10.1177/0018726707076698

Golden, T. D., & Fromen, A. (2011). Does it matter where your manager works? Comparing managerial work mode (traditional, telework, virtual) across subordinate work experiences and outcomes. *Human Relations*, *64*(11), 1451–1475. doi:10.1177/0018726711418387

Greckhamer, T. (2011). Cross-cultural Differences in Compensation Level and Inequality across Occupations: A Set-theoretic Analysis. *Organization Studies*, *32*(1), 85–115. doi:10.1177/0170840610380806

Griffith, T. L. (2012). Technology and teams: The next ten years. In M. A. Neale & E. A. Mannix (Eds.), *Looking back, moving forward: A review of group and team-based research* (pp. 245–278). Bingley, UK: Emerald Group Publishing Ltd. doi:10.1108/S1534-0856(2012)0000015013

Haas, M. R. (2006). Acquiring and applying knowledge in transnational teams: The roles of cosmopolitans and locals. *Organization Science*, *17*(3), 367–384. doi:10.1287/orsc.1060.0187

Hatch, M. J. (2002). Essai: Doing time in organization theory. *Organization Studies*, *23*(6), 869–875. doi:10.1177/0170840602236008

Helin, J., Hernes, T., Hjorth, D., & Holt, R. (2014). Process is how process does. In J. Helin, T. Hernes, D. Hjorth, & R. Holt (Eds.), *The Oxford Handbook of Process Philosophy and Organization Studies* (pp. 1–16). Oxford, UK: Oxford University Press.

Henfridsson, O., & Bygstad, B. (2013). The generative mechanisms of digital infrastructure evolution. *Management Information Systems Quarterly*, *37*(3), 907–931. doi:10.25300/MISQ/2013/37.3.11

Hernes, T., Simpson, B., & Söderlund, J. (2013). Managing and temporality. *Scandinavian Journal of Management*, *29*(1), 1–6. doi:10.1016/j.scaman.2012.11.008

Hinds, P., Liu, L., & Lyon, J. (2011). Putting the global in global work: An intercultural lens on the practice of cross-national collaboration. *The Academy of Management Annals*, *5*(1), 135–188. doi:10.1080/19416520.2011.586108

Hinds, P. J., & Bailey, D. E. (2003). Out of sight, out of sync: Understanding conflict in distributed teams. *Organization Science*, *14*(6), 615–632. doi:10.1287/orsc.14.6.615.24872

Hosking, D. M. (2006). Discourses of relations and relational processes. In O. Kyriakidou & M. Ozbilgin (Eds.), *Relational Perspectives in Organizational Studies: A Research Companion* (pp. 265–277). Cheltenham, UK: Edward Elgar Publishing Ltd. doi:10.4337/9781781950548.00020

Hosking, D. M. (2011). Telling tales of relations: Appreciating relational constructionism. *Organization Studies*, *32*(1), 47–65. doi:10.1177/0170840610394296

Hosking, D. M., Dachler, H. P., & Gergen, K. J. (1995). *Management and Organization: Relational alternatives to individualisation.* Aldershot, UK: Avenbury.

Hoyer, P., & Steyaert, C. (2015). Narrative identity construction in times of career change: Taking note of unconscious desires. *Human Relations*, *68*(12), 1837–1863. doi:10.1177/0018726715570383

Hsiao, R.-L., Tsai, D.-H., & Lee, C.-F. (2012). Collaborative knowing: The adaptive nature of cross-boundary spanning. *Journal of Management Studies*, *49*(3), 463–491. doi:10.1111/j.1467-6486.2011.01024.x

Hwang, E. H., Singh, P. V., & Argote, L. (2015). Knowledge sharing in online communities: Learning to cross geographic and hierarchical boundaries. *Organization Science*, *26*(6), 1593–1611. doi:10.1287/orsc.2015.1009

Jarvenpaa, S. L., & Leidner, D. E. (1999). Communication and trust in global virtual teams. *Organization Science*, *10*(6), 791–815. doi:10.1287/orsc.10.6.791

Jonsen, K., Butler, C. L., Makela, K., Piekkari, R., Drogendijk, R., Lauring, J., ... Zander, L. (2013). Processes of international collaboration in management research: A reflexive, autoethnographic approach. *Journal of Management Inquiry*, *22*(4), 394–413. doi:10.1177/1056492612464429

Josefy, M., Kuban, S., Ireland, R. D., & Hitt, M. A. (2015). All things great and small: Organizational size, boundaries of the firm, and a changing environment. *The Academy of Management Annals*, *9*(1), 715–802. doi:10.1080/19416520.2015.1027086

Joshi, A., Lazarova, M. B., & Liao, H. (2009). Getting everyone on board: The role of inspirational leadership in geographically dispersed teams. *Organization Science*, *20*(1), 240–252. doi:10.1287/orsc.1080.0383

Kasper-Fuehrer, E. C., & Ashkanasy, N. M. (2001). Communicating trustworthiness and building trust in interorganizational virtual organizations. *Journal of Management*, *27*(3), 235–254. doi:10.1016/S0149-2063(01)00090-3

Kelley, E. (2001). Keys to effective virtual global teams. *The Academy of Management Executive*, *15*(2), 132–133. doi:10.5465/AME.2001.4615036

Kirkman, B. L., Cordery, J. L., Mathieu, J., Rosen, B., & Kukenberger, M. (2013). Global organizational communities of practice: The effects of nationality diversity, psychological safety, and media richness on community performance. *Human Relations*, *66*(3), 333–362. doi:10.1177/0018726712464076

Kirkman, B. L., & Mathieu, J. E. (2005). The dimensions and antecedents of team virtuality. *Journal of Management*, *31*(5), 700–718. doi:10.1177/0149206305279113

Kirkman, B. L., Rosen, B., Gibson, C. B., Tesluk, P. E., & McPherson, S. O. (2002). Five challenges to virtual team success: Lessons from Sabre, Inc. *The Academy of Management Executive*, *16*(3), 67–79. doi:10.5465/AME.2002.8540322

Kolb, D. G., Caza, A., & Collins, P. D. (2012). States of connectivity: New questions and new directions. *Organization Studies*, *33*(2), 267–273. doi:10.1177/0170840611431653

Latour, B. (2005). *Reassembling the social: An introduction to actor-network-theory*. Oxford, UK: Oxford University Press.

Lee, H., & Liebenau, J. (1999). Time in organizational studies: Towards a new research direction. *Organization Studies*, *20*(6), 1035–1058. doi:10.1177/0170840699206006

Leonardi, P. M. (2011). Innovation blindness: Culture, frames, and cross-boundary problem construction in the development of new technology concepts. *Organization Science*, *22*(2), 347–369. doi:10.1287/orsc.1100.0529

Leung, K., & Wang, J. (2015). Social processes and team creativity in multicultural teams: A socio-technical framework. *Journal of Organizational Behavior*, *36*(7), 1008–1025. doi:10.1002/job.2021

Levi, D. (2015). *Group dynamics for teams* (5th ed.). Thousand Oaks, CA: Sage Publications Inc.

Mailhot, C., Gagnon, S., Langley, A., & Binette, L.-F. (2016). Distributing leadership across people and objects in a collaborative research project. *Leadership*, *12*(1), 53–85. doi:10.1177/1742715014543578

Majchrzak, A., Rice, R. E., Malhotra, A., King, N., & Ba, S. L. (2000). Technology adaptation: The case of a computer-supported inter-organizational virtual team. *Management Information Systems Quarterly*, *24*(4), 569–600. doi:10.2307/3250948

Malhotra, A., & Majchrzak, A. (2014). Enhancing performance of geographically distributed teams through targeted use of information and communication technologies. *Human Relations*, *67*(4), 389–411. doi:10.1177/0018726713495284

Malhotra, A., Majchrzak, A., Carman, R., & Lott, V. (2001). Radical innovation without collocation: A case study at Boeing-Rocketdyne. *Management Information Systems Quarterly*, *25*(2), 229–249. doi:10.2307/3250930

Mäntymäki, M., & Riemer, K. (2014). Digital natives in social virtual worlds: A multi-method study of gratifications and social influences in Habbo Hotel. *International Journal of Information Management*, *34*(2), 210–220. doi:10.1016/j.ijinfomgt.2013.12.010

Martins, L. L., Gilson, L. L., & Maynard, M. T. (2004). Virtual teams: What do we know and where do we go from here? *Journal of Management*, *30*(6), 805–835. doi:10.1016/j.jm.2004.05.002

Maynard, M. T., Mathieu, J. E., Rapp, T. L., & Gilson, L. L. (2012). Something(s) old and something(s) new: Modeling drivers of global virtual team effectiveness. *Journal of Organizational Behavior*, *33*(3), 342–365. doi:10.1002/job.1772

Maznevski, M. L., & Chudoba, K. M. (2000). Bridging space over time: Global virtual team dynamics and effectiveness. *Organization Science*, *11*(5), 473–492. doi:10.1287/orsc.11.5.473.15200

McKnight, D. H., Cummings, L. L., & Chervany, N. L. (1998). Initial trust formation in new organizational relationships. *Academy of Management Review*, *23*, 473–490.

McNamee, S. (1998). Re-inscribing organizational wisdom and courage: The relationally engaged organization. In S. Srivastva & D. L. Cooperrider (Eds.), *Organizational wisdom and executive courage* (pp. 101–117). San Francisco: New Lexington Press.

Michailova, S., & Hutchings, K. (2006). National cultural influences on knowledge sharing: A comparison of China and Russia. *Journal of Management Studies*, *43*(3), 383–405. doi:10.1111/j.1467-6486.2006.00595.x

Mohrman, S. A. (1999). The contexts for geographically dispersed teams and networks. *Journal of Organizational Behavior*, *6*, 63–80.

Montoya-Weiss, M. M., Massey, A. P., & Song, M. (2001). Getting it together: Temporal coordination and conflict management in global virtual teams. *Academy of Management Journal*, *44*(6), 1251–1262. doi:10.2307/3069399

Neeley, T. B., Hinds, P. J., & Cramton, C. D. (2012). The (un)hidden turmoil of language in global collaboration. *Organizational Dynamics*, *41*(3), 236–244. doi:10.1016/j.orgdyn.2012.03.008

Nyberg, D. (2009). Computers, customer service operatives and cyborgs: Intra-actions in call centres. *Organization Studies*, *30*(11), 1181–1199. doi:10.1177/0170840609337955

O'Leary, M. B., & Cummings, J. N. (2007). The spatial, temporal, and configurational characteristics of geographic dispersion in teams. *Management Information Systems Quarterly*, *31*(3), 433–452. doi:10.2307/25148802

O'Leary, M. B., & Mortensen, M. (2010). Go (con)figure: Subgroups, imbalance, and isolates in geographically dispersed teams. *Organization Science*, *21*(1), 115–131. doi:10.1287/orsc.1090.0434

Orlikowski, W. J. (1992). The duality of technology: Rethinking the concept of technology in organizations. *Organization Science*, *3*(3), 398–427. doi:10.1287/orsc.3.3.398

Orlikowski, W. J. (2000). Using technology and constituting structures: A practice lens for studying technology in organizations. *Organization Science*, *11*(4), 404–428. doi:10.1287/orsc.11.4.404.14600

Orlikowski, W. J., & Scott, S. V. (2008). Sociomateriality: Challenging the separation of technology, work and organization. *The Academy of Management Annals*, *2*(1), 433–474. doi:10.1080/19416520802211644

Pines, A. M., & Zaidman, N. (2014). Stress and burnout in bicultural teams in hi-tech industry. *British Journal of Management*, *25*(4), 819–832. doi:10.1111/1467-8551.12056

Puranam, P., Alexy, O., & Reitzig, M. (2014). What's "new" about new forms of organizing? *Academy of Management Review*, *39*(2), 162–180. doi:10.5465/amr.2011.0436

Raelin, J. A. (2016). Imagine there are no leaders: Reframing leadership as collaborative agency. *Leadership*, *12*(2), 131–158. doi:10.1177/1742715014558076

Raghuram, S., Garud, R., Wiesenfeld, B., & Gupta, V. (2001). Factors contributing to virtual work adjustment. *Journal of Management*, *27*(3), 383–405. doi:10.1177/014920630102700309

Ravlin, E. C., Ward, A.-K., & Thomas, D. C. (2014). Exchanging social information across cultural boundaries. *Journal of Management*, *40*(5), 1437–1465. doi:10.1177/0149206314531623

Saunders, C., Van Slyke, C., & Vogel, D. R. (2004). My time or yours? Managing time visions in global virtual teams. *The Academy of Management Executive*, *18*(1), 19–31. doi:10.5465/AME.2004.12691177

Sewell, G., & Taskin, L. (2015). Out of sight, out of mind in a new world of work? Autonomy, control, and spatiotemporal scaling in telework. *Organization Studies*, *36*(11), 1507–1529. doi:10.1177/0170840615593587

Shapira, Z., Lewin, A. W., Schoonhoven, C. B., Argote, L., & Levinthal, D. (2014). Editorial: Organization Science at 25. *Organization Science*, *25*(6), 1573–1578. doi:10.1287/orsc.2014.0943

Sieben, B. (2007). Doing research on emotion and virtual work: A compass to assist orientation. *Human Relations*, *60*(4), 561–580. doi:10.1177/0018726707078347

Sole, D., & Edmondson, A. (2002). Situated knowledge and learning in dispersed teams. *British Journal of Management*, *13*(S2), 17–34. doi:10.1111/1467-8551.13.s2.3

Van Maanen, J. (1979). The fact of fiction in organizational ethnography. *Administrative Science Quarterly*, *24*(4), 539–550. doi:10.2307/2392360

Vesa, M., & Franck, H. (2013). Bringing strategy to time, studying strategy as experiential vectors. *Scandinavian Journal of Management*, *29*(1), 23–34. doi:10.1016/j.scaman.2012.11.005

Wageman, R., Gardner, H., & Mortensen, M. (2012). The changing ecology of teams: New directions for teams research. *Journal of Organizational Behavior*, *33*(3), 301–315. doi:10.1002/job.1775

Wai Fong, B., Yuqing, R., Kiesler, S., & Bussjaeger, R. (2007). Expertise and collaboration in the geographically dispersed organization. *Organization Science*, *18*(4), 595–612. doi:10.1287/orsc.1070.0263

Weick, K. E., & Roberts, K. H. (1993). Collective mind in organizations - heedful interrelating on flight decks. *Administrative Science Quarterly*, *38*(3), 357–381. doi:10.2307/2393372

Wiesenfeld, B. M., Raghuram, S., & Garud, R. (1999). Communication patterns as determinants of organizational identification in a virtual organization. *Organization Science*, *10*(6), 777–790. doi:10.1287/orsc.10.6.777

Wilson, J., Crisp, C. B., & Mortensen, M. (2013). Extending construal-level theory to distributed groups: Understanding the effects of virtuality. *Organization Science*, *24*(2), 629–644. doi:10.1287/orsc.1120.0750

Wilson, J. M., O'Leary, M. B., Metiu, A., & Jett, Q. R. (2008). Perceived proximity in virtual work: Explaining the paradox of far-but-close. *Organization Studies*, *29*(7), 979–1002. doi:10.1177/0170840607083105

Yoo, Y., Boland, R. J. Jr, Lyytinen, K., & Majchrzak, A. (2012). Organizing for innovation in the digitized world. *Organization Science*, *23*(5), 1398–1408. doi:10.1287/orsc.1120.0771

Zammuto, R. F., Griffith, T. L., Majchrzak, A., Dougherty, D. J., & Faraj, S. (2007). Information technology and the changing fabric of organization. *Organization Science*, *18*(5), 749–762. doi:10.1287/orsc.1070.0307

Zimmermann, P., Wit, A., & Gill, R. (2008). The relative importance of leadership behaviours in virtual and face-to-face communication settings. *Leadership*, *4*(3), 321–337. doi:10.1177/1742715008092388

KEY TERMS AND DEFINITIONS

E-Leadership: The understanding of leadership in virtual and online contexts, characterised by a-temporal and a-spatial orientations.

Relationality: Points to a relational ontology of becoming from a process perspective. The conceptual focus is on the relatedness to one another, rather than individual action.

Swift Trust: An a priori bestowment of trustworthiness.

Technological Integration: Highlights the changes in infrastructure that have occurred across many organisations, leading to new ways of conducting and knowing work life.

Time-Space: The domain in and through which a team achieves its rhythm of collaboration. The physical parameters of time (chronology) and space (container of action) are irrelevant, important is immersion into work.

Virtual Teams: A dynamic work arrangement characterised by people who collaborate across a set of boundaries in pursuit of common goals, using technology to communicate and coordinate work.

Virtuality: The context *in use* of virtual teams.

Chapter 2
Design-Based Research as a Methodology for Studying Learning in the Context of Work:
Suggestions for Guidelines

Ditte Kolbæk
Aalborg University, Denmark

ABSTRACT

This chapter presents a suggestion for design-based research (DBR) as a methodology for organizational studies. Although DBR was developed for investigating classroom training, this chapter discusses the methodological issues that are involved when DBR is employed for investigating learning in the context of work. DBR seems to be suitable in this complex context as it is an authentic learning environment. The purpose of this chapter is to provide new perspectives on DBR, including suggestions for guidelines regarding how to conduct DBR for studying learning from experience in the context of work. The research question is: What is needed to utilize DBR to explore learning from experience in the context of work? The theoretical framework is based on Yrjö Engeström and John Dewey's work on learning, and the exploration of DBR is based on a literature review and findings in a seven-year DBR study on learning from experience in the context of work.

INTRODUCTION

There is growing evidence that organizations produce superior results when they facilitate and support learning within the organization (Aktthar & Khan, 2011; Holsapple & Wu, 2011). Consequently, it is important to find methodologies that focus on learning in the context of work. "Learning from experience in the context of work" refers to situations in which employees learn from their practices, as opposed to learning from courses or work-related texts. Rather than improve competencies in defined courses,

DOI: 10.4018/978-1-5225-4094-6.ch002

employees should improve products, services, and work practices (Brandi & Elkjaer, 2011) to improve business results, competitive advantages, and revenue (Nonaka & Takeuchi, 1995).

This chapter explores design-based research (DBR) as a methodology for studying learning from experience in the context of work. This is a new setting, as DBR was developed to study learning in educational settings (Anderson & Shattuck, 2012; Andriessen, 2007; Brown, 1992; Dede, 2004; Design-Based Research Collective, 2003; Kolbaek, 2017; Wang & Hannafin, 2005).

The research question is as follows: What is needed to utilize DBR to explore learning from experience in the context of work?

The chapter is structured as follows: First, "learning from experience in the context of work" is examined, based on John Dewey's (Dewey & Boydston, 1976) explanation of learning from experience and Engeström's view on learning in the context of work (Engeström, 2001). Next, Kolbaek's (2014) educational design for learning in the context of work is briefly presented. Then, DBR is presented in its original format, followed by the research setting and four iterations in the development of the educational design of Proactive Reviews (PRs). The analysis focuses on the weaknesses of DBR when utilized in the context of learning from experience in the context of work. The analysis leads to suggestions for new guidelines for DBR as a methodology for studying learning from experience in the context of work. The suggested guidelines are called "DBR at work." Finally, the conclusion presents theoretical, methodological, and practical contributions from this study.

BACKGROUND

Authors use different terms to describe DBR, including "design research," "development research," "design experiments," and "design studies" (Anderson & Shattuck, 2012; Andriessen, 2007). However, this chapter only will use the term "design-based research" (DBR), which was introduced by Ann Brown in 1992. DBR is different from Design Science Research (DSR), which focuses on the design of objects, processes (making and using objects), and reflection and evaluation of object design (Winter, 2008). The "object" of DBR is "learning" in the classroom, as DBR is used for solving educational design problems, not design problems in general (Collins 2010).

DBR is not a well-defined methodology; rather, it is a combination of processes leading to knowledge about learning designs, in the form of qualitative and quantitative data (Collins, Joseph, & Bielaczyc, 2004). Wang and Hannafin (2005) define DBR this way: "DBR is a systematic, but flexible, methodology aimed to improve educational practices through iterative analysis, design, development, and implementation, based on collaboration among researchers and practitioners in real-world settings, and leading to contextually sensitive design principles and theories" (Wang & Hannafin 2005, p. 6).

According to Zheng (2015), there is a lack of long-term DBR studies that include several iterations of intervention on educational designs. This chapter includes a longitudinal DBR study of learning from experience in the context of work that follows four iterations of an educational design called Proactive Review.

THEORETICAL APPROACH

This section presents a brief introduction to theories and an educational design for learning in the context of work.

Learning in the Context of Work

Learning in the context of work is different from learning in the classroom, most visibly because of the lack of both a teacher and an educational design, but also because employees are often not conscious of participating in a learning process. Furthermore, neither the employees nor their managers define what is to be learned (Engeström, 2001).

When learning from experience takes place in a specific organization, the organization defines the framework for learning. This framework is the context of the learning. The context may be defined by various factors, such as physical surroundings, historical background, and sociological aspects. The *physical surroundings* include geography, buildings, and tools, and they are part of the resources available for the learning process (Engeström et al., 2013; Gherardi et al., 1998; Nonaka et al., 2000; Wenger 1998). As the context develops over time, the *history* of the context is necessary for understanding the context itself.

The *historical background* provides insights about what has happened, which may lead to an understanding of the present situation (Engeström et al. 2013; Gherardi et al. 1998; Nonaka et al. 2000).

Social aspects of the context may include organizational structure and power, division of labor, critical organizational competencies, rules, channels of communication, language, and time for learning purposes (Engeström et al., 2013; Gherardi et al., 1998; Nonaka et al., 2000; Wenger, 1998).

Dewey and Boydston (1976) perceive knowledge as a result of experimentation, reflection, and conversation. Consequently, knowledge is not a commodity, but a process that evolves, which may be explained with the words "knowledge" and "knowing." Knowledge is an embedded capability, whereas knowing is a process of negotiation between individuals and is grounded in everyday practices (Gherardi as cited in Easterby-Smith et al., 2011; Shipton & DeFillippi as cited in Easterby-Smith et al., 2011). Furthermore, knowing cannot be passively received, as knowing is action, and inquiry leads to the development of knowing (Dewey, 1916). Knowing is constructed in collaboration with peers, and conversation is the means of learning that begins with perplexity, confusion, doubt (Dewey & Boydston, 1976), disturbance, contradictions, or conflicts (Engeström, 2001).

Actions and interactions with the world may lead to experience, provided that the individuals reflect on what they tried to do and what happened as a consequence; thus, experience is based on activities and involves emotions (Dewey & Boydston, 1976). Experience may be achieved by trying and trying again. Often, experience is not achieved in one attempt; therefore, experience is constantly under construction as the context changes over time, and individuals adapt accordingly while integrating previous experiences into new ones (Elkjaer, 2003). Experience is based in the past, utilized in the present, and maintained for the future (Dewey & Boydston, 1976).

Learning is, thus, perceived as being dependent on the context. Those who are subject to learning in the context of work are employees (Engeström, 2001; Engeström et al., 2013; Kolbaek, 2015), and they may contribute to the design of learning situations, provide feedback to the researcher, and even take part in the analysis. This collaboration enables the researcher to take advantage of their expertise. Hence, reflecting, questioning, and analyzing are important activities for learning in the context of work, and these activities are based on the conversations among those who learn (Engeström, 2001; Kolbaek, 2015). Learning from experience in the context of work implies that employees become aware of the realities of work; thus, learning in a work context should aim to enable employees to influence these realities (Engeström, 2001, 2013).

The employee is simultaneously an individual, a participant in a team, and a member of the organization---the three ontological dimensions that are not strictly divided. Instead, the ontological dimensions are a way to describe different approaches to learning in an organization (Nonaka & Takeuchi, 1995). These three dimensions are included in DBR (Collins et al. 2004).

Proactive Review (PR)

When teachers plan for a learning situation, this planning is called an educational design, which includes resources and practices, as well as considering the way they are utilized in the learning context (Dohn et al., 2016). Proactive Review (PR) is an educational design for learning from experience in the context of work (Kolbaek 2014), developed over a period of seven years at a global IT company, here called Cleo. PR includes 13 iterations (see Appendix), starting with a pilot based on the four questions from After Action Review (Kolbaek 2014), then adding more questions according to the requirements of Cleo the (Kolbaek, 2014).

The seven questions are:

1. What is the purpose of this Proactive Review?
2. What was our goal?
3. What happened and why?
4. What should we do next time?
5. What are we going to report, to whom, when, and how?
6. What management challenge may be addressed from this Proactive Review?
7. What is your personal highlight from this Proactive Review?

These seven open-ended questions should be asked in this specific sequence. PRs reveal experience from employees who have solved a task together or are going to achieve a complex task. They are conducted by a trained facilitator. The physical surroundings for PRs include face-to-face meetings or online tools. The sociological aspects -- such as power, rules, and time keeping -- are taken care of by a trained facilitator, and the historical background is identified in the question "What happened and why?"

The tangible results of the PR are a communication and action plan and, sometimes, a management challenge (Kolbaek, 2014). Individual learning is achieved by participating in the PR, and team learning is achieved when the participants share their new knowing with peers, and when the manager initiates new work processes as suggested in the communication and action plan. Organizational learning is achieved when participants share their new insights with colleagues outside their team, e.g., in a community of practice, when managers discuss PR results with peers, or when upper management initiates organizational changes due to addressed management challenges (Kolbaek, 2014).

METHODOLOGY AND DATA COLLECTION

A methodology for studying learning from experience in the context of work should be based in a real-world setting, and it should include individual, team, and organizational learning. The methodology should define the context by the physical surroundings, historical background, and social aspects, and it should include iterations on the educational design that may uncover development in knowledge and knowing. Furthermore, the methodology should include the analysis of conversations as a means of learning. In the following section, the use of DBR for studying learning from experience in the context of work is described and exemplified. This section presents four iterations on the development of a PR.

Design-Based Research

DBR was first developed and used for studying learning in classroom settings (Andersen & Shattuck, 2012; Brown, 1992; Collins, 2010; Collins et al., 2004; Design-Based Research Collective, 2003). Thus, DBR embraces the messiness and complexity existing in real-world settings that are suitable for studying learning in the context of work.

The starting point of DBR is the identification of critical elements, such as failures and problems, to improve learning in a classroom setting. DBR allows an educational design to be tested and refined several times. A new iteration of the educational design requires a description of the failures or problems that initiate the iteration.

Table 1 shows the process of DBR in a classroom setting.

DBR includes the following three roles (Wang & Hannafin, 2005): researcher; practitioners, i.e., the teacher and learners; and other stakeholders, including administrators or managers in the organization (Collins, 2010). Researchers collaborate closely with teachers and students/learners to maintain a fine

Table 1. Original guidelines for design-based research

1	Implementing an educational design, including critical elements and how these elements interact.
2	Modifying the educational design; each modification is an intervention that starts a new iteration of the educational design.
3	Analyzing the intervention, including the perspectives of the cognitive development of the individual learner; the resources available for the learning processes; the interpersonal relationships between the learners and the teacher, as well as between the learners; the social aspects of the group involved; and the institutional aspects, such as the stakeholders' engagement and support.
4	Dependent variables consist of "climate variables," which indicate engagement, cooperation, and risk taking, and "outcome variables," which indicate learners' new skills. The teacher influences the dependent variables. Independent variables consist of "setting," "nature of the learner," "required resources and support," "professional development," "financial requirements," and "implementation path." The teacher cannot influence or select independent variables.
5	DBR reporting should include the goals and elements of the educational design, the research setting, a description of the iterations and the findings/outcomes, and lessons learned.

(Based on *Collins, 2010; Collins et al., 2004)*

balance between the theoretical and the practical (Wang & Hannafin, 2005). Thus, the teacher and researcher are two different people -- but that's not always the case. DBR provides variables for change in the dispositions of the individual learner and the development of his or her skills and learning strategies (Collins, 2010). DBR examines the social interaction between learners (Collins, 2010) and uncovers group dynamics and interactions between students and teachers (Collins, 2010). DBR also includes analysis at the institutional and resource levels; consequently, DBR includes three ontological dimensions in the methodology: individual learning, team learning, and organizational learning (Collins et al., 2004, p. 35).

Collins has determined a number of variables for researching the complexity in learning situations in real life, and he categorizes them as dependent variables and independent variables (Collins, 2010). Dependent variables are those that the teacher may influence directly, whereas independent variables are contextual (Collins et al., 2004). The variables are not to be held constant. *Climate variables* describe the following participant behaviors: engagement, cooperation, risk-taking, efforts to understand others' points of view, and participants' control (Collins, 2010). *Outcome variables* show how learners become more knowledgeable about content, and they include content knowledge, skills, dispositions, and learning strategies (Collins, 2010). *Resources* include the materials available to participants and the ease with which participants can find and use materials; information and communications technology (ICT) becomes an increasingly important resource that plays an important role when technology frames the learning situation (Collins et al., 2004). A lack of resources can hinder an educational design, so the researcher should explore what is needed to carry out the design in terms of costs, management support, materials, and technical support. Lack of *professional development* could be an obstacle to success because of a lack of skills. *Financial requirements* consist of the expenses needed for the educational design, e.g., time, equipment, support, and education. Collins et al. (2004) argue that it is important to report the process and the results of any DBR.

Research Setting

The empirical setting is Cleo - a world-class IT company that only works business-to-business and delivers hardware, middleware, and software. In 2011, the company had approximately 108,000 employees worldwide and 30,000 employees in Europe, the Middle East, and Africa (EMEA). In 2002 its revenue was 36.7 billion U.S. dollars over a fiscal year (Rao, 2003). From 2005 to 2011, the company changed its strategy from being a multinational company with local organizational entities in multiple countries to being a global company with global lines of business. More than 60 companies were merged into Cleo, and these mergers and acquisitions involved a vast number of new employees who worked online with managers and colleagues from other countries. Thus, it became increasingly important to learn from experience. The author served for seven years as manager of organizational learning at Cleo in EMEA. DBR came into play when she developed and implemented an educational design for learning from experience in the context of work, called Proactive Review (Kolbaek, 2014) in Cleo, EMEA.

Data Collection

The seven years of PR development used DBR as a methodology for the research and represented a longitudinal study of DBR in the context of work. Table 2 presents four of the 13 iterations of DBR described as required by Collins (2010) using four criteria: critical elements in the educational design, intervention, data collection, and the reporting of DBR results. The next section, "Analysis and Find-

Table 2. Data collection for four iterations of the development of DBR at work

	First Iteration 2005	Second Iteration 2005	Third Iteration 2008	Fourth Iteration 2009
Critical elements in the educational design	Pilot: From After Action Review to Proactive Review	Train PR facilitators	Management challenges	Online PR
Intervention	Establish an educational design for learning in the context of work	Define facilitator role Provide training of facilitators	New process for reporting bottom-up	Move from face-to-face meetings to online meetings
Data collection	Feedback from participants in two diaries (diaries) Two interviews with managers (stories in internal newsletters) Two interviews with participants (stories in internal newsletters) Observations of After Action Review (field notes)	Three e-mail interviews with facilitators (stories in internal newsletters) Observation of four PRs (field notes) Observations of two conference calls among facilitators (slides from the calls and field notes)	One interview with a manager (internal story in a newsletter) One literature review One future workshop with a task force (field notes) Two interviews with facilitators (stories in the PR blog) Discussion forum for facilitators lasting for three months (slides and field notes) One discussion between facilitators in a web conference (slides and field notes)	One focus-group interview with a task force in a web conference (slides and field notes) One future workshop (slides and field notes for later presentation) Discussion forum lasting for four months Shared folders on the Intranet
Reporting on DBR results	Internal newsletters Web seminars Presentation on management meetings	Approval from top management Presentation at web conferences for knowledge managers Internal newsletters PR web page	Internal newsletter PR blog PR web page Presentation at the PR community web conference	Internal newsletters Presentations at web conferences PR web page PR blog Presented at PR web conference

ings," includes an analysis of the intervention. This study focuses on weaknesses in the dependent and independent variables when utilized in the context of work. The section also will examine reporting of DBR iterations internally at Cleo. The primary focus of the analysis is on DBR, not on the actual educational design. The people used as participants in this study were employees of Cleo.

The initial iteration started in 2005 with a format called *After Action Review*, which was developed to learn from combat situations among U.S. soldiers in the Vietnam War (U.S. Army, 1993). The first iteration incorporated observations of the After Action Review and feedback from participants in After Action Reviews (Kolbaek, 2014), using two diaries and interviews with two managers and two participants that were published in internal newsletters. The data uncovered the need for trained facilitators, which will be described in the second iteration.

The objective for the second iteration of the educational design was the need for trained facilitators who could conduct PRs. The data collection included interviews with three facilitators that were developed into three stories in internal newsletters, observations of four PRs recorded in field notes, and observations of two conference calls for facilitators, including slides from the calls and field notes. This training was an educational design developed by the Manager of Organizational Learning, who also taught the training course and is a PR-development researcher, as well as the author of this chapter. Obviously, she had to consider her role(s) and ethical issues. During the development of the training course, she collaborated with peers to distance her management role from her role as a developer and teacher. She ensured that the pilot of the training course also was run by peers, primarily with two teachers, so that one teacher could observe and take notes on what happened. The researcher was not a neutral scientist, but an engaged manager in a specific organization who was aware of her bias. Her ethical considerations included protection of confidentiality, so that no one was exposed in the organization in a way that he or she did not approve of. The ethical considerations led to the decision to interview the facilitators through emails, so that she could not influence their responses by facial expressions or body language (Bowden & Galindo-Gonzalez, 2015). The feedback from the peer trainers was gathered in a web conference, in which they met to exchange experiences to refine the educational design of the training course. The interviews were changed into "stories," and the interviewees consented to the publication of these stories in internal newsletters and on the new internal PR web page. The reporting on this new intervention addressed different stakeholders: top management who had provided resources for the training course, knowledge managers throughout EMEA, and participants and potential participants in the facilitator-training program. The trained facilitators became members of a community of practice called the "PR community," in which they exchanged experiences to develop their skills continuously, and to be visible and available for conducting PRs in their home countries when required. Also, the EMEA knowledge managers participated in regular web conferences in which the training program was presented. The top managers each received an email that included a thank-you note for the opportunity to develop and run the training course, as well as a link to the new PR web page, where they could read the stories based on the interviews with participants.

The third iteration was initiated by top management who realized the potential of PR and wanted to gain insights into "what happens at the grassroots," i.e., the issues identified by the participants at any hierarchical level. The top management wanted to know when employees at any level encountered issues they could not solve with their manager because the problem then needed to be pushed upward in the organization to be solved. Thus, a critical element in the educational design was the lack of information directly from PR participants to top management. This third iteration required theoretical considerations, including learning from peers and from bottom-up or top-down in the organization. The researcher reflected on her role: In this intervention, she would be a superior over the participants in a pilot test. She could choose to be a peer in the task force she intended to establish, and she would be a subordinate to the top management regarding the final educational design. Her ethical considerations concerned the risk of the "grassroots," as no one would risk getting in "bad standing" with his or her superior by "sneaking" around behind the manager's back and going to top management. The stakeholder analysis made her involve top management and the PR facilitator community. As she knew this was a new process, she identified participants for a task force that voluntarily contributed to this intervention. She interviewed a top manager over the phone to fully understand the requirements, and she invited very experienced facilitators of a high organizational status to participate in the task force that created the intervention. First, the task force studied theories from Engeström (2001), Nonaka and Takeuchi (1995), and Wenger (1998)—texts chosen by the Manager of Organizational Learning. Next, they met for a PR conducted by a very experienced facilitator. The PR enabled the task force to discuss the past and to sketch the preferred future. Over the following weeks, they worked on the inputs, discussed various suggestions, and recognized the deficiencies in the available tools in an internal discussion forum. The proposal was discussed with the PR community in a web conference, and the inputs from the facilitators were taken into consideration by the task force. The final proposal was presented to a top manager, who approved the design and the budget for the development of a PR application. The result seemed simple—just the addition of a question to the original PR format, followed up by the development of a new application to capture the inputs from the "grassroots" and the development of a process to deliver the inputs to top management (Kolbaek, 2014). The additional question was: "What management challenge may be addressed from this Proactive Review?"

The new version of the PR had to be aggressively communicated, as hundreds of facilitators in various countries needed to apply the new format and update the PR application. The new PR format was presented in the PR community web conference, on the PR web page, and as a story on the PR blog. Additionally, managers at all levels needed to be aware of this new process, in which issues could be addressed directly to top management without managerial interference. This was explained in various internal newsletters.

The fourth iteration was initiated as Cleo changed its strategy from having local organizational entities in multiple countries to becoming a global entity, with 60 companies having merged into it. These mergers and acquisitions included a vast number of new employees. Therefore, it became increasingly important to collaborate online and learn from experience. The critical element in the educational design was the lack of online conversations and facilitation. The objective of this iteration was the creation of an online educational PR design that would enable geographically dispersed colleagues to attend the PR without travel costs and the time involved in travel. Technology available in 2009 did not support videoconferencing, so bodily cues were not available. When people cannot see each other, the conversation format changes, and it may be more difficult to build the trust needed for sharing good or bad experiences. Consequently, the theoretical considerations included trust building online (Hinds & Mortensen, 2005). The stakeholders in this intervention were, literally, all the employees. Very important stakeholders would be the facilitators, who needed new skills for conducting the PRs online, and specialists in using new social tools. The researcher reflected on her role. As the Manager of Organizational Learning, she would be responsible for the results. She would be heavily involved in the development of the new format as the main "teacher," and she would be the researcher who gathers the data and analyzes the process and results. Her ethical considerations included concern for the facilitators, who wanted to appear professional, and the risks they might feel they were taking when they had to facilitate online—frequently using tools that were new to them. Again, she established a small task force of very experienced facilitators who met in a web conference, in which she presented the requirements of the intervention. The methods included a focus-group interview to create a common understanding of the task and a future workshop in which the task force agreed on available tools for online collaboration, with the intervention divided into "before," "during," and "after" segments of the online PR. They created a discussion forum in which they could discuss their proposals independent of time and space, and placed all materials in a shared folder. The result was an online training course that included videos and multiple-choice questionnaires that the trained facilitators could access when needed. The researcher and the task-force members discussed the conditions for success and agreed that more than 80% of the attendees in the online training course should finish the course with 85% correct answers on the multiple-choice questionnaires.

As an integrated part of the development of this intervention, the task force developed a communication plan to ensure that all employees were aware that the PRs could be run online and that the facilitators could attend additional training for online facilitation. The online PRs were published as stories in internal newsletters and presented at management meetings and in various communities of practice. The online training course was placed on the PR web page and presented at the PR community web conference several months in a row, and stories about the development of the training course, as well as interviews with facilitators who had finalized the course, were published on the PR blog. Many facilitators were involved in the development of this intervention, and the researcher/teacher's role primarily involved project management. The fine result indicates an excellent collaboration between the most important stakeholders (the facilitators and the specialists in the task force) and the researcher.

ANALYSIS AND FINDINGS

The analysis looks at climate variables, outcome variables, and resources available in the development of educational design, as suggested by Collins et al. (2004) and Collins (2010). The analysis also focuses on weaknesses in the original DBR guidelines and suggests others to use in the context of work.

DBR Weaknesses Identified in the First Iteration

The first iteration uncovered the following DBR weaknesses when utilized in the context of work: methods and data collection, variables, interpersonal level/group level, role of the researcher, and ethical considerations.

Collins et al. (2004) suggest that the researcher looks for available resources that are easy to understand and use, but they do not point out the implications of any methods. As methods may influence those who are researched, as well as the research environment, and as these methods may influence what the researcher "sees," it seems important to carefully consider applicable methods. DBR allows for a messy setting that enables research in authentic learning situations. At the same time, the messy setting can make it difficult to control the interventions and what happens during the learning situation (Dede, 2004). While researchers try to examine various challenges in the learning situation, they easily end up with overwhelming amounts of data with no guarantee of effectiveness (Collins, 2010; Dede, 2004). DBR provides variables for the context, including the physical surroundings and sociological aspects, but lacks historical background (Collins, 2010; Collins et al., 2004). DBR includes analysis of collaborations between peers, but lacks analyses of conversations as the means of learning. To obtain sufficient data, *considerations on methods and data collection* should be added to the DBR guidelines at work.

DBR provides *variables* that the teacher influences and variables that the teacher does not influence; this doesn't seem to fit into the context of work, as there is no teacher. Furthermore, the extent to which the researcher may or may not influence the variables depends on the specific organization. Therefore, there should be no distinction between dependent and independent variables in DBR at work.

The many variables were taken into consideration in the pilot. The *climate variables* include engagement, cooperation, risk-taking, and the effort to understand others' points of view, which could be used as an inspiration for researchers and practitioners when they collaboratively develop an intervention.

Table 3. Themes of analysis based on the original DBR guidelines

	First Iteration 2005	Second Iteration 2005	Third Iteration 2008	Fourth Iteration 2009
Analysis: Dependent and independent variables and resources	Climate variables Interaction between researcher and the people under study Resources	Literature review Stakeholder analysis	Outcome variables Resources Reporting the results Collaboration between researcher, teacher, and learners	Outcome variables Climate variables Resources

Outcome variables show how learners become more knowledgeable about the content. In the context of work, the content to be learned is related to the vision and objectives of the organization, e.g., the improvement of products, services, and work practices (Brandi & Elkjaer, 2011). *Resources* include accessibility to learning materials, professional development, and financial requirements, which must be specified when DBR is applied for learning in the context of work.

According to Dede (2004) and Engeström (2011), the variables appear to be too numerous. In terms of costs, DBR at work should focus on *a limited number of variables* for the iteration.

Moreover, DBR calls for analyses of the learning environment, at the individual, group, and institutional levels (Collins, 2010). At the individual level, the researcher is supposed to understand what the participant learned (Collins et al., 2004), but in the context of work, learners are often unaware that they are learning, so learning objectives are likely missing. However, observations and interviews could be used to explore the development of individual learning; both methods were utilized in the first iteration of the PR.

Collins (2010) recommends analysis of interactions between the teacher and students, which he calls the "interpersonal level"; he also calls for analysis of interactions between learners, which he calls the "group level." When DBR is employed for learning in the context of work, the teacher is lacking. Instead, there may be a facilitator present to conduct the learning situation (Engeström, 2011; Kolbaek, 2014). To simplify the analysis, "group level" can be an umbrella term for "interpersonal level" and "group level."

In the first iteration, the institutional level was not included in the inquiry. However, the second iteration includes the institutional level.

There is not a strong distinction between the researcher and those to be studied, as DBR allows for interaction between the researcher and those under study (Collins, 2010). The upside of the double role as manager and "scientist" was the easy access to those sought for study and the advantage of their expertise (Barab & Squire, 2004). The downside was the challenge of staying neutral and refraining from investigating the motivations and intentions of those under study (Brinkmann & Tanggaard, 2010). There was an organizational power distance between the researcher and her colleagues, who may easily have provided less feedback on occurrences or experiences that they found problematic or negative and, by doing so, may have delivered less-useful information (Kvale, 2002, p. 170). The experience of work is personal and political, and it is sometimes dangerous to speak out. Therefore, the researcher must think carefully about how to encourage employees to share their experiences while bearing in mind the inherent difficulties. Consequently, the *researcher needs to be aware of her role* and the possible consequences that the research may pose for those under study.

This requirement includes ethical considerations (Andriessen, 2007), but ethical considerations regarding DBR are not thoroughly described by Brown (1992), the Design-Based Research Collective (2003), Collins, et al. (2004), Wang and Hannafin (2005), Collins (2010), or Anderson and Shattuck (2012). The researcher must consider informants' desire to appear professional at work and the risks they may fear from sharing their emotions, thoughts, and considerations with colleagues. In competitive environments, it may be very tempting to be visible in newsletters or other communication channels. When the researcher offers such an opportunity to informants, she may jeopardize their honesty and spontaneity. Consequently, *ethical considerations* should be added to the guidelines for DBR at work.

DBR Weaknesses Identified in the Second Iteration

The suggestion for facilitator training was presented to the top management as a business case to get financial support. The second iteration of the educational design for learning in the context of work uncovered the following three DBR weaknesses: non-numbered iterations, lack of theoretical considerations, and stakeholder analyses.

New iterations are prompted by a problem or an issue discovered while running the educational design, but the original guidelines do not track the iterations. Thus, the researcher cannot return and look for changes and results in the iteration and compare them with later iterations. This historical aspect is missing. Consequently, it would be beneficial to provide the iteration of the intervention with a specific number and call it "Modifying the intervention – iteration version X."

The second PR iteration included knowledge about "facilitation," "training," and "teaching"; hence, theories on these topics were crucial. In general, DBR lacks a strong theoretical foundation, which may mean the results reported are simply common sense instead of a refinement of the theories (Anderson & Shattuck, 2012; Dede, 2004; Engeström, 2011). Furthermore, theories behind an intervention are important for identifying critical elements (Collins, 2010), and Anderson and Shattuck (2012) recommend conducting a literature review before the creation of a new intervention. To ensure a proper theoretical foundation for the iteration, *theoretical considerations* should be added to the guidelines for DBR at work.

The lack of trained facilitators needed attention. In such cases, the top managers should provide resources for the development of the training, the managers should allow employees to attend the training, and the individual employee should learn facilitation skills. The trained facilitators became members of the PR community where they shared experiences and learned new interventions. Hence, the analysis of the PR learning environment includes the individual, the group, and the organizational levels, which are comparable to the three ontological dimensions of learning: individual, team, and organizational learning (Nonaka & Takeuchi, 1995). According to Collins (2010), the organizational level includes stakeholders from inside and outside the organization, where the learning is studied. This is called *stakeholder analysis* in the guidelines for DBR at work.

DBR Weaknesses Identified in the Third Iteration

The objective of the third intervention was to create a process that allowed participants to identify, formulate, and address issues that could not be solved by the manager and the team. Also, it was essential that the process avoid to hurt participants, their manager, or their cooperation.

Top management wanted to know about issues that could not be solved at a given hierarchical spot in the organization; they wanted to know about employee dispositions toward changes in working practices, products, or services. Consequently, this intervention included *outcome variables*. The intervention required minor professional development of the trained facilitators and financial investment in a new PR application; therefore, it included *resources*.

The identification and delivery of management challenges enabled learning from the "grassroots" level to be considered by top management, who initiated organizational changes as a result of their new insights. The new process implied vertical learning bottom-up, or grassroots to top management, as well as top-down, when top management initiated organizational changes that all relevant employees had to apply. In this way, the educational design of the PR strengthened learning at the organizational level.

The third iteration uncovered the following DBR weaknesses when utilized in the context of work: lack of a communication plan, lack of reflections on the collaboration between stakeholders and the researcher, and lack of a definition of conditions for success.

Collins (2010) stresses the importance of reporting the results from the intervention. The report is meant to present all important findings from the intervention and, if possible, provide an improved intervention for learning. However, Collins (2010) does not suggest when or how to report the findings or the improved intervention for learning; therefore, "create a communication plan and report the changes to relevant stakeholders" should be added to the guidelines for DBR at work.

This intervention drew heavily on various stakeholders, including top management, very experienced facilitators, facilitator community, and developers of the new PR application. The collaboration between the researcher/teacher and these stakeholders was a prerequisite for success. DBR stresses the importance of the collaboration between the researcher, teacher, and learners, as they have a huge influence on the implementation of results (Anderson & Shattuck, 2012; Collins, 2010; Collins et al., 2004; Dede, 2004). However, collaborations with other practitioners, including other parts of the organization, are not part of the DBR guidelines provided by Collins et al. (2004). To raise awareness of collaborations between the researcher and stakeholders, reflections on the "cooperation between the stakeholders and the researcher" should be added to the DBR guidelines for workplaces.

This intervention brings up consideration of the conditions for success in an intervention. Is it enough that top management has approved the intervention, given that their expertise does not include learning, or is the success factor achieved when the facilitators and the task force find the new format useful?

Dede (2004) criticizes DBR for being "endless" because of the lack of conditions for success, contending that the research design is differentiated from its conditions for success and that these conditions should not be changed between DBR iterations. These conditions for success may be used as standards for deciding when to cancel iterations of the educational design or when to finish the development of new iterations. Thus, "conditions for success" should be added to the DBR guidelines for workplaces.

DBR Weaknesses Identified in the Fourth Iteration

The fourth intervention is an example of significant engagement and support, and the development of online PR would not have been possible without the facilitators and specialists' willingness to volunteer for this difficult work. It is important to support practitioners' influence on the implementation of an intervention because they will adapt to the intervention according to their needs, abilities, and interests (Barab & Squire, 2004; Collins et al., 2004). Consequently, the researcher needs to identify important stakeholders and allow them to be actively involved in the implementation of the new intervention. This involvement is especially important in the context of work, where there is no teacher to make decisions and conduct the implementation process. Therefore, "implementation in the organization" should be added to the DBR guidelines for workplaces.

In this fourth intervention, the researcher served mainly as the project manager, and the online training course on PRs was developed by very skilled facilitators and specialists. The learning and development did not come from the teacher/researcher—it was a process that occurred among the members of the task force, who maintained the process and delivered the results. In this case, the researcher should remember whether she explained her objectives and intentions to those who collaborated so intensively, and she should reflect on her ability to maintain a positive environment during the long working process and consider how feedback was given.

The researcher should allow herself to reflect on unexpected results and dwell on positive experiences during the DBR at work. Because of this, "self-reflection" should be added to the DBR guidelines for workplaces.

The analysis uncovered recommendations for DBR guidelines at work that may be divided into three phases: A) Planning for DBR; B) Executing the DBR; and C) After the DBR.

DISCUSSION AND RECOMMENDATIONS

The examples of interventions showed that learning from experience in the context of work enables practitioners to exchange ideas and solve problems. New interventions supported the practitioners in building expertise at the individual, team, and organizational levels. Solving problems at work made the practitioners aware of the realities and enabled them to influence realities (Engeström, 2001, 2013). The interventions improved the educational design of the PR that is used for improving products, services, and work practices (Brandi & Elkjaer, 2011; Kolbaek, 2014) to enhance business results, competitive advantages, and revenue (Nonaka & Takeuchi, 1995; Kolbaek, 2014).

DBR at Work: Recommendations for New Guidelines

New guidelines for DBR at work are based on the theory and the analysis of the four iterations described above.

- **Planning for DBR at Work**
  - Context of research
  - Role of the researcher
  - Ethical considerations
  - Modification of the intervention – iteration X
- **Executing the DBR at Work**
  - Problem identification
  - Theoretical considerations
  - Stakeholder analysis
  - Conditions for success
  - Consideration of methods and data collection
  - Mixed methods
  - Analyzing the intervention
  - Identifying a limited number of variables
  - Cooperation between stakeholders and the researcher
- **After the DBR at Work**
  - Creating a communication plan and reporting the changes to relevant recipients
  - Achievements regarding conditions for success
  - Implementation in the organization
  - Self-reflection

Planning for DBR at Work

The analysis showed that before the researcher begins DBR in the context of work, she needs to prepare by familiarizing herself with the organization under study, considering her role, and numbering the iterations of the interventions. The context of the DBR at work can be described by the physical surroundings, i.e., the geography, buildings, and tools, including machines and ICT, and working practices (Engeström 2001, 2011, Gherardi et al. 1998). The historical background uncovers the past that led the organization to where it is today. The social aspects include organizational structure and power, division of labor, critical organizational competencies, rules, channels of communication, and language (Engeström 2001, 2011, Gherardi et al. 1998). Lack of resources could stop the development of an intervention (Andersen & Shattuck, 2012, Collins 2010), so the researcher should explore what is needed to carry out the intervention in terms of costs, management support, materials, and technical support. Lack of skills could be an obstacle to success; thus, professional development is an important resource. Financial requirements cover costs in terms of time, equipment, support, and education needed for the intervention (Andersen & Shattuck, 2012, Collins 2010).

The role of the researcher should not be at a distance from those under study in DBR at work (Wang & Hannafin, 2005). The engagement of the researcher—placed within the context being studied—leads to the best understanding of what's going on (Flyvbjerg, 2005). This involvement does not necessarily cause problems, as long as the researcher is aware of her position and role. The researcher decides the place, the informants, the actions to research, the time to spend in the field, and the way to build relationships with the people to be studied (Brinkmann & Tanggaard 2010; Flyvbjerg 2005). Learning in the context of work is not a neutral arena, but an area of research influenced by power, ambitions, and emotions that may be difficult for the researcher to perceive. Therefore, the researcher must think carefully about how to enable employees to share their experiences, while bearing in mind the difficulties associated with doing this.

Consequently, ethical considerations are needed in DBR at work (Brinkmann &Tanggaard 2010; Pedersen & Klitmøller, 2012). Because the researcher is close to those under study, the researcher should use methods that allow those under study to express themselves with small interference from the researcher, e.g., ask them to write diaries or interview them via email. She must create a positive atmosphere during interviews, ensure confidentiality, consider to what extent the informants should participate in the analysis, and consider the possible consequences to interviewees when results are published. She could minimize bias by being aware of her pre-conceived notions and trying to make them explicit, e.g., in her research log (Kvale, 2002).

DBR at work should distinguish the steps of development by assigning each intervention a number, starting with zero and adding one to each intervention. This way, the historical progression of the interventions will be recorded.

Executing DBR at Work

Any intervention starts with *problem identification* (Engeström 2001, Collins, 2010; Collins et al., 2004; Wang & Hannafin, 2005). The problem needs to be described to provide sustainable solutions. The problem may be perplexity, confusion, doubt or disturbance, contradictions, conflicts, or critical elements in the intervention. The starting point may be initiated by those who experience the problem—which could be stakeholders or the researcher.

The analysis suggests that the researcher should familiarize herself with relevant theories and decide which theoretical approaches to apply. These are called *theoretical considerations*. The researcher should look for relevant theories before she plans the intervention, or iteration X. The theoretical considerations may add new views, enable the development of a new version of the educational design, or improve the way the design operates in practice. This provides a new lens to explore the problem, leading to a working hypothesis or suggestions for solutions.

The researcher can employ a *stakeholder analysis* to identify people or groups that either impact the intervention or are impacted by the intervention (Anderson & Shattuck, 2012; Collins, 2010; Collins et al., 2004; Dede, 2004). The stakeholders may be actively involved in the new intervention, and they may adapt to the intervention according to their needs, abilities, and interests. The collaboration between the researcher and stakeholders ensures there are many points of view in the development process and enables stakeholders to influence future realities in the workplace.

The analysis uncovered that the researcher should define the *conditions for success* (Dede, 2004) of an intervention in cooperation with relevant stakeholders; this ensures a common understanding of the objective for the intervention and that the achievements will be appreciated by important stakeholders. The conditions should be stable throughout an intervention and only change when a new intervention begins (Dede, 2004, Engeström, 2011). The conditions for success may be used as the standard for deciding when to cancel an intervention or when to finish the development of an intervention. Conditions for success should be specific and easy to understand and communicate, and they should be based on data that are implied in the research of the intervention so as not to waste time and energy gathering less-relevant data. The time frame must be explicit, i.e., the date when the condition for success is supposed to be achieved (Dede, 2004, Engeström, 2011).

DBR at work places the research in a messy setting that requires considerations on applicable *methods and data collection* (Anderson & Shattuck, 2012; Collins, 2010; Brinkmann & Tanggaard 2010). The methods should be used to gather data about the variables that the researcher has defined beforehand, and the methods should imply the physical setting, historical background, and sociological aspects of the context (Engeström 2001, Gherardi et al. 1998). It is recommended that the researcher consider methods that distance her from the informants to avoid influencing the people under study.

The theory and analysis revealed that to analyze the intervention, the researcher identifies a *limited number of variables,* these being climate and outcome variables, and resources (Dede, 2004, Engeström, 2011). *Climate variables* include behaviors such as engagement, cooperation, risk-taking, and the effort to understand others' points of view (Collins et al. 2004). *Outcome variables* show how learners become more knowledgeable about content, e.g., improvements in products, services, or work practices (Brandi & Elkjaer, 2011; Nonaka & Takeuchi, 1995). *Resources* include accessibility to learning materials, professional development, and financial requirements (Collins et al. 2004). The analysis should include the three ontological dimensions of learning: individual learning, team learning, and organizational learning. The analysis also should look for the individual's ability to take risks or understand alternative points of view. Considerations about taking risks include the possibility of the intervention collapsing and whether some participants feel offended or scared. The analysis at the group level reveals group dynamics, such as authority or domination and participation or lack of participation, as well as the development of relationships among participants. Learning at the organizational level involves stakeholder engagement and support.

Cooperation between the researcher and stakeholders is a prerequisite for success. It is critical that stakeholders contribute to the intervention, provide feedback, and take part in the analysis. Stakeholders' active involvement is an advantage, as the research benefits from their expertise (Flyvbjerg 2005). It is important to support stakeholders' influence on the implementation of an intervention because they will adapt to the intervention according to their needs, abilities, and interests. Consequently, the researcher should reflect on the cooperation between herself and stakeholders to ensure commitment and engagement.

After DBR at Work

When the research is completed, the results should be reported, and the researcher should reflect on the process and results (Collins 2010).

The analysis uncovered that the researcher should create *a communication plan* to ensure that the changes to the educational design are reported to relevant recipients. DBR reporting may consist of four elements: goals and elements of the intervention, research setting, a description of the iterations, and changes. The report is supposed to present all important changes from the intervention and, if possible, provide an improved educational design.

The analysis suggested that the researcher should reflect on the *achievements* regarding the conditions for success to determine whether she should change direction or stop the intervention. When she looks back, she may consider whether she lived up to the conditions of success and whether the methods and data collection actually delivered the information she needed.

In the context of work, the researcher may not be responsible for the *implementation of the intervention* in the organization. Consequently, the researcher needs to identify important stakeholders and support them in the implementation of the new intervention. Otherwise, the results of the DBR at work are at risk. The implementation itself depends on the given context and stakeholders, which are often out of the researcher's control.

The final activity in DBR at work is the researcher's *self-reflection.* The researcher is encouraged to reflect on unexpected results, dwell on good experiences during the DBR at work, and consider her choices and decisions to learn from her own experiences.

FUTURE RESEARCH DIRECTIONS

This chapter focused on DBR and learning in the context of work. Future research may include DBR in other organizations, as well as online settings, and it should explore how learning from experience as part of online teams may be studied from a DBR-at-work perspective.

CONCLUSION

This chapter aimed to answer the question: What is needed to utilize DBR to explore learning from experience in the context of work?

It appears as if DBR may be a suitable approach for studying learning from experience in the context of work because the methodology is based in a real-world setting and includes three ontological dimensions for learning: the individual, the team, and the organization. DBR at work is based on iterations of interventions in educational designs that may uncover developments in knowledge and knowing or dispositions for the future. Furthermore, DBR includes possibilities for exploring the context defined by physical surroundings, historical background, and sociological aspects.

This study makes a theoretical contribution by exploring learning in the context of work through the lens of DBR. Since the literature contains few references to DBR in the context of work, this study provides new insights into this subject. Furthermore, this study provides an example of a longitudinal study on DBR, which is rare, based on the literature review.

This study makes a methodological contribution by exploring DBR as a methodology for studying a workplace learning environment, as opposed to traditional educational settings.

Finally, this study provides practical contributions, namely guidelines for DBR at work. These guidelines illustrate how to prepare for and complete DBR in the context of work.

REFERENCES

Anderson, T., & Shattuck, J. (2012). Design-based research: A decade of progress in education research? *Educational Researcher, 41*(1), 16–25. doi:10.3102/0013189X11428813

Andriessen, D. (2007). Combining design-based research and action research to test management solutions. Paper presented at the 7th World Congress on Action Learning, Action Research and Process Management, Groningen.

Barab, S., & Squire, K. (2004). Design-based research: Putting a stake in the ground. *Journal of the Learning Sciences, 13*(1), 1–14. doi:10.1207/s15327809jls1301_1

Bowden, C., & Galindo-Gonzalez, S. (2015). Interviewing when you're not face-to-face: The use of email interviews in a phenomenological study. *International Journal of Doctoral Studies, 10*, 79–92. doi:10.28945/2104

Brandi, U., & Elkjaer, B. (2011). Tidsskrift for arbejdsliv. *Organisatorisk læring i organisatorisk forandring, 13*(2), 73.

Brinkmann, S., & Tanggaard, L. (2010). *Kvalitative metoder: En grundbog*. København: Hans Reitzels Forlag.

Brown, A. L. (1992). Design experiments: Theoretical and methodological challenges in creating complex interventions in classroom settings. *Journal of the Learning Sciences, 2*(2), 141–178. doi:10.1207/s15327809jls0202_2

Brown, A. L. (1992). Design experiments: Theoretical and methodological challenges in creating complex interventions in classroom settings. *Journal of the Learning Sciences, 2*(2), 141–178. doi:10.1207/s15327809jls0202_2

Collins, A. (2010). *Design experiments. In International Encyclopedia of Education.* Elsevier Ltd.

Collins, A., Joseph, D., & Bielaczyc, K. (2004). Design Research: Theoretical and methodological issues. *Journal of the Learning Sciences, 13*(1), 15–42. doi:10.1207/s15327809jls1301_2

Dede, C. (2004). If design-based research is the answer, what is the question? A commentary on Collins, Joseph, and Bielaczyc; diSessa and Cobb; and Fishman, Marx, Blumenthal, Krajcik, and Soloway in the JLS Special Issue on design-based research. *Journal of the Learning Sciences, 13*(1), 105–114. doi:10.1207/s15327809jls1301_5

Design-Based Research Collective. (2003). Design-based research: An emerging paradigm for educational inquiry. *Educational Researcher, 32*(1), 5–8. doi:10.3102/0013189X032001005

Dewey, J., & Boydston, J. A. (1976). *The Middle Works: 1899–1924* (Vol. 9). Carbondale, IL: Southern Illinois Univ. Press.

Dohn, N. B., Hansen, J. J., Jensen, J., Johnsen, L., Lauridsen, E., Møller, M. H., & Yepez-Reyes, V. (2016). *Digital educational design - process, product, and practice*. Paper presented at the Designs for Learning, Fifth International Conference, København. Retrieved from http://www.designsforlearning2016.aau.dk/digitalAssets/198/19 8492_dfl_short-papers-final-2.pdf

Easterby-Smith, M., & Lyles, M. A. (2011). *Handbook of Organizational Learning and Knowledge Management* (2nd ed.). Chichester, UK: Wiley.

Easterby-Smith, M., & Prieto, I. M. (2008). Dynamic capabilities and knowledge management: An integrative role for learning? *British Journal of Management, 19*(3), 235–249. doi:10.1111/j.1467-8551.2007.00543.x

Elkjaer, B. (2003). Organizational learning with a pragmatic slant. *International Journal of Lifelong Education, 22*(5), 481–494. doi:10.1080/02601370320000102841

Engeström, Y. (2001). Expansive learning at work: Toward an activity theoretical reconceptualization. *Journal of Education and Work, 14*(1), 133–156. doi:10.1080/13639080020028747

Engeström, Y. (2011). From design experiments to formative interventions. *Theory & Psychology, 21*(5), 598–628. doi:10.1177/0959354311419252

Engeström, Y., Rantavuori, J., & Kerosuo, H. (2013). Expansive learning in a library: Actions, cycles, and deviations from instructional intentions. *Vocations and Learning, 6*(1), 81–106. doi:10.1007/s12186-012-9089-6

Flyvbjerg, B. (2006). Five misunderstandings about case-study research. *Qualitative Inquiry, 12*(2), 219–245. doi:10.1177/1077800405284363

Gherardi, S., Nicolini, D., & Odella, F. (1998). Toward a social understanding of how people learn in organizations: The notion of situated curriculum. *Management Learning, 29*(3), 273–297. doi:10.1177/1350507698293002

Hinds, P. J., & Mortensen, M. (2005). Understanding conflict in geographically distributed teams: The moderating effects of shared identity, shared context, and spontaneous communication. *Organization Science, 16*(3), 290–307. doi:10.1287/orsc.1050.0122

Holsapple, C. W., & Wu, J. (2011). An elusive antecedent of superior firm performance: The knowledge management factor. *Decision Support Systems*, *52*(1), 271–283. doi:10.1016/j.dss.2011.08.003

Kolbaek, D. (2014). *Learning From Experience in the Context of Work*. Aalborg: Aalborg University Press.

Kolbaek, D. (2015). Proactive Review. In S. McIntyre, K. Dalkir, P. Paul, & I. C. Kitimbo (Eds.), *Utilizing Evidence-Based Lessons Learned for Enhanced Organizational Innovation and Change* (pp. 82–100). Hershey, PA: IGI Global. doi:10.4018/978-1-4666-6453-1.ch005

Kolbaek, D. (2017). Design-based research as a "smart" methodology for studying learning in the context of work. *Proceedings of IRIS40, Information Systems Research Seminar in Scandinavia*. Retrieved from https://pdfs.semanticscholar.org/26b5/411610955317c57761f7902dd6ed704b0671.pdf

Kvale, S. (2002). *Interview: En introduction til det kvalitative forskningsinterview*. København: Hans Reitzels Forlag.

Nonaka, I., & Takeuchi, H. (1995). *The Knowledge-Creating Company*. New York: Oxford University Press, Inc.

Nonaka, I., Toyama, R., & Konno, N. (2000). ECI, Ba and leadership: A unified model of dynamic knowledge creation. *Long Range Planning*, *33*(1), 5–34. doi:10.1016/S0024-6301(99)00115-6

Pedersen, M., Klitmøller, J., & Nielsen, K. (2012). *Deltagerobservation*. København: Hans Reitzels Forlag.

Qvortrup, L. (2000). *Det hyperkomplekse samfund: 14 fortællinger om informationssamfundet* (2nd ed.). København: Gyldendal.

Rao, M. (2003). *Leading with knowledge*. Delhi: TATA McGraw-Hill.

US Army, . (1993). *A leader's guide to after-action reviews*. Washington, DC: Department of the Army. Retrieved from http://www.au.af.mil/au/awc/awcgate/army/tc_25-20/tc25-20.pdf

Wang, F., & Hannafin, M. J. (2005). Design-based research and technology-enhanced learning environments. *Educational Technology Research and Development*, *53*(4), 5–23. doi:10.1007/BF02504682

Wenger, E. (1998). *Communities of Practice: Learning, Meaning, and Identity*. New York: Cambridge University Press. doi:10.1017/CBO9780511803932

Winter, R. (2008). Design Science Research in Europe. *European Journal of Information Systems*, *17*(5), 470–475. doi:10.1057/ejis.2008.44

Zheng, L. (2015). A systematic literature review of design-based research from 2004 to 2013. *Journal of Computers in Education*, *2*(4), 399–420. doi:10.1007/s40692-015-0036-z

KEY TERMS AND DEFINITIONS

Design-Based Research: A methodology for researching and improving educational designs in classroom and work settings.

Educational Design: Designs dedicated to learning environments for enhancing learning processes.

Facilitator: The person who conducts proactive reviews. The facilitator is neutral and does not provide answers to the problems to be solved.

Management Challenge: An issue or a problem that cannot be solved at the hierarchical level, where it is identified in the organization, but needs to be pushed upward to higher hierarchical levels to be solved.

Organizational Learning: Learning in three ontological dimensions: individual learning, team learning, and learning throughout the organization. May be initiated top-down or bottom-up.

Proactive Review: An educational design for learning from experience in the context of work.

APPENDIX

Table 4. 13 iterations in the development of Proactive Review

#	Iteration	Initiator	Year
1	From AAR to PR	Top management	2005
2	Training facilitators	Manager org. learning	2005
3	PR community	Manager org. learning	2006
4	Managers' responsibilities	Manager org. learning	2007
5	Management challenges	Top management	2008
6	PR application	Top manager	2008
7	Online PR	Facilitators	2009
8	Online facilitator training	Manager org. learning	2009
9	Online connection between managers and facilitators	Managers	2010
10	PR triggers When to run PRs	External Audit	2010
11	PR recognition - stamp	Facilitators	2011
12	Inter-organizational PR	Manager org. learning	2011
13	PR light for sales	Sales manager	2012

Section 2
Challenges in Online Leadership and Online Collaboration

Distance matters when people work together. The lack of bodily clues along with high diversity are challenges that must be overcome to create a successful online team. This second section explores challenges in online leadership and online collaboration by linking theoretical considerations of online collaboration and online leadership with descriptions and cases that illustrate practitioners' ways of handling the online work environment. The researches apply different methodologies and methods for studying online collaboration and leadership in different settings, and Section 2 provides insights into different approaches to understand and to handle online collaborarion and online leadership including cultural understanding, perceptions of trust, perceptions of contributions to the organization, perceptions of being present, as well as recommendations for overcoming the challenges in online collaboration.

Chapter 3
Is *Urgent Evoke* a Digital Ba?
How a Game Can Make Space for Knowledge Creation

Mette Wichmand
Roskilde University, Denmark

ABSTRACT

Web 2.0 brought a participatory potential with it, prompting organisations to ask themselves how to design social media platforms that can engage external stakeholders in the organisation's process of knowledge creation and innovation. This chapter presents an analysis of the data from a case study of such a platform, the World Bank's online game Urgent Evoke, which has been designed with the purpose of engaging citizens in developing innovative solutions for sociopolitical problems like poverty. The analysis is based on Nonaka's concept of Ba, which means "place" and is described as a platform for advancing the creation of knowledge. The analysis suggests that, in order to create a digital Ba, the design should not only facilitate the four characteristics of Ba that Nonaka has described –socialization, externalization, combination, and internalization. The design should also allow multiple forms of participation, as well as recognise external stakeholders' contributions to the process of knowledge creation and innovation.

INTRODUCTION

When Web 2.0 was born, the social possibilities online were strengthened. This development enabled the creation of a new media landscape, which, since then, has been dominated by social media platforms, such as Facebook, YouTube, and Wikipedia.

In this landscape, a new communication culture, characterized by open access and co-creation, emerged, and a participatory optimism was ignited. An optimism that, among other things, was based on the idea that social media would enable organisations not just to communicate to or with their external stakeholders, but to turn their clients, customers, and audiences into an ever-ready crowd of cocreators and prod-users (Bruns, 2008). A crowd expected to be equipped with permanent Internet access via their smart phones, tablets, and laptops eager to contribute to the shared pool of the organization's collective

DOI: 10.4018/978-1-5225-4094-6.ch003

intelligence with the ultimate goal of creating better services and products (Benkler 2006; Jenkins 1992; Jenkins, Ford, & Green, 2013; Leadbeater 2008; Lévy & Bonomo 1999; McGonigal 2011; Shirkey 2008).

However, over the years, the participatory optimism has become more tempered, as it has become clear that having access to social media does not guarantee that Internet users at large are willing, feeling qualified or having the time to engage in a process of collective knowledge creation online (Fuchs 2015; McGonigal 2008; Preece, Nonnecke, & Andrews 2004; van Dijck 2013).

Already in 1996, in the early Internet days, Kollock and Smith (1996) pointed out that, even though a growing number of people found their way into the cyberspace, most Internet users were just looking at the content, which was created by others online. Kollock and Smith found the behaviour of this group of users problematic and named them 'lurkers'.

Ten years later, the 'lurkers' were still prevailing, as the bloggers Ben McConnell and Jakie Huba underlined *by coining the popular term 'the 1% rule'*[1]. The term referred to the fact that in most social media platforms the original content was produced by 1% of the users. Another 9% spent their time curating the content others produced, whereas the remaining 90% were still keeping to the role of spectators.

In 2012, research made by BBC Future Media showed that the 1% rule had been broken, at least in the UK, and only 23% of the British population online could now be defined as pure spectators. Still, it was just 17% of the online population who could be characterized as 'intense users' producing original content for the world to see. Thus, even though the majority of the online population had started to use the creative opportunities of social media, they were only sharing their creations with a closed circle of existing friends and family (Goodier, 2012). This short historical view on the use of social media illustrates that making people engage online as creators in a process of collective knowledge creation is maybe not as effortlessly done as the discourse surrounding social media implies. As a result, new ways of designing social media platforms need to be developed, analysed, and discussed if organisations are to fully enjoy the participatory potential of Web 2.0.

In this chapter, the Wold Bank's social network game Urgent Evoke will be presented as an example of a new approach to social media design that can facilitate collective knowledge creation. The game was developed and played for the first time in 2010 and the aim of the game was

to empower players to: investigate the most pressing challenges around the world, collaborate to generate innovative and creative solutions, and act to turn ideas into reality within their own communities and beyond (Gaible & Dabla 2010, p. 9).

Urgent Evoke is a social network game intended to be played by a massive number of players together simultaneously, and it therefore belongs under the umbrella term 'social media'. The game makes an interesting case when one is discussing online collaboration and communication in contemporary organizations, because games like Urgent Evoke – designed by a large international organization to facilitate player-driven knowledge creation and positive social change - are relatively new and few. Furthermore, even though knowledge creation in virtual spaces has received extensive attention in the academic literature (von Krogh & Geilinger, 2014) very little research has been done on the type of social space that games form and their effect (Connolly, Boyle, MacArthur, Hainey, & Boyle, 2012). Therefore, this chapter will present an analysis of the game design and the players' experience of playing the game.

The analysis is based on the Japanese organizational researcher Ikujiro Nonaka's concept of Ba and his heuristic model of knowledge creation, called the 'SECI model' (Nonaka, Hirose & Takeda 2016; Nonaka & Konno, 1998; Nonaka & Takeuchi, 1995; Nonaka, Toyama, & Konno, 2000). Nonaka is chosen because he was one of the first organizational researchers to introduce space as a relevant category when discussing organisational communication and knowledge creation. During the years many researcher have been working with his concept and model (von Krogh & Geilinger, 2014), but the original concept and model are still intact and relevant.

'Ba' means space, and the concept describes a physical, digital or even a mental space which is intended to enable social learning and collective knowledge creation in organisations. Nonaka's ideas are part of the literature on organisational knowledge management and could be defined as a learning and innovation theory. In this chapter, Nonaka's ideas are used as a framework for a design analysis. In particular, Nonaka's understanding of how to facilitate organisational learning and knowledge creation has been chosen as the analytical framework because it enables a discussion of how a game can function as a Ba. With Nonaka, it is possible to define *Urgent Evoke* as an organisational learning space where the World Bank can collaborate with its external stakeholders on creating innovative solutions to the sociopolitical problems that the World Bank aims to tackle.

In the following, the first section introduces the methodology and the methods, which have been applied, the second section presents Nonaka's understanding of organisational learning, and the third section describes the *Urgent Evoke* game play. Hereafter, the *Discussion* section provides an analysis of the game design and the players' experience of how the game play enabled them to create knowledge in the form of social innovations.

The chapter aims at analysing if and how Urgent Evoke, as a state-of-the-art example of a game design, functions as a Ba, that is an organisational space for knowledge creation with external stakeholders. Furthermore, the chapter aims at discussing the strengths and weaknesses of Nonaka's theory as a framework for a design analysis of social network games.

METHODOLOGY AND METHODS

The data, which are presented in this chapter stem from an extensive single case study of *Urgent Evoke*, based on a mixed method approach. The case study has been chosen as methodology because games like *Urgent Evoke* – which was designed to facilitate player-driven knowledge creation and social innovation - are, as mentioned above, relatively new and few (Connolly et al., 2012). When a new real-life phenomena such as *Urgent Evoke* emerges, Bent Flyvbjerg (2006) says it is "worth repeating the insight of Thomas Kuhn that a discipline without a large number of thoroughly executed case studies is a discipline without systematic production of exemplars, and that a discipline without exemplars is an ineffective one" (p. 242).

The aim of the case study is to produce a rich, context-dependent description of the game that provides a first-hand understanding of how the game design works as a social space for knowledge creation in an organisational context. Furthermore, the case can help depict the complexity of the relation between the game design and game players' process of knowledge creation (Yin, 2009).

The mixed method approach has been chosen in order to enable an analysis of the relation between the organizational intentions behind the design, the affordances of the game design, as well as the users' experiences of playing the game.

The ontological starting point for this research is that it is impossible to capture *one* 'true' social reality because the social world only shows itself through representations or interpretations, and because no one experiences or interprets things in the same way. It is thus necessary to operate with a pluralistic concept of reality (Denzin, 2012; Greene, 2012; Pearce, 2012). The epistemological consequence of this pluralism is that the knowledge created about the social world should be seen as an interpretation, i.e., a construction, or one story among several possible. In such a constructivist ontological and epistemological framework, the aim of the researcher is not to produce 'truth', but to bring "breadth complexity, richness, and depth to any inquiry" (Denzin, 2012, p. 82). Such an imperative calls for a methodology that makes room for multiple voices, differences, and dialogue. The methodology does not try to reduce complexity by ruling out certain types of methods or data *a priori*, but tries to move between paradigms and mix different approaches, in order to produce new perspectives and reflections on the social world. The mix of methods, which were chosen for the case study consists of qualitative and quantitative methods. Together, they represent different ways of looking at the same phenomenon and different voices (Denzin, 2012). The methods are:

- Non-participatory observations on the site of *Urgent Evoke* after the game ended. The observations were performed on the game site containing written and visual materials from *Urgent Evoke* 2010, such as chat, sound files, picture files, 10 chapters of a graphic novel, and data from blog posts and social networks.
- A document analysis of an evaluation report on *Urgent Evoke*, written by the Natoma Group for the World Bank.
- Open and semi-structured qualitative interviews, made via e-mail and Skype with 12 winners of *Urgent Evoke*, as well as one face-to-face interview made with Robert Hawkins, senior advisor at the World Bank and executive producer of *Urgent Evoke*. The interviews were constructed using Brenda Dervin's (2008) sense-making methodology in the design and carrying out.
- Last but not least, a social network analysis of the players' in-game social network, made using NodeXL software. The data from the social network analysis are too extensive to be included in this chapter and will therefore not be presented.

The creation of data happened post-game during a period of 24 months, starting in 2012. The observations and the document analysis were performed first. This was chosen to build up a first-hand understanding of the game design and the history of the game, before entering into the interviews. At the time of the observations, the game was not being played and was mostly functioning as an archival site.

The interviews were made with only the winners of the game because they had played the game to the end for sure.

In the analysis, the informants are given pseudonyms when they are quoted. This is done not so much to cover their real identity, but to underline that the analysis should be seen as the author's interpretation and not necessarily as the informants' opinion.

BA: A FRAMEWORK TO UNDERSTAND KNOWLEDGE CREATION

In the following, Ikujiro Nonaka's concept of Ba and the SECI model will be introduced and the argument for why the concept is interesting in relation to an analysis of a social network game like *Urgent Evoke* will be unfolded.

The Concept of Ba

Ba means place, a place for the creation of new knowledge. A Ba makes it possible for humans to meet, interact, and form relations with the purpose of sharing their knowledge, experiences, values, and ideas, and ultimately create new knowledge together. Ba is thus a spatial concept concerned with how to establish a context for social learning and higher-order thinking (Nonaka, Hirose & Takeda 2016; Nonaka & Konno, 1998; Nonaka & Takeuchi, 1995; Nonaka, Toyama, & Konno, 2000).

The concept of Ba is originally Japanese and has a longer history. In the 1990's, it was revitalised by the organizational researcher Ikujiro Nonaka, who saw in the concept an opportunity to discuss and provide a model for how organisations could focus not just on sharing information and existing knowledge, but facilitate the creation of new knowledge (Nonaka, Hirose & Takeda 2016; Nonaka & Konno, 1998; Nonaka, Toyama, & Konno, 2000). Nonaka's definition of the concept is applied in this chapter.

In Nonaka's perspective, Ba can be a physical, virtual as well as mental space, and most often it is a combination. Ba should be thought of as a resourceful place where the participants are challenged as well as given the freedom to engage with a variety of experiences, perspectives, and ideas present in the space itself and brought in by the users of the space. It is in the process of engaging with the resources, which are present and of starting to combine them in new ways that the opportunity for knowledge creation emerges.

The concept of Ba includes a dialectical relation between the users of the space and the space itself, as the space only becomes a Ba when the users acknowledge and share the collective and creative intention of the space. The space is, thus, a context that '*harbours meaning*' (Nonaka & Konno, 1998, p.40), a meaning that must be perceived and recognized in the use of the space.

Knowledge Creation

As knowledge creation is central to the concept of Ba, Nonaka also provides a model for how knowledge is created. This model is called the SECI model, an acronym formed from the initials of the four words: socialization, externalization, combination, and internalization.

The SECI model is founded on the understanding that two types of knowledge exist: explicit and tacit. *Explicit knowledge* is transmittable. It is the knowledge that people are aware that they have, can talk about, and communicate to others in words, numbers, drawings, and similar. *Tacit knowledge* is personal, embodied, and difficult to communicate and share because it has become an integral part of how people operate and think, thus they do not notice it any more. Tacit knowledge consists of, for example, subjective insights, intuitions, and hunches (Nonaka & Konno, 1998; Nonaka et al., 2000).

In the SECI model, the creation of knowledge is seen as the result of a continuous spiral process where the participants move to a higher level by going through the four phases of socialization, externalization, combination, and internalization. While going through the four phases, participants make their internalized tacit knowledge explicit and they internalize the newly gained and created knowledge.

The four phases in the SECI model could be described in the following way:

1. **Socialization:** This phase is about meeting, interacting with, and being close to other people 'on the job'. Socialization is thus connected to working or being active together, and it is in the process of 'doing' that the participants connect and get to know each other. The socialization process is important in order to enable the externalization and sharing of personal knowledge.

2. **Externalization:** It is about expressing an individual's personal tacit knowledge as well as experiencing that of others. The focus is on the articulation of ideas, experiences, know-how, and values. Dialogue is central in the externalization process. Moreover, the dialogical perspective underlines that in the Ba it is as important to listen to and acknowledge the contributions of others as it is to raise one's own voice.
3. **Combination:** It is about capturing, editing, sampling, and processing the knowledge that is present in the context and shared in the process of socialization and externalization. It is in the phase of combination that new knowledge is created.
4. **Internalization:** This phase is about converting the new knowledge into embodied tacit knowledge. It is in this process, which is associated with learning-by-doing, training, exercises, etc. that the new knowledge is accepted as relevant for the participants and influences - if not radically changes - their ideas and actions, but also their intuitions and values.

Nonaka visualizes the SECI model in the following way (Figure 1).

The spiral, which is depicted in the centre of the model indicates that the participants can keep going through the four-step process of socialization, externalization, combination, and internalization, as the creation of knowledge is an endless process. Actually, the four steps in the process happen in a more meshed and overlapping way than the model indicates.

What Makes the Concept of Ba Interesting?

Because Nonaka's Ba can be either a digital or a physical space, his concept and ideas become relevant in a time where Web 2.0 and social media enhance organisations' possibilities to become more fluid and networked entities that are not limited by the walls of the office buildings they occupy, but operating in the physical as well as the virtual world.

Figure 1. Reproduction of Nonaka's visualization of the SECI model (Nonaka, Toyama & Byosière 2001)

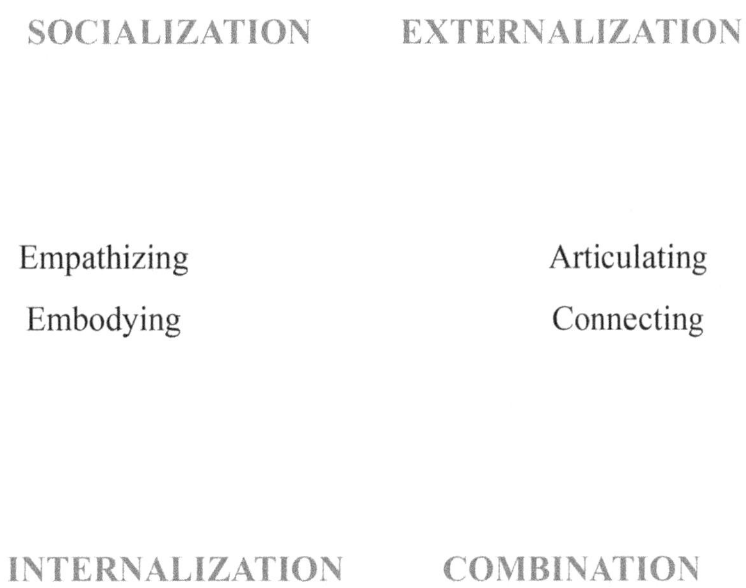

Together, Nonaka's concept of Ba and his SECI model provide a framework to analyse and understand social media platforms, not just as communication and entertainment tools, but as digital spaces for organisational knowledge creation.

Drawing on Nonaka also helps anchoring the meaning of the ambiguous notion of the 'participatory potential', which is tied to social media. Most often, when the participatory potential of social media is discussed, it is not clear in what people should participate–it could be anything from discussing the secrets of tomato growing on a horticultural digital forum to setting-up an online petition to close the Guantanamo Bay detention camp. With Nonaka, the 'participatory potential' of social media must be understood as the ability of the medium to facilitate users' participation in the creation of new knowledge in an organisational setting.

URGENT EVOKE

This section presents *Urgent Evoke*. First, it provides a description of the game design and central functionalities. Subsequently, it offers an analysis of the game design and the players' relevant experience using Nonaka's SECI model. Moreover, the section establishes an understanding of how the design enabled the creation of a digital Ba.

The Game Play

Urgent Evoke is a social network game, which means that the game is designed to form a network of multiple players who will play together online simultaneously. The game was played for the first time in 2010, during a period of ten weeks. As the introduction described, the World Bank's goal with the game is to facilitate collaborative player-driven social innovation.

The game is meant to attract ordinary citizens from all over the world with an interest in development issues. The game is free, but access to the Internet via a computer, tablet, or mobile phone[2] is needed.

Urgent Evoke is based on a 2D interface–it is a simple, but visually strong design. The main opening page leads to a series of subpages and a mix of digital artefacts (e.g., links to external sites, blogs, jpegs, and Mp3 recordings) incorporated into the site. The design of the interface is to a large extent based on images and sound, bringing the use of text back to a minimum. For example, the instructions about how to play the game are delivered as five video files.

The game is constructed around a graphic novel consisting of ten chapters. The narrative tells the story of a secret international network of agents with innovative superpowers who are capable of solving some of the most complex problems facing the world. The graphic novel follows the work of a team of three agents– Eureka, Ember, and Quinn–who travel to all corners of the world to solve problems like famine, epidemics, and natural catastrophes. Each member of the group has a defining background story and a set of characteristics, skills, and knowledge which set them apart as individuals, but makes them function as a resourceful team of social innovators. The players of *Urgent Evoke* are playing the role of an agent in the secret network.

The graphic novel drives the game forward. Every week a new chapter is released and with it a new mission and quest that the players are asked to tackle. The missions are connected to the socio-political problems, which the graphic novel treats. If the agents in the novel are fighting a famine crisis, so are the players. The quests function as personal development questions. For example, players are asked to describe what motivates them to play the game and to outline the competences of a social innovator they admire. The questions should help the player get a sense of his/her strengths as well as development goals in regard to being or becoming a social innovator.

In order to complete a mission, the players have to go through three steps: Learn, imagine, and act. 'Learn' is about doing research and finding out more about the problem, which is treated in the mission. 'Imagine' is about formulating possible solutions to the problem. 'Act' is about taking on actions, in the physical world, that can have a small positive impact on the problems, which are related to the mission.

Urgent Evoke has only a few rules, but, in order to win the game, the players have to complete the missions and quests and create an 'Evokation'. An Evokation is a detailed plan for how the player will tackle a sociopolitical challenge of his/her choice after the game is over.

The winners receive one or a mix of the following awards: Seed money for further development of their Evokation; the opportunity to participate in a crowd-funding campaign on Kickstarter; the opportunity to participate in a World Bank Summit; or an experienced social innovator as mentor.

In 2010 the game received 258,219 visits by 171,958 individuals. 19,386 of the visitors registered as players and 6,618 of them completed at least one mission or quest. 142 players completed all ten missions and quests and 73 people submitted the final Evokation. The World Bank awarded 32 out of the 73 Evokations (Gaible & Dabla, 2010, p. 5). The players represented 150 countries, with the majority living in Western Europe or North America (Gaible & Dabla, 2010, p.16).

Urgent Evoke: A Digital Ba?

The formulation of the goal of *Urgent Evoke*–which was described in the introduction-makes it relevant to look at the game as an attempt to design a digital Ba. The formulation underlines that the game is not primarily designed to entertain, but to engage ordinary citizens in developing new knowledge in the form of innovative solutions to the sociopolitical problems, which the World Banks aims to eliminate.

In the following, an analysis of the game is made using Nonaka's SECI model. The analysis will focus on how the design facilitates the players' knowledge creation by allowing a process of socialization, externalization, combination, and internalization. Also, the analysis is constructed around the four phases, which will be treated individually, even though they are intertwined and overlapping. The analysis of every phase will include both the game design and the players' experience of the game.

Socialization

Nonaka describes how a Ba should support the socialization among the users of the space. Socialization is important because it is through the interaction and getting to know each other that the following phase of externalization and sharing of knowledge can take place. In the following, five of the most essential ways in which the design of *Urgent Evoke* facilitates the players' interaction and socializing will be described.

- **One:** In order to play the game the players must create a personal page. The personal page functions in many ways, like a personal page on Facebook. Here the players can create an online representation of themselves. For example, they can place a photo of themselves, make a short description of who they are, and collect and list their in-game friends. During the game, on their personal page players will also store and exhibit their answers to the missions and quests of the game. Players can visit each other's pages and leave comments and feedback. Therefore, the page functions as a tool for interaction between players.

- **Two:** Players are invited to discuss the missions and quests with each other in the chat room of the game. The chat room is a collective space that enables the players to get to know more about each other's experiences, existing knowledge, and values. The chat room enables the players to listen to each other and to ask questions.
- **Three:** The missions of the game are based on what would be defined as wicked problems. A wicked problem is a complex problem that does not have one right solution, but is asking for a multi-facetted approach. An example of a wicked problem could be poverty. Poverty can be the result of social, cultural, political, geographical, and historical challenges or a mix of these. Working with wicked problems in the game means that none of the players is likely to have 'the right answer'. Instead, the players will benefit from entering into a dialogue and collaboration, and thereby help each other understand, analyse, and develop answers to the problems, which are presented in the game.
- **Four:** *Urgent Evoke* knows no levels, and players can therefore follow and participate in the game without completing any missions or quests, for example. This means that players who are only interested in looking at the work of others, providing some feedback or sharing some ideas or experiences can still participate in the game right to the end.
- **Five:** Players are encouraged to be social and interact. This is communicated through the graphic novel that shows the players that social innovation is a team effort that demands team players with different competences. Prosocial behaviour is also promoted through the point system of the game. In *Urgent Evoke* it is possible to earn points for ten abilities[3] that are promoted as the core abilities of a successful social innovator. Two of the ten abilities have a social focus: *collaboration* and *knowledge share*. Working together and sharing knowledge are thus stressed in the game design as something for which the players should strive.

The five points above illustrate how the game design both communicates to the players that collaboration is important and enables players to socialize as a group (e.g., in the chat room) and as individuals, via the personal pages.

In the interviews, the winners were asked how they experienced the social interaction in the game and what it meant for their game play.

One of the winners characterized the in-game network in the following way: "We actually developed some of the network with people around the world, who were actually working for the same objective or who had same kind of mission or goal to contribute for overall humanity and international development" (David).

The quote shows that the game design facilitated the creation of an international network among the players. The quote highlights that what brought the players together was that all of them shared an interest in the goal of the game. It can therefore be argued that part of the reason why *Urgent Evoke* can be defined as a Ba is because the World Bank with their game design have been able to create a 'space that harbours meaning', a meaning that players can unite around.

The positive impact of finding a space where a social network of people sharing a goal can emerge is something that is repeated in several of the interviews. For example, another player said:

I think it is very empowering in the sense that it connects you with people that are really thinking in the same terms as you are thinking, at least the issues, the solutions might be completely different, but the awareness and the willingness to change certain subjects. I think that it is very empowering to meet them. (Jonas)

Jonas feels empowered from getting connected to like-minded people. Other players are expressing similar feelings from receiving feedback from co-players in the game:

But the feedback that I got from people was – 'you are really good at that'. [...] That was the biggest thing that I got out of the whole experience is that feedback of things that I was good at (Allan).

I think I actually got encouragement from fellow players who said,' you know we think you should do an Evokation', and I was thinking to myself 'okay' [...] I live in a community where I see very many challenges [...] which one should I choose [...] Eventually I went back, looked all the posts I had been making, and I realized I probably did well on food security and it might be a good place to start [....] But it was basically due to the initial comments and people giving me extra motivation to turn in a Evokation. (William)

In Allan and William's words, getting positive feedback from other players encouraged them. According to William, it even made him go all the way in the game and turn in an Evokation.

Nevertheless, according to Nonaka, socializing in a Ba is more than meeting and greeting or liking, hearting, and commenting each other's work. It is also about collaborating and working together. In relation to this, one of the players says (Nonaka & Konno, 1998):

I was always looking for a community to be a part of. One that focused on globally inspired activism in a positive way. It's hard to find on the Internet. Certainly, you can find other activists, but they generally aren't in a community working together and sharing goals, and it's even rarer to find a place that's encouraging learning and exploring while also focusing on doing practical things. (Anni)

In the quote, Anni describes how *Urgent Evoke* in her eyes is a unique space because the socializing that takes place has the form of a collaborative learning and exploration, something that she has not been able to find on any other social platform. Furthermore, she is also talking about how other players helped her with practical stuff at the end of the game in relation to realizing the Evokation that she had produced:

Evoke also offered me many of my higher needs: not only a community of people who were similarly interested in making the world a better place, so that I didn't feel so alone, but it also offered me that practical support that helped me in actually getting some material things accomplished-a Web site, some classes, postcards to promote the organization and its ideas, and some references for applying for grants for more workshops. (Anni)

The data above show how the design of *Urgent Evoke* has succeeded in creating a social space where the interviewees experience that the players can interact in a positive way, share their goals and ideas, think, learn, and work together. The data also tell how the interviewees felt empowered and encouraged by meeting like-minded people and interacting with them (e.g., though receiving or giving feedback and practical help).

Externalization

A Ba should enable the creation of new knowledge by facilitating the externalization of the existing knowledge, so that it can be shared. This means that, in order for *Urgent Evoke* to function as a Ba, the design must not only facilitate the socializing among players; it must also help players externalize and communicate their explicit as well as tacit knowledge.

In *Urgent Evoke*, the process of communicating–both tacit and explicit–knowledge is facilitated in different ways. As the author mentioned earlier, players are invited to use the chat room to discuss existing ideas, experiences, values as well as the findings they have obtained during the game, and they are asked to communicate their 'conclusions' on the missions and quests on their personal page. Indeed, it is the questions raised in the missions and quests that are meant to ignite the explication of knowledge, because they are open questions allowing for many possible answers. Especially the missions that are related to wicked problems such as poverty, hunger, and migration are, because of their complexity, open to many different interpretations and answers. The complexity and openendedness compel the players to work together. The complexity asks for many different ways of looking at and approaching an analysis, and, therefore, the players need each other and each other's knowledge in order to secure that multifaceted perspective. This helps players enter a dialogue, because they have a shared interest in listening to each other and collaborate. The dialogue is a characteristic of Ba, and an important prerequisite for the externalization and sharing of knowledge.

Nonaka characterizes Ba as a resourceful space. In *Urgent Evoke* it is not only the players' knowledge and ideas that are externalized and turned into a shared resource, but the World Bank's knowledge as well.

The whole game design reflects the knowledge that the organisation has accumulated over the years, which is now externalized and used to create a resourceful space for the players to explore. Undoubtedly the World Bank's existing knowledge, experience, and values have provided input to the graphic novel, shaped the missions and quests, and guided the choice of links to external sites, blogs, jpegs, and Mp3 recordings which are curated in the game.

Thus, it is through the game design that the bank interacts and collaborates with the players; it is a way for the World Bank to make their knowledge available and accessible to the players. By doing so, the World Bank helps facilitate the players' process of knowledge creation.

In the interviews with the winners, they express how the sharing of knowledge in the game helped them in their game play. William, one of winners, is talking about how the game gave him access to an international network:

Because you know the quests were the same for everybody, but from someone in Mexico and someone in India and someone in Australia and the UK and South Africa and comparing all these different, you know, opinions and experiences with mine, was a treasure. It was something that you can't get in one day and yet during the game we had access to everybody's knowledge at the same time. (William)

This quote stresses how the international network of *Urgent Evoke* helps the players get multiple perspectives on the issues with which the game deals. The ideas of the individual players are challenged by the ideas of others living very different lives and having very different experiences. This means that the social learning environment of *Urgent Evoke* is a diversified space that allows players to access different types of knowledge in a fast and easy way.

In the interviews, one of the winners reflects on how having access to other players thoughts influenced her game play:

Sometimes, before I submitted my answer or case study, I'd go and visit other popular leading players' input...with that I learnt a lot ...I was impressed with the diversity of achievements around the world. [...] I was taken by people's miracles and to what extent some went into remote areas to help others [...] all of the stories were impressive and it took me by surprise what humans can do and how they can break the barriers and go and try where other people failed. I used to feel special and talented, but when I looked at what others were doing, I felt that there was so much power in others and that it was beautiful to make the changes that I or others are not strong enough to do. (Alice)

The player in the quote above is telling how she is learning from looking at others' work. She is impressed and taken by what people are capable of and by the diversity in the thoughts and ideas that she encounters.

As the author stated above, dialogue is an important element of Ba. The analysis of the game design pointed out how the game design meant to facilitate that *Urgent Evoke* becomes a dialogical space, where players interact with each other's ideas and knowledge, both in direct discussions and in a more indirect way, by spending time looking at, reflecting on, and giving feedback to the content on other players' personal page. Two players are talking about entering into a dialogue with others' work by looking at it and giving feedback: "I think, as the game went on, you also spent more time reading other people's work. That was something else that was important to me. You didn't want to be just someone who just created content, producing, but actually giving that same feedback" (Allan). "I think every day to see other responses, because the interesting thing about the game was not how you were contributing. The interesting thing was to see others, and learn from their approaches and to challenge them, and you know, give them feedback" (David).

In both the quotes, the players find educational to see others' works. In the first quote, the player is telling how looking at other players' work in the game is something that is taking up more and more of his time as the game progresses. In the second quote, the player is saying that the opportunity to see others' work is actually the most interesting thing about the game.

Both players are talking about how they, too, when visiting other players' sites, often provide feedback. The players frame this both as a way of challenging others and a way of giving something back to the community. In this perspective, the focus on the creation of content is exchanged with a focus on interaction and contribution to a shared goal.

The analysis has showed how the game design facilitates the externalization of knowledge through the missions and quests and by providing players with tools that support the dialogical communication (e.g., a chat room and feedback possibilities on the personal pages) among players. Furthermore, the analysis has showed how looking at others' work is something that the players think of as productive and educational.

Combination

Combination is the third phase in Nonaka's SECI model. In order for *Urgent Evoke* to function as a Ba the game should facilitate that the externalized knowledge is (re)combined, so new knowledge can be created.

In *Urgent Evoke,* the combination phase is also driven by the missions and quests prompting players not only to discuss and research the questions asked, but also to try to produce original answers to the questions – ultimately in the form of an Evokation.

The combination phase depends on that a trustful, dialogical, and collective atmosphere has been established during the two first phases. In the process of socializing and externalizing, the phase of combination has already started. This can be recognized in the two last quotes above, where players tell how they provide feedback to others during the game. In order to be able to give feedback, the players have to reflect on how their own knowledge connects to that of the other. As a result, the process of combination starts and a possibility for knowledge creation occurs.

As the chapter pointed out earlier, an important element in the design of *Urgent Evoke* is the open-endedness of the questions asked in the missions and quests. None of the questions has one correct answer rather they could be answered in many possible ways. When players are not competing to come up with 'the right answer', sharing knowledge does not become a threat, and, thereby, a space where externalization and combination is possible can emerge.

One of the winners describes how the social space of *Urgent Evoke* influences his individual process of knowledge creation: "I was exposed to a lot of ideas I had never thought out and then out of that I generated some ideas that I had never thought about and that really energized me" (Peter).

The interviewee describes how he is surprised about the ideas he is capable of creating in the game, something that he sees as the result of being exposed to many different and, to him, new ideas.

The analysis has illustrated how the game design created a dialogical space where the players can share their own knowledge and ideas as well as listen to each other, and, thereby, gain the necessary resources to start combining the knowledge in new ways. In addition, Peter-above-told how being exposed to others' ideas enables him to produce new ideas that he had never believed he would be capable of producing.

Internalization

In a Ba, tacit knowledge has to be externalized, in order to be shared and combined in new ways. However, when this happens, the new knowledge also has to be internalized again and become a tacit resource that supports on-the-job actions, approaches, and values.

According to Nonaka (Nonaka & Konno, 1998), the process of internalization is associated with learning-by-doing, training, and exercises. It is thus through participating in work-related activities that the new knowledge is internalized and becomes an integrated part of the individuals' way of thinking and acting. In *Urgent Evoke,* the design supports the process of internalization in several ways.

First of all, the design is intended to provide the players with the experience of being social innovators for real. The World Bank calls their game an 'imagination-infused reality' (Gaible & Dabla, 2010, p.36) and thereby signifies how the game aims to offer experiences that equal experiences in the physical world. The fact that the game should be accepted as real-as in opposition to make-believe-is reflected in the game design, where players are asked to tackle real world problems in the missions and to use their real name instead of an alias; also, they are instructed that what takes place in the game is 'for real':

This is not a simulation. You are about to tackle real problems [...] when we evoke, we look for creative solutions. We use whatever resources we have. We get as many people involved as possible. We take risks. We come up with ideas that have never been tried before [...] your mission is to practice your world-changing skills, right now, wherever you are. (How to play, 2010)

Second, in order to complete a mission, players are asked to go through the three steps *learn, imagine,* and *act*. Especially the last step of acting is prompting players to practice some of the things that they have learned from playing the game in the physical world. In the mission on sustainable energy, the players are asked to find another way to power a device that they use every day, rather than using electricity.

Third, the players are asked to produce an Evokation, a detailed plan for how they would tackle a sociopolitical problem postgame. The Evokation is encouraging players to look at their own local community and start thinking about how they could generate positive change there. Thus, they have to reflect on what steps they would take in order to make the change happen and, thereby, both implement the knowledge they gained from the game and start to see themselves as social innovators.

Ten of the interviewed winners stated that they succeeded in implementing their Evokations postgame. The author has been able to find evidence of a postgame implementation in seven of the ten cases. For example, one of the winners gave up his university job and started a production company producing documentaries about the sociopolitical problems that he had described in his Evokation. Other players are building respectively a new type of library in Africa and an agricultural school for children in a local community in the Philippines. One player set up a collaboration with a group of female farmers in Uganda, helping them build up a knowledge sharing system combined with microloans.

These players were all busy with other jobs or education before the game, and, during the game, all of them changed their view of their own capacities and their role in creating positive sociopolitical change. In order to be able to do this, players have not only built up knowledge of how to tackle real-life problems during the game, but they have also started to see themselves as social innovators. This change of self-perception is reflected in the following two quotes:

So it [the game] has taken me from how I saw myself there [...] to someone that can have a larger impact. [...] I think that is so nice, [...] I haven't said this out loud, but I am thinking about it, and it is such a nice change [...] in my own perception of myself. So, now I do believe I am a social innovator [...] Maybe, I knew before that I had an impact on a smaller scale [...], but not in such a large scale and with subjects that are not necessarily related to what I went to school for. (Jonas)

As I said, personally I found it builds up your confidence that you do have a voice and you can actually make a change. I don't think I would have had the confidence to approach the prisons and do the work projects [implementing his Evokation] that I am doing now, if I hadn't been through Evoke. (Axel)

In both quotes, the players report how playing the game gave them confidence in themselves as social innovators. In the game, the players played the role of a social innovator, an experience that they have been able to internalize to an extent where they start to see themselves and behave as social innovators postgame.

This last part of the analysis shows how the game design through meta-communication ('This is not a simulation'), through the subjects of missions, the 'act' component in the missions, and the formulation of the Evokation, facilitates the internalization of the new knowledge, which was produced in and gained from the game.

DISCUSSION

The analysis which has been conducted in this chapter illustrates how the World Bank, with the design of *Urgent Evoke,* has succeeded in creating a digital Ba: a resourceful and meaningful space that can attract and engage organisations' external stakeholders in a collective and participatory process of developing answers to the sociopolitical problems the World Bank is fighting. This process that resulted in the creation of 73 Evokations, some of which were even implemented postgame.

However, if on one side the 73 Evokations indicate some kind of success, they are equivalent to a little more than 1% of the registered players. Therefore, the 73 Evokations are also a reminder of the 1%-rule and the difficulties of making people create original content online. Based on the numbers alone–showing that *Urgent Evoke* resulted in 73 of the 6,618 registered players sending in an Evokation-, it is impossible to argue that games like *Urgent Evoke* are better at engaging an organisation's external stakeholders in collective knowledge creation online than other kinds of social media.

Nevertheless, the contrast between the player data and the picture of *Urgent Evoke* as a collaborative, creative, and productive community which the players provided in the interviews is interesting because it raises questions about how to understand and measure participation.

The World Bank has chosen to measure participation by counting the number of completed missions and quests, but this seems to leave them blind to other forms of participation because the interviews indicate that a large number of players participated in the game in other ways than by producing the missions and the quests (e.g., by liking and hearting the works of others, providing feedback, and participating in the discussions).

Therefore, the analysis of *Urgent Evoke* also raises the question if the participatory hope attached to social media is, and maybe has always been, too big. Maybe designers of social media platforms should be more interested in how to make designs that enable people to unite around a shared interest and contribute to the community and the collective production of new knowledge in many different ways, instead of looking for ways to turn lurkers into producers of original content.

Further, even though it cannot be argued that *Urgent Evoke* provided a higher turnout than other types of social media, it can be argued that the game did provide the winners with an empowering experience of being part of a well-designed Ba.

In the analysis of *Urgent Evoke,* Nonaka's concept of Ba and the SECI model have proven to be a good framework for a design analysis aimed at understanding how the design facilitates the collective process of knowledge creation in the game. However, the analysis also challenges Nonaka's concept and ideas about knowledge creation.

Nonaka's description of the Ba and the SECI model provides no differentiation between the participants: everybody entering the Ba is there to participate in the creation of knowledge and supposed to go through the four phases of the SECI model. The analysis of *Urgent Evoke*, though, shows that the players use and interact with the game design in different ways. A large number of people are participating in the phase of socializing by visiting the game and registering as players. Many take part in the more demanding phase of externalizing knowledge and combining it in a new way by completing some or all of the games missions and quests, but only a few got all the way and internalized the knowledge which they produced by creating an Evokation and implementing it postgame.

Nonaka's ideas about what constitutes a Ba and how knowledge is created through socialization, externalization, combination, and internalization makes it difficult to grasp and account for the role of the different forms of participation which, as the analysis showed, took place in *Urgent Evoke*. The analysis of *Urgent Evoke* therefore calls for a development of Nonaka's ideas that includes a more multi-facetted understanding of how participation in a Ba could be performed in different meaningful ways.

FUTURE RESEARCH DIRECTIONS

The research presented in this chapter has shown that, games like *Urgent Evoke* have a lot to offer as digital spaces for organizational communication and knowledge creation together with external stakeholders. Future research could build on and extend these findings by documenting, describing and analysing similar games, their design and game play. This research would provide a valuable understanding of the strength and weakness of different design choices.

Furthermore, the research made ought to followed up by future studies of and experiments on how to design games that can support the players' capacity to transfer knowledge from the digital Ba and act as social innovators and knowledge creators in the post-game situation

CONCLUSION

With *Urgent Evoke* the World Bank set out to create a digital, free-to-use game that could enable ordinary citizens to collaborate on developing knowledge in the form of social innovations, and thereby help the World Bank generate positive change in relation to some of the world's most wicked problems. By making the game open and free for people from all over the world, the World Bank connected to the communitarian optimism coupled with the Web 2.0 and the birth of social media. In addition, the organisation thereby positioned itself in an on-going discussion of whether social media in general can actually engage the crowd in collective creation as they promised.

The World Bank's intention with the game design of enabling people to co-create knowledge online, by providing them with a social space where they were asked to work and innovate together, made it interesting to analyse the game design using Nonaka's concept of Ba and SECI model.

The aim of the chapter was to provide an analysis of if and how the design of Urgent Evoke could enable the game to function as a Ba that is an organisational space for knowledge creation with external stakeholders. Furthermore, the aim was to discuss the strengths and weaknesses of Nonaka's concept of Ba and the SECI model as a unified framework for a design analysis of social network games similar to Urgent Evoke.

Nonaka describes a Ba as a social learning space. It is a resourceful place that harbours meaning, which provides the users with a tight framework and the freedom to explore, at the same time. A Ba should be a dialogical place, where the users not only talk, but also listen to each other. Further, a Ba should facilitate the users' externalization and internalization of knowledge while they move through the phases of socialization, externalization, combination, and internalization.

The analysis shows how the game design facilities the players' movement through all the four phases in the SECI model, although more players take part in the socialization and externalization phases, and in the combination and internalization.

The design forms a social learning space that provides the players with access to a vast number of knowledge resources–both those brought in by the players and those accumulated by the World Bank and implemented in the design. The clear communication of the goal of the game through, for example, the narrative of the game, the missions and quests, the point system, and the powers that the players were encouraged to develop created a meaningful context that provided the players with a sense of direction and shared purpose. At the same time, the game design also offered players freedom to explore and develop their own ideas by setting up very few rules and by basing the missions and quests on open-ended questions.

Throughout the analysis, the missions and the quests are central to the knowledge creation in the game. Together the missions and the quests drive the game forward and support the players' externalization as well as implementation of the knowledge they created in the game. The openness and complexity of the missions and the quests permit the game to function as a dialogical space that compel players to socialize, share knowledge, and collaborate.

The analysis of *Urgent Evoke* does not show that games are better at engaging people in organisational knowledge creation than other types of social media. Instead, the analysis raises questions about whether the focus on making people create and share original content online has been too narrow, and that it has made it impossible to see other forms of meaningful participation that have contributed to the creation of new knowledge. In other words, when designing social media platforms including games, the focus should not be on facilitating the 'creators' as a special group, but on creating creative communities where many forms of participation can thrive and the created knowledge is seen as the result of a collective process.

The analysis has showed that Nonaka's concept of Ba and his SECI model functions very well as an analytical framework for a design analysis of a game such as *Urgent Evoke,* aimed at facilitating player-driven knowledge creation. Nonaka's concept of Ba makes it possible to understand and define a game such as *Urgent Evoke* as an organisational space for social learning, and not merely as a platform for entertainment or as a tool for disseminating information.

Both the concept of Ba and the SECI model create a focus in the analysis on how the game design with its graphical expression, functionalities, story line, point system, and other features enables the creation of knowledge by facilitating the externalization and internalization of knowledge as the players move through the four phases of the SECI model.

REFERENCES

Benkler, Y. (2006). *The Wealth of Networks: How social production transforms markets and freedom.* Retrieved November 2, 2016, from http://www.benkler.org/Benkler_Wealth_Of_Networks.pdf

Bruns, A. (2008). *Blogs, Wikipedia, Second Life, and beyond: From production to produsage.* Bern, Switzerland: Peter Lang.

Connolly, T. M., Boyle, E. A., MacArthur, E., Hainey, T., & Boyle, J. M. (2012). A systematic literature review of empirical evidence on computer games and serious games. *Computers & Education, 59*(2), 661–686. doi:10.1016/j.compedu.2012.03.004

Denzin, N. K. (2012). Triangulation 2.0. *Journal of Mixed Methods Research, 6*(2), 80–88. doi:10.1177/1558689812437186

Dervin, B. (2008). Interviewing as dialectical practice: Sense-Making methodology as exemplar. In *Proceedings of the International Association for Media and Communication Research (IAMCR) 2008 Annual Meeting* (pp. 20-25). Stockholm: Academic Press.

Flyvbjerg, B. (2006). Five misunderstandings about case-study research. *Qualitative Inquiry, 12*(2), 219–245. doi:10.1177/1077800405284363

Fuchs, C. (2015). *Culture and economy in the age of social media.* New York, NY: Routledge.

Gaible, E., & Dabla, A. (2010). *EVOKE project evaluation.* Retrieved from www.siteresources.worldbank.org/EDUCATION/Resources/ProjectEvoke-evaluation-final-16oct11.pdf

Goodier, H. (2012). The Participation choice. *BBC Marketing and Audience.* Retrieved from http://www.bbc.co.uk/blogs/bbcinternet/2012/05/bbc_online_ briefing_spring2012_participation.pdf

Greene, J. C. (2012). Engaging critical issues in social inquiry by mixing methods. *The American Behavioral Scientist, 56*(6), 755–773. doi:10.1177/0002764211433794

How to play. (2010, May 12). Retrieved from http://www.urgentevoke.com/page/how-to-play

Jenkins, H. (1992). *Textual poachers: Television fans and participatory culture.* New York, NY: Routledge.

Jenkins, H., Ford, S., & Green, J. (2013). *Spreadable media: Creating value and meaning in a networked culture.* New York: New York University Press.

Kollock, P., & Smith, M. (1996). Managing the Virtual Commons: Cooperation and conflict in computer communities. In S. Herring (Ed.), *Computer-Mediated Communication: Linguistic, Social, and Cross-Cultural Perspectives* (pp. 109–128). Amsterdam: John Benjamins. doi:10.1075/pbns.39.10kol

Leadbeater, C. (2008). *We-think.* London, UK: Profile Books.

Lévy, P., & Bonomo, R. (1999). *Collective intelligence: Mankind's emerging world in cyberspace*. Cambridge, MA: Perseus Publishing.

McGonigal, J. (2008). Engagement economy: The future of massively scaled collaboration and participation. *Institute for the Future*. Retrieved from www.iftf.org/uploads/media/Engagement_Economy_sm_0.pdf

McGonigal, J. (2011). *Reality is broken: Why games make us better and how they can change the world*. London, UK: Jonathan Cape.

Nonaka, I., Hirose, A., & Takeda, Y. (2016). 'Meso'-foundations of dynamic capabilities: Team-Level synthesis and distributed leadership as the source of dynamic creativity. *Global Strategy Journal*, *6*(3), 168–182. doi:10.1002/gsj.1125

Nonaka, I., & Konno, N. (1998). The concept of "ba": Building a foundation for knowledge creation. *California Management Review*, *40*(3), 40–54. doi:10.2307/41165942

Nonaka, I., & Takeuchi, H. (1995). *The knowledge creation company: How Japanese companies create the dynamics of innovation*. New York: Oxford University Press.

Nonaka, I., Toyama, R., & Byosière, P. (2001). A theory of organizational knowledge creation: Understanding the dynamic process of creating knowledge. In M. Dierkes, A. B. Antal, J. Child, & I. Nonaka (Eds.), *Handbook of organizational learning and knowledge* (pp. 491–517). Oxford, UK: Oxford University Press.

Nonaka, I., Toyama, R., & Konno, N. (2000). SECI, Ba and leadership: A unified model of dynamic knowledge creation. *Long Range Planning*, *33*(1), 5–34. doi:10.1016/S0024-6301(99)00115-6

Pearce, L. D. (2012). Mixed methods inquiry in sociology. *The American Behavioral Scientist*, *56*(6), 829–848. doi:10.1177/0002764211433798

Preece, J., Nonnecke, B., & Andrews, D. (2004). The top five reasons for lurking: Improving community experiences for everyone. *Computers in Human Behavior*, *20*(2), 201–223. doi:10.1016/j.chb.2003.10.015

Shirkey, C. (2008). *Here comes everybody: The power of organizing without organizations*. London: The Penguin Press.

Van Dijck, J. (2013). *The culture of connectivity: A critical history of social media*. New York: Oxford University Press. doi:10.1093/acprof:oso/9780199970773.001.0001

Von Krogh, G., & Geilinger, N. (2014). Knowledge creation in the eco-system: Research imperatives. *European Management Journal*, *32*(1), 155–163. doi:10.1016/j.emj.2013.04.002

Yin, R. K. (2009). *Case study research: Design and methods*. Thousand Oaks, CA: Sage.

KEY TERMS AND DEFINITIONS

Collective Knowledge: A concept built on the idea that distributed knowledge can be pulled together and shared through collaboration and collective efforts. Social media is seen as a technology that can support the process of creating collective knowledge.

Game Play: It denotes the players' interaction with and experience of the game while playing.

Higher-Order Learning: It refers to the idea that there are different types of learning demanding more or less cognitive efforts. An example of higher-order learning skills is the ability to analyse, evaluate, and create new knowledge.

Prod-Users: A concept created by drawing the words *production* and *usage* together. The concept communicates the idea that social media today enable users and buyers to play an active role in the production of the products, services, etc. that they normally would consume passively.

Social Learning: The idea that learning takes place in a social context and is a social activity.

Social Network: A structural understanding of human relations. A social network is made up by the connections between social actors. An example of a social network could be the links between the members of a group of friends.

Wicked Problems: Problems characterized by their complexity. Wicked problems are difficult to solve, often change over time, and call for multiple "solutions."

ENDNOTES

1. Blogger Ben McConnell and Jackie Huba posted on May 3, 2006 on their blog: "The 1% Rule: Charting citizen participation". The post was retrieved 2017-1-6.
2. Today, the game is only open to teachers who want to use it as part of the curriculum.
3. The ten abilities are: Collaboration, courage, creativity, entrepreneurship, local insight, knowledge share, resourcefulness, spark, sustainability, and vision.

Chapter 4
Online Leadership and Communication Across Cultures:
Developing an Interdisciplinary Approach

Iben Jensen
Aalborg University, Denmark

ABSTRACT

Cross-cultural management is challenging. However, online leadership across cultures demands even more intercultural competence, as the actors involved neither share physical space nor everyday experiences. This chapter offers a new interdisciplinary perspective on global collaboration, which addresses the complexity in online cross-cultural leadership. The perspective is formed as a post-cultural prism comprising practice theory, positioning theory, and postcolonial theory. It is argued that this interdisciplinary approach is capable of filling the gap in the field of intercultural competence, as it offers tools for reflection on online interactions in everyday situations. The aim of this chapter is to contribute to a conceptualization in cross-cultural leadership and to introduce basic intercultural practices for practitioners.

INTRODUCTION

Online leadership across cultures requires theoretical and practical rethinking, as digital technologies and globalization has changed former practices in cross-cultural management. Hence, the situation today is complex; on the one hand, employees have qualified cultural presuppositions of each other across cultures (Tange & Jensen, 2012), created through mass media, social media, and experiences from everyday life in multicultural societies. On the other hand, global leadership today (and especially online global leadership) is carried out in relation to employees who do not share a physical environment in their everyday practices. Furthermore, unlike previous global management, only the most dedicated employee went abroad or worked internationally10 years ago. Today more staff members are involved

DOI: 10.4018/978-1-5225-4094-6.ch004

internationally, and expatriates are abroad for shorter periods (Gertsen et al., 2012). It is argued that this kind of complexity calls for theories that are capable of conceptualizing how culture is part of specific practices in cultural encounters. How does culture, for example, influence performance among employees from Germany, India, and USA during a meeting? What is at stake, apart from language and time zones? While former theories in cross-cultural management have been devoted to studies of individual adaption, the complexity of the situation today asks for an organizational approach that envisions the workplace as creating space for intercultural encounters in organizations.

Will a Postcultural Perspective be Productive?

In order to contribute to an organizational approach, this chapter explores whether online leadership across cultures will benefit from an interdisciplinary postcultural perspective. The postcultural perspective is developed by the author and is comprised of three theoretical perspectives adapted to a global frame: *practice theory,* by which it is argued that culture can be seen as a series of practices, which are shared among members across cultures. Hence, the task for the global leader is to get access to knowledge about the degree to which the employees share understandings of relevant practices at the workplace; *positioning theory*, by which it is argued that cross-cultural online leadership ought to be built on an awareness and sensitivity of which positions the employees are offered in the company; *postcolonial theory,* by which it is argued that European identity was developed in the colonial era, which must be considered when cross-cultural managing crosses the lines between the center and the periphery. Postcolonial theory offers important perspectives on why conflicts in cross-cultural management sometimes escalate rapidly and why arguments often draw on racial or ethnic differences rather than on individual or political differences. Hence, the aim of this article is to discuss whether this postcultural approach is beneficial for developing an organizational approach to cross-cultural online leadership.

The article is structured in four parts. The first part introduces the three fields that cross-cultural management draws on. The second part presents the theoretical perspective of the article and an introduction and discussion of the postcultural perspective. The third part is the analysis conducted from two vignettes. The last part presents a discussion of the findings, future research, and the conclusion.

BACKGROUND

Online leadership and communication across cultures are related to at least three research fields: cross-cultural management (Hofstede, 1980; Gertsen et al., 2012), intercultural competence (Deardorff, 2006; Martin, 2015; Chen & Starosta, 2008; Byram, 1997; 2000; Arasaranam, 2013), and intercultural communication (Hall, 1976; Collier, 1998; Jensen, 2006a, 2006b; Dahl, 2016; Dervin, 2011; Halualani, 2009; Hall, 1996).

The Field of Cross-Cultural Management

The most dominant research in the field of cross-cultural management is still the work of Geert Hofstede (1980), who pointed out six cultural dimensions. Hofstede's cultural dimensions have established themselves as a benchmark that all researchers are forced to reflect upon. Quantitative methods dominate this field. However, some authors, such as Gertsen, Søderberg, and Zølner (2012), have argued for the importance of using qualitative methods and take their point of departure in an understanding of organizations in line with Sackmann and Phillips (2004) and with Brannen and Salk (2000), in terms of the multiplicity of cultures and of culture as contextually negotiated content. Working with the d'Iribarne's (2009) definition, they see culture as shared, explicit, or implicit assumptions and meanings having a degree of stability but being interactively redefined over time. In this way, culture is both the pace of continuity and change (Gertsen et al., 2012, p. 5–6).

The Field of Intercultural Communication

Edward Hall is one of the most influential researchers in the field of intercultural communication. As with the other prominent researchers in this field, the aim of his work was to establish taxonomies to predict intercultural communication. One of Hall's most well-known concepts is the concept of high- and low-context societies. A high-context society is, for example, the Japanese culture, where an important part of a message's information is to be found in the context. A low-context society could be the American culture, where most meaning is explicitly expressed in verbal or written conversation (Hall, 1976).

After many years of total domination of the field by Gudykunst, Hall, Hofstede, and a few insistently critical researchers, such as Tanno, Gonzales, Jandt, and Collier, a new critical intercultural handbook was published in 2010 by Nakayama and Halualani. With an article on junctures in the field of intercultural communication, Halualani et al. also contributed importantly to the field. First and foremost, they raised a critique of the "embedded assumptions of race-less, gender-less, and class-absent individualism" (Halualani et al., 2009). Secondly, a) they argued for the importance of context and history in theorizing culture, b) they criticized the unitary conceptualization of culture as a nation, c) they argued for the importance of seeing culture as a field of struggle, and d) they suggested the field should be open to and inspired by other disciplines.

The Field of Intercultural Competence

Arasaratnam, conducted a survey (2015) of the field of intercultural competence by comparing the themes published in three scientific journals. She found that intercultural competence was surprisingly seldom represented. Intercultural competence is developed and used in international cooperation and collaboration and has especially focused upon an individual's ability to interact with people from other cultures. In business, intercultural competence is often described by three dimensions: the cognitive dimension (knowledge about other cultures), the affective dimension (the ability to accept another positioning, e.g., a woman in a negotiation where a female CEO might leave the negotiation to a lower-ranked male colleague), and communication competence (the ability to meta-communicate about, e.g., conflicts) (Gertsen, 1990). In the last 10 years, the field of intercultural competence has been dominated by the work of Darla Deardorff, who conducted a study in which she approached 20 prominent researchers on intercultural competence and asked the researchers to agree on one common definition for intercultural competence. The researchers ended up with a definition of *intercultural competence* as "effective and appropriate behavior and communication in intercultural situations" (Deardorff, 2006). Based upon this definition, Deardorff developed a model for intercultural competence, which emphasizes a process approach. Although the model mediates the previous researchers' diverse positions and visualizes some kind of a process, the content of the model has residues of the positivist paradigm that influences these fields.

Representing the constructivist paradigm, Michael Byram (also one of the invited researchers in Deardorff's project) works with intercultural communicative competence, with the aim of addressing how intercultural communicative competences can best be taught and evaluated. Byram defined *intercultural competence* as follows:

- ***Attitudes:*** *Curiosity and openness, readiness to suspend disbelief about other cultures and belief about one's own.*
- ***Knowledge:*** *Of social groups and their products and practices in one's own and in one's interlocutor's country, and of the general processes of societal and individual interaction.*
- ***Skills of Interpreting and Relating:*** *The ability to interpret a document or event from another culture, to explain it and relate it to documents from one's own.*
- ***Skills of Discovery and Interaction:*** *The ability to acquire new knowledge of a culture and cultural practices and the ability to operate knowledge, attitudes, and skills under the constraints of real–time communication and interaction.*
- ***Critical Cultural Awareness/Political Education:*** *An ability to evaluate critically and on the basis of explicit criteria perspectives, practices and products in one's own and other cultures and countries.* (Byram, 2000, p. 9)

"Revisiting the Field of Intercultural Competence" (2015) is a short and sharp essay by Judith Martin. First, she states that most models still reflect ABC (affect, behavior, and cognition/knowledge), which seems to contrast with the theoretical clarifications and synthesis developed in the same period. Her suggestions for future research are to a) move beyond individually focused, reductionist models to frameworks that capture a more holistic, relational, and spiritual view of intercultural communication competence; b) move from a focus on national culture groups that are presumed to be homogenous to conceptualizations that acknowledge multiple cultural identities, intercultural interactions, and the fluid, dynamic, and contested nature of cultures; and c) acknowledge that power relations are part of every intercultural encounter and that all encounters (and notions of competence) are impacted (and constrained) by larger societal, historical, and political forces (Martin, 2015, p. 6).

Summing up, the background for developing a new perspective for cross-cultural encounters, which takes its departure from an organizational approach, is the demand for theories and practices that are capable of moving 1) beyond individualism, 2) away from the concept of national cultures, and 3) toward applicable theories that have attracted massive interest. Hence, the aim of this chapter is to develop a theoretical prism that is capable of contributing to the research to fill this gap. Practice theory challenges the concept of national cultures and offers a specific awareness of practical leadership. Positioning theory offers an awareness of the situated communication and the psychological impact of categorization in an organization, and postcolonial theory addresses the historical and present geopolitical power relationship in specific global settings.

THEORY

Practice Theory

First and foremost, the postcultural perspective draws on practice theory. In this section, practice theory therefore is introduced, and furthermore, how the concept of culture is to be seen from a practice theoretical approach is discussed. A relevant start might be to ask, what a practice is. A practice can be to "take a seat," "to participate in an online meeting," or "to greet a friend." A practice is characterized by the following: a) it is recognizable for persons familiar with the practice (Reckwitz, 2002), and b) it is related to normativity (Rouse, 2007), which means that all practices are the result of social negotiations regarding what is deemed as appropriate in a specific practice. Practice theory is based upon the assumption that social action (the practice) is a precondition of all existence. Practices are the starting point, not the subject. Thereby, it follows that it is the practice that concurrently constitutes both the subject and the object. Practices train, so to speak, subjects to develop certain ways to act and to handle objects (materiality), and the objects also form the subject (Schatzki, 1996; Reckwitz, 2002; Kemmis & Mutton, 2012).

According to Theodore Schatzki, practices are

organized nexuses of actions. This means that the doings and sayings composing them hang together. More specifically the doings and sayings that compose a given practice are linked through 1) practical understanding, 2) rules, 3) a teleoaffective structure, and 4) general understandings. (Schatzki, 2002, p. 77)

Furthermore, materiality and bodies are part of all practices as well (Reckwitz, 2002). If "online leadership across cultures" is taken as an example of a practice, the concepts would be as follows: "Doings and sayings" are to be understood literally as what is done and said related to online leadership. An understanding of the practice is to know what to expect of the situation; it is a kind of practical knowledge and expectation of the situation. In online leadership across cultures, a *practical understanding* of what such a praxis might mean could vary significantly across cultures, as it is a relatively new practice, which probably is done quite differently. *Rules* are part of every practice; rules are defined as formal rules, which, for example, could be a language policy in the company or rules for breaks during the day. A teleoaffective structure might be the most relevant concept in cross-cultural settings, as it comprises the informal rules in a practice or how it is appropriate to act to obtain one's goal and which emotions it seems appropriate to show in this process (Schatzki, 2002). In online leadership practices, the *teleoaffective* structure is related to which emotions the team leader and the team members find appropriate to invest or show in a meeting. Finally, a practice is tied together by general understandings. *General understandings* are defined as a common way of thinking in a social setting. This could be a general understanding of education as important for a good life or an understanding of one's parents as one's responsibility. As mentioned, body and materiality are part of all practices as well. As materiality has an increasing influence on how subjects are conducting communication, this factor is relevant for cross-cultural leadership.

This description of practice is furthermore useful as a framework for its analysis. Which understandings of the practice do the employees share? Which rules must be followed in a specific practice? What is appropriate to do to reach your goal? Additionally, which general understandings are silenced, and which are dominating?

Culture as Practice

Defining culture as practice is not a new idea. The English anthropologist Ann Swidler (2001) argued that seeing culture as practice answers to one of the major challenges in sociology, which is to address which activities are related to abstract concepts as values or ideas. Swidler writes, "Both discourses and practices are concretely observable in a way that meaning, idealism and values never really were" (Swidler 2001, p. 76).

Taking a practice theoretical approach to culture means that practice constitutes the social. The logical consequence must be that culture can be seen as comprised of practices, which means that it is possible to focus on specific practices. Practice theory offers a way to conceptualize cultural encounters by their practices rather than from national belonging. This is crucial, as cultural theories for many years have argued that culture is distributed (Barth, 1967) and members of a society do not share all values or practices. Practices are distributed as well, which means that we share some practices with one another—but we do not share the same practices with everyone in a society. In a global context, it is even more transparent to see that many practices are shared with members of other societies. A good example is our use of IT, where an increasing number of people in the world are trained by Microsoft or Apple and thereby share IT practices.

Developing New Practices?

According to Shove and Pantzar (2005), practices of consumption are always "homegrown." Parts of the practice circulate, but the way the practices are pieced together is informed "by previous and related practices. What looks like a diffusion of Nordic Walking is therefore better understood as its successive, but necessarily localized, (re)invention" (Shove & Pantzar, 2005, p. 43). A practical example of this is seen in Japan, where a previous interface on cell phones has worked with a double click. Consequently, the use of a double click was easy to implement on cell phones in a Japanese context but turned out to be problematic in an American context. Seeing practices in context is inherent in the work of the Australian researcher Stephen Kemmis, who argued that all practices must be understood in a practice archaeology (Kemmis & Mutton 2012). Hence, developing new practices must always be negotiated with the existing practices. In an organizational context, this is of interest for two reasons. First, if the existing practices are disregarded, a new practice would either create conflict or be ignored. Secondly, an actor builds upon his or her former practice while developing and negotiating new practices (Shove, Pantzar, & Watson, 2012). In global leadership, it is therefore a key issue for a leader to know which practices the employees are drawing from while the employees are learning a new practice and how new practices will be negotiated with the practice architecture.

Positioning Theory

The concept of subject positioning draws on the work of subjectification developed by Michel Foucault (1979). According to Bronwyn Davies and Rom Harré (1990) *subject position* is defined as the positions that subjects are offered and interact with. Such a position can either be *confirmed* by the subject if the subject answers as expected, or it can be *rejected* if the subject denies it or ignores it. Alternatively, the position can be *recepted* by the subject, while the subject tries to *redefine* the offered subject position

(Davies & Harré, 1990; Harré & Langenhove, 1998). From the perspective of positioning theory, it is impossible to avoid taking a position. Accordingly, partners are always a part of communication and are always positioning themselves and others both verbally and nonverbally. "Generally speaking, positions are relational, in that for one to be positioned as powerful other must be positioned as powerless" (Harré & Langenhove, 1998 p. 1–2). Harré and Langenhove offered three different forms of positioning: individual, in-group, and categorial.

Individual positioning can, as an analytical tool, be divided into three forms of positioning: *superior* positioning (giving one the right to decide for others), *equal* positioning (everyone having the same rights to discuss and decide), and *inferior* positioning (inviting others to make decisions for oneself) (Jensen, 2011b). In a Scandinavian context, the ideal seems to be equal positioning among staff, while an inferior position is seen as weak and too dependent on others (Jensen, 2011b).

In-group positioning is defined as a description of a social group in which the actor either positions oneself as a part of or as a group that the actor has a certain knowledge about. It can take forms of identification or of distance. An in-group position would be: "In our department, we are good team members." It seems to be normal to include oneself when the narrative is positive and create some distance when the narrative is less positive.

A *categorial positioning* is defined as a positioning that refers to social categories, such as gender, age, ethnicity, or national group, as an explanation of vital importance. "Women are better car drivers than men" would be a statement that works as categorial positioning related to gender. Hence, categorial positioning revolves around stereotypes.

Regarding cultural encounters, positioning is a key concept, because many misunderstandings in cultural encounters can be related to different uses of positioning and—when always related to power—a negotiation of the positions that partners offer each other. In cross-cultural online leadership, awareness of these patterns of positioning is crucial, as the online practice is dependent on relatively shorter spaces with fewer opportunities to meet on a daily basis. Misunderstandings related to ill-considered or even clumsy positioning takes a while to repair online. These concepts of positioning are all useful as analytical tools, as they create an awareness of the importance of positioning in online communication.

Postcolonial Theory

"Colonialism continues to live on in ways that perhaps we have only begun to recognize" (Dirks, 1992: 23). In postcolonial theory, it is argued that the colonial epoch and colonial discourses have had an immense impact on both the colonizer and colonized. The colonial epoch was not just about economic dominance and a boost for the colonizers that made the Industrial Revolution possible. The colonial past is also related to the fact that European identity and ideology have been developed through the encounter with and construction of "the other." It was a requisite for being capable of colonizing others that a systematic discourse of the other was established in which the other was described as less civilized and less human (Smith, 1999). One of the most important researchers in postcolonial theory is Franz Fanon, who was born in French-colonized Martinique in the Caribbean. Fanon was educated in Paris as a psychiatrist and wrote the influential book *Black Skin White Mask* (1967). Fanon argued that the colonized subject is not capable of being in sound contact with himself or herself, as that person throughout childhood has existed as an invisible subject. In addition, the subject has only existed as a general negative representation (Fanon, 1967).

In both Norway and Sweden, researchers have pointed out the importance of their mutual colonial history. The Norwegian anthropologist Marianne Gullestad, who conducted a historical study of Norwegian missionaries in Africa, showed how discourses constructed by the missionaries are still part of presuppositions about people from Africa. Gullestad argued that this part of Norwegian history plays a role in the way Norwegians relate to nonwhite Norwegians or refugees from the Global South (Gullestad, 2007, p. 5).

The Swedish development researcher Maria Baaz examined how Scandinavian nongovernmental (NGO) workers relate to collaboration with people from the Global South. Moreover, she pointed at a clear connection between the discourse of the colonial period and the discourse used by NGO workers. Baaz documented that the presupposition of the lazy, irrational African is still an active myth. According to Baaz, nonwhite citizens were constructed as the "subordinated other," which set up the NGO to be superior in the situation (Baaz, 2002). Inspired by these studies, it has been suggested that postcoloniality also plays a central role in cross-cultural management. As an analytical tool, it can teach us how we are all part of our descriptions of "the other." The missionaries' pictures of the "wild ones" also gave a picture of the missionaries as civilized and rational. Telling a story about employees from Brazil who took no initiative constructs the non-Brazilian leader as right on time, well prepared, and logical.

Summing up, a postcultural perspective offers a prism that is capable of exploring the actual practice in its context, examining positions produced in interaction and communication, and allowing a reflection on how historical images of "the other" are still alive.

METHODS AND DATA

Inspired by the vignette method (Langer, 2016; Jenkins et al., 2010) and practice theory (Schatzki, 2002), the method in this chapter takes its departure in two short vignettes. While most vignette methods use vignettes in their interview, posing questions to an interviewee about the content of the vignette (Langer, 2016), this study uses the vignette the other way around, letting the vignette work as text material for a postcultural analysis. The vignettes are examples of sayings or discourses. The first vignette is quite short, as it is a condensed interview represented by a single utterance. The other vignette is a broader description of Scandinavians in a cross-cultural relation. According to practice theory, sayings (the utterances) are always related to doings, to rules and teleoaffective structure, and to general understandings. Likewise, sayings are also related to body and materiality, as they are part of all practices. This assumption offers this method its validity.

The objective of the analysis is to examine and discuss to which extent a postcultural prism is capable as an analytical tool in the situations the vignettes represent. The strength of this method is that it provides the possibility to bridge between empirical and philosophical data production. By focusing on specific situations experienced in lived life, it is possible to discuss the theoretical prism in a specific context. Contrary to this, the weakness of the method is that it can be mistaken as empirical data rather than empirical examples.

ANALYSIS

Vignette 1

The first vignette descends from fieldwork conducted by the author in January 2017 on a course on "Intercultural Awareness." The course was organized by a Danish municipality, their local job center, and Aperian Global, an international consultancy. The course was free of charge for the participants, as the objective of the course was also to invite local companies to offer traineeships to refugees from Syria. Aperian Global has developed its own "cultural dimensions," heavily inspired by the work of Geert Hofstede (1980). According to the invitation, the course offered knowledge about a) the national Danish profile, b) the participant's individual cultural profile, and c) strategies and practical tools for collaborating and communicating most effectively with new employees, regardless of whether their cultural heritage was Danish or foreign. The author was invited as a participant, although introduced as a professor in the field of intercultural communication and learning. During the course, the participants were asked to interview each other on their experiences with cross-cultural management. The participant next to the author was a woman aged 50+, with many experiences who was also employed in online cross-cultural management.

We were really challenged when we outsourced some of our activities to India. For example, they did not appreciate our privileges as employees: having a free lunch, a nice meeting room and those kinds of luxuries. They preferred money. It took us some time to figure that one out. (Jensen, 2017)

This utterance, in which the Danish manager explains how it took them by surprise that the Indian employees did not appreciate the same employee privileges (fringe benefits), such as having a free lunch, a nice meeting room and "those kinds of luxuries," is analyzed in the following section from a postcultural perspective. Accordingly, a practical understanding, rules, a teleoaffective structure, general understandings, and body and materiality are examined, and furthermore, positioning and a postcolonial perspective are added.

The practice under analysis is a writing-a-contract practice.

The *practical understanding* of contract writing is recognizable, and this includes the negotiation of salaries and privileges for both partners. The disagreement seems to be the value of a fringe benefit. *Rules* for "writing-a-contract practice" are related to the whole setup in writing and negotiating a contract. The rules involved are presumably directed by national rules, company rules, and local HR rules. While developing a contract, all these rules are interwoven. The *teleoaffective* structure, the aim of the practice—to write and negotiate a contract—literally involves negotiation. As seen in the *saying*, the disagreement in the negotiation is the fringe benefits. Due to the manager's "saying," fringe privileges are naturalized in such a way that she has taken it for granted that a free lunch is better than a higher salary. Regarding *general understanding* in this quote, it is primarily related to work as an identity, which can be shown as a fringe benefit. *Body and materiality* are part of this practice, as the fringe benefits are a luxury related to body and materiality; a free lunch and a nice meeting room are related to both.

According to positioning theory, the manager primarily uses an in-group position and categorial positionings: "*We* were challenged"; "*we* had outsourced some of *our* activities. It took *us* (in-group position) a while to figure that one out." The in-group positioning emphasizes a common understanding of this break with normativity that the Indian employees had brought forth in their negotiation.

"They" (categorial positioning) is used as a positioning of difference: "*They* did not appreciate our privileges as employees." The geopolitical position and history might be part of the relation between a Danish manager and staff from India. Denmark was part of the colonizing period and was active in India, having occupied the colony of Trankebar. Denmark was at the time the sixth most aggressive colonial power. Trade had priority in the relation between Denmark and India. However, Indians represented the Orient, being read as sensitive, childish, and so on (Said, 1978). Besides, India has had the status of an underdeveloped country and has received development aid from Denmark since decolonization.

The residues of colonialism are in this utterance, to be found in the way the managers explain that "they" do not appreciate the same things as "we" do—they do not appreciate luxury. However, as it is an underlying assumption in a postcultural perspective that the researcher has no access to the inside of the actor's mind, the analysis can only be based upon their sayings and doings and on the interpreted pattern from history and concurrent geopolitical positions. Consequently, an analysis of an utterance such as this is mainly to create an awareness of how sayings and doings are carried out in cross-cultural management. How to work with these insights in a practical manner is discussed in the last part of this chapter.

Vignette 2

Vignette 2 descends from an article written by Jan Selmer and Jakob Lauring (2012). The article is called "Globalization Denmark: Open-minded while ethnocentric." The authors explored a paradox that is experienced when Danish management goes abroad. It seems that while managers are in Denmark, they act in line with the ideals from the work of Hofstede, emphasizing feminine values and individualism. While abroad, "when confronted with individuals that have not been socialized and educated in similar surroundings, Danish managers change their management style to be less tolerant or inclusive (Bjørn 1977; Lauring, Bjerregaard, & Søderberg, 2008)" (Selmer & Lauring, 2012, p. 10).

The study is based upon 30 semi-structured interviews conducted with 16 expatriate managers and 13 immigrants from third countries working in Saudi. These data were produced by 3 months of full-time participant observation. This vignette consists of quotes from the empirical findings presented in the article. The vignette will be analyzed through the postcultural prism.

The practice at stake is the Scandinavian management practice, which seems to be a recognizable practice. The analysis will be based upon the quotes below:

Manager I, about his intercultural manager practice:

What characterizes us is that we speak to our middle managers as equal partners. We don't care whether they have light or dark skin. We don't have presumptions about them being idiots. (Selmer & Lauring, 2012, p. 18).

Manager II, about his experiences with Scandinavian management practice:

I tried to be flexible, but after two months I found that I really distrusted some of my coworkers. Therefore, it is by no means possible to be soft-hearted. Then you can choose to call it a racist attitude, but what is racism actually? Am I a racist just because I distrust a black person? I think there are some concepts about which you have become more flexible since I first arrived (Danish manager). (Selmer & Lauring, 2012, p. 19)

Employee I, about his experience with Scandinavian management practice:

We always just have to say yes, yes; We can never say "no" to anything, If there is a manager saying it is night when it is day and we object, we will be fired.

Employee II, about his experience with Scandinavian management practice:

Europeans are very racist, they just don't know it.

As indicated in these quotes, Selmer and Lauring showed a paradox: although Danish international managers define themselves as open minded, they seem to create a counter reaction and become very dominating and exclusive (Selmer & Lauring, 2013, p. 15). The authors concluded that their study confirmed several studies that suggesting "that Scandinavians overcompensate for their initial, democratic approach by assuming an overly authoritarian management style." (Selmer & Lauring, 2012, p. 22)

Analysis of Vignette 2

As was the aim in the first vignette, the objective of this analysis is to explore whether a postcultural prism is productive in creating awareness in online cross-cultural management. In this vignette, the practice chosen is Scandinavian management practice.

A practical understanding of a Scandinavian management practice is shared by the two Danish managers. The shared understanding is that Danes manage as equal partners; they have equal relations to employees, independent of their color of skin, and they have no negative presumptions. This saying is contradicted by the sayings of both a Danish manager and employee I and II in the Danish company. It seems that manager II is unable to carry out his Scandinavian practice when the employees are not carriers of the same practice. This illustrates how practices are homegrown and the importance of getting the expectations out in the light, which is even more crucial in cross-cultural settings.

Regarding *rules*, it is presumable that the company does have some CSR rules, and working with equal rights for all individuals is written in their HR policy. The discriminating utterances from manager II do not follow the official rules for Scandinavian management practice.

The *Teleoaffective structure* is in this case divided between the two positions; the first position argues that an equal relationship creates good relations. The second position takes a conflictual position and breaks with the UN declaration on CSR and discrimination. Quite unusually, this manager positions himself as responsible for direct discrimination, while he carries out actions on the basis of the color of the skin.

The *body/materiality* at stake in this practice is mainly the color of skin, which Manager I does not ascribe meaning, while Manager II does. Regarding positioning, when Manager I uses an in-group positioning, it is related to positive statements about the superiority of Danish managers. A categorial positioning is used by Manager II when he describes the employees as untrustworthy. Individual positioning is only taken in one case, while Manager II describes how he has been forced to just give orders. Last but not least, from the perspective of the employees, they are mainly offered inferior positions.

This vignette addresses the postcolonial perspective in an unusually directly way. Mentioned already is the case with the Danish manager who just accepts his distrust of his employees—while this is corroborated by the harsh remark from Employee II, who explains that European are racists but not aware of it. This could be seen as a residue from a colonial past and as sedimentation of a postcolonial present.

FINDINGS

Analyzed through a postcultural prism, the findings are first and foremost that the Danish manager from Vignette 1 does not seem aware of her own practices and therefore appears as if she is ethnocentric, although she is in fact just doing what most of us do in everyday life, expecting that most of our practices are negotiated as usual. The analysis also demonstrated that the manager's short narrative created cultural categories rather than curiosity about another relationship between job/salary/fringe-privileges. Furthermore, the analysis indicates that the negotiation of the contract puzzled the manager, but there was no sign of mutual intercultural learning in this example. What if the manager had brought a postcultural prism when they had their negotiation? First, she would have been capable of focusing on the practice: the contract negotiation. In doing so, the company could have started the process by analyzing their own contract practice; that is, the formal rules they normally followed. More importantly, which informal rules would they follow? What would be seen as an inappropriate demand? Why? Which general understandings did they follow themselves, for example, in the relation between personal identity and corporate belonging? After such a discussion, they could have asked someone who knew the specific company or another comparable company about the usual practices related to contract writing in terms of formal rules are to be followed, but more importantly, which teleoaffective structures would be dominant. This could also include a discussion on general understandings. The informant could be an Indian or an expatriate.

The analysis of Vignette 2 shows the strength of the postcultural prism, as it is capable of creating awareness of practices that violate HR rules and human rights. The prism is able to shed light on three different perspectives, which provides richness in the analysis.

DISCUSSION

Most handbooks in the intercultural field include a chapter on politeness, as well as do's and don'ts in different situations. However, what is even more important in intercultural competence is a genuine interest in the logic of the other. Behind every question in cross-cultural management and intercultural competence should be a basic assumption that "their" social order is of equal importance to ours, and if it does not seem so, it is due to a lack in our knowledge or sense making. To be respectful is to search for the advanced or unknown explanations of others instead of making a shortcut to postcolonial residues such as Manager II did. Showing respect in a cultural encounter is also intended to create awareness of our own cultural presuppositions.

Collaboration Across Cultures

Collaboration across cultures follows the same mechanisms and forms as other teamwork. The only difference is that in-group positioning and categorial positioning are part of our nationalized or colonized past, which means that it is part of our cultural presupposition of the other. When we search for a reason for a problematic relation, we are provided with a postcolonial discourse. Many conflicts in intercultural communication arise due to an idealization of in-group practices and an ethnocentric reading of others' practices. "The other" is often described from within a deficit approach, and accordingly, they seem to act irrationally and to lack the qualities that the majority has (Børhaug, 2016). This is addressed in both vignettes.

SOLUTIONS AND RECOMMENDATIONS

It is argued throughout this chapter that a postcultural prism is beneficial for online leadership across cultures because it embraces the ability to point at and talk about what is normally hidden in routines and naturalizations. One of the most crucial reasons behind misunderstandings and conflicts in intercultural communication is a lack of information about what the intention with a question or an activity is (Gumperz et al., 1979)—the prism can shed different light on diverse practices, which opens up a rich pool of information about expectations and unwritten rules. The prism is hereby beneficial because it manages to put in words what we are doing, how we carry out our practices, which positions we offer each other, and how colonial residues might remain in both positions and practices. Furthermore, the prism manages to take an organizational position rather than focusing on individual competences. Using the prism, it is possible to work with practices instead of national differences. Regarding social change, the prism invites us to dialogue, as the intention is to pose questions about practices, positioning, and postcolonial residues and not to predict their existence or a general solution for global leadership. In this way, the postcultural prism offers questions that can function as tools in everyday intercultural collaboration. With this, the prism underpins an organizational initiative for making practices that create mutual intercultural learning.

Space for Cross-Cultural Learning at the Workplace

How does a leader get access to knowledge about all the practices? One possibility is to hire someone to explore the practices, but it is more beneficial for intercultural learning to start a shared process at the workplace. An online meeting could be one opportunity.

Practically, the process can start at the workplace quite simply by asking a participant to be an observer at an online meeting across national departments in the company. The observer can registrate the actual practices in order to explore what seems to be appropriate in the online-meeting-practice. Is it appropriate to read the agenda on the computer? Is it appropriate to answer emails before the meeting starts? Is it inappropriate to take a call on the cell phone? Or to leave the office due to a call? After the first observation session, whether the participant actually found the meeting practice effective, pleasant, boring, or a waste of time could be discussed. This could start another process, in which all participants would be asked to give their opinions, for example, to break the norms between the most and the least speaking participants. The same activity could be conducted in other groups to make them aware of their own practices and to determine whether they could negotiate some changes to it.

Online Leadership and Cultural Diversity

Most of the examples in this chapter have been related to cultural encounters between the Global North and Global South—but the encounters can also be south-south and north-north or a combination of these.

If the encounter is north-north, a classic postcolonial perspective might not be relevant in relation to an entire group of staff members, but the perspective could be relevant in individual positioning represented by nonwhite persons. It might be relevant to have a look at segregation in the company regarding ethnicity or color of skin. It might be relevant to assess equality and career opportunities in the company, for example, to determine whether the company practice establishes a glass ceiling that is a hindrance for employees' career regarding class and ethnicity. It might also be relevant to examine which in-group positionings are expressed and which categorial positionings are made. Are minorities "singled out" or are they categorized as a group with many similarities?

FUTURE RESEARCH DIRECTIONS

While the vignette method provides sufficient data to discuss the benefits from a postcultural prism, further empirical studies are necessary. As mentioned several times, it might be quite easy to establish an online analysis, either by inviting staff members as co-researchers or by making an analysis of the virtual part online.

CONCLUSION

Online leadership across cultures demands new theoretical and practical takes to avoid reproducing stereotypical imaginations of "the other." The existing theories in the fields of interculturality offer mainly abstract concepts of culture, which presupposes the influence of monocultural national subjects. There is a demand for studies of specific work-related practices and for reflective tools for both staff and leaders. In this chapter, the author has introduced three different theories, which together form a prism that both researchers and leaders can bring to their work.

The contribution to the research field is a cross-disciplinary approach that has both conceptual and analytical advantages. The interdisciplinary approach offers theoretical advantages in that practice theory is capable of pointing out and exploring existing practices—and it offers a systematic way to qualify the development of new practices in consideration of the practice archaeology.

As culture is seen as a series of practices, it is possible to conceptualize how practices are "done" differently but also to point specifically to all the similarities across carriers of practices. Last but not least interesting is the capability of practice theory to specify how practices that look identical are constructed from different experiences, which resembles the description of all practices as being "homegrown." A postcultural prism is capable illuminating the process of developing new common practices across cultures.

The authors' exploration of the postcultural prism is similar to holding up a magnifying glass to see patterns and differences to seek an explanation of how practices are understood and interpreted locally and how they relate to each other. Gaining access to this knowledge resembles classic anthropological fieldwork in many ways, but the difference is in the prism; not all practices are of equal interest. The researcher or the practitioner decides which practices are seen as the most relevant for exploration.

Positioning theory contributes a psychological/sociological awareness of the relations between colleagues and leaders in different positions in a company. Contrary to discourse analysis, the operationalized form as presented in this chapter is quite easy to integrate into everyday language. The challenge in the prism might be the postcolonial approach because it on the one hand can be awkward to speak out on former brutality and enslavement, as it might not be productive for an equal relationship. On the other hand, silencing the past can easily turn into the unspoken taboos in the room. Remembering the words of Fanon, it is crucial for both partners in a cultural encounter to face the history or the present geopolitical situation. If the Danish manager had been expatriated to India, it would have been appropriate to seek insight into the country's historical and social development. Although expatriation occurs less often and the periods of expatriation are shortened with online leadership, it is crucial that leaders show their employees abroad the respect of knowing at least part of their history, especially if the mother country has a common colonial past.

REFERENCES

Arasaratnam, L. A. (2015). Research in intercultural communication: Reviewing the past decade. *Journal of International and Intercultural Communication*, *8*(4), 290–310. doi:10.1080/17513057.2015.1087096

Baaz, M. E. (2002). *The White Wo/Man's Burden in the Age of Partnership – A Postcolonial Reading of Identity in Development Aid*. Göteborg: Department of Peace and Development Research, Göteborg University.

Barth, F. (1994). *Manifestasjon og prosess* [Manifestation and process]. Oslo: Universitetsforlaget.

Børhaug, F. B. (2016). How to challenge a culturalization of human existence? Promoting interculturalism and ethical thinking in education. FLEKS. *Scandinavian Journal of Intercultural Theory and Practice*, *3*(1), 1–15.

Brannen, M. Y., & Salk, J. E. (2000). Partnering across borders: Negotiating organizational culture in a German-Japanese joint venture. *Human Relations*, *53*(4), 451–487. doi:10.1177/0018726700534001

Byram, M. (1997). *Teaching and Assessing Intercultural Communicative Competence*. Clevedon, UK: Multilingual Matters LTD.

Byram, M. (2000). Assessing Intercultural Competence in Language Teaching. *Sprogforum*, *6*(18), 8–13.

Chen, G., & Starosta, W. J. (2008). Intercultural communication competence. A Synthesis. In M. K. Asante, Y. Miike, & J. Yin (Eds.), *The global intercultural communication reader* (pp. 215–237). New York: Routledge.

Collier, M. J. (1998). Researching cultural identity: Reconciling interpretive and postcolonial perspectives. In D. V. Tanno & A. Gonzales (Eds.), *Communication and Identity across Cultures* (pp. 122–147). Thousand Oaks, CA: Sage.

d'Iribarne, P. (2009). National cultures and organizations in search of a theory: An interpretative approach. *International Journal of Cross Cultural Management*, *9*(3), 309–321. doi:10.1177/1470595809346601

Dahl, Ø. (2016). *Human encounters. Introduction to intercultural communication*. Oxford, UK: Peter Lang.

Davies, B., & Harré, R. (1990). Positioning: The discursive production of selves. *Journal for the Theory of Social Behaviour*, *20*(1), 43–63. doi:10.1111/j.1468-5914.1990.tb00174.x

Deardorff, D. K. (2006). Identification and assessment of intercultural competence as a student outcome of internationalization. *Journal of Studies in International Education*, *10*(3), 241–266. doi:10.1177/1028315306287002

Dervin, F. (2011). A plea for change in research on intercultural discourses: A "liquid" approach to the study of the acculturation of Chinese students. *Journal of Multicultural Discourses*, *6*(1), 37–52. doi:10.1080/17447143.2010.532218

Fanon, F. (1967). *Black Skin, White Masks*. New York: Grove Press.

Foucault, M. (1979). *Discipline and punish: the birth of the prison*. New York: Vintage Books.

Gertsen, M. C. (1990). Intercultural competence and expatries. *International Journal of Human Resource Management*, *1*(3), 341–362. doi:10.1080/09585199000000054

Gertsen, M. C., Søderberg, A., & Zølner, M. (2012). Introduction and overview. In M. C. Gertsen, A. Søderberg, & M. Zølner (Eds.), *Global collaboration: Intercultural experiences and learning* (pp. 1–11). London: Palgrave Macmillan. doi:10.1057/9781137026064_1

Gullestad, M. (2007). *Misjonsbilder – Bidrag til norsk selvforståelse* [Mission pictures – Contributions to Norwegian self-understanding]. Oslo: Universitetsforlaget.

Gumperz, J., Jupp, T. C., & Roberts, C. (1979). *Crosstalk – A study of cross-cultural communication: Background material and notes to accompany the BBC film*. London: The National Centre for Industrial Language Training.

Hall, E. (1976). *Beyond Culture*. New York, NY: Doubleday.

Hall, S. (1996). When was the "postcolonial?" Thinking at the limit. In I. Chambers & L. Curti (Eds.), *The postcolonial question: Common skies, divided horizons* (pp. 242–260). London: Routledge.

Halualani, R. T., Mendoza, S. L., & Drzewiecka, J. A. (2009). Critical junctures in intercultural communication studies: A review. *The Review of Communication Journal*, *9*(1), 17–35. doi:10.1080/15358590802169504

Harré, R., & Langenhove, L. (1998). *Positioning Theory*. Oxford, UK: Blackwell.

Hofstede, G. H. (1980). *Culture's consequences: International differences in workrelated values*. London: Sage.

Jensen, I. (2006a). The practice of intercultural communication: Reflections for professionals in cultural meetings. In L. A. Samovar, R. E. Porter, & E. R. McDaniel (Eds.), *Intercultural communication – a reader* (pp. 39–48). Wadsworth.

Jensen, I. (2006b). The Aspect of Power in Intercultural Communication Practice. In Ø. Dahl, I. Jensen, & P. Nynäs (Eds.), *Bridges of Understanding* (pp. 85–100). Oslo: Unipub.

Jensen, I. (2011a). If culture is practice …? A practice theoretical perspective on intercultural communication and mediation. In J. Finch & P. Nynäs (Eds.), *Transforming Otherness* (pp. 9–33). Transaction.

Jensen, I. (2011b). *Jobsamtaler med etniske minoriteter* [Job interviews with ethnic minorities]. Frederiksberg: Roskilde University Press.

Jensen, I. (2017). *Unpublished fieldnotes from Course in Intercultural awareness in Kolding Municipality*. Academic Press.

Kemmis, S., & Mutton, R. (2012). Education for sustainability (EfS): Practice and practice architectures. *Environmental Education Research, 18*(2), 187–207. doi:10.1080/13504622.2011.596929

Langer, P. (2016). The Research Vignette: Reflexive Writing as Interpretative Representation of Qualitative Inquiry—A Methodological Proposition. *Qualitative Inquiry, 22*(9), 735–744. Retrieved from qix.sagepub.com

Lauring, J. (2008). Rethinking social identity theory in international encounters: Languages as a negotiated object for identity making. *International Journal of Cross Cultural Management, 8*(3), 343–361. doi:10.1177/1470595808096673

Martin, J. N. (2015). Revisiting intercultural communication competence: Where to go from he- re. *International Journal of Intercultural Relations, 48*, 6–8. doi:10.1016/j.ijintrel.2015.03.008

Nakayama, T. K., & Halualani, R. T. (Eds.). (2010). *The handbook of critical intercultural communication*. West Sussex, UK: Wiley-Blackwell. doi:10.1002/9781444390681

Reckwitz, A. (2002). Toward a theory of social practices. A development in culturalist theorizing. *European Journal of Social Theory, 5*(2), 243–263. doi:10.1177/13684310222225432

Rouse, J. (2007). Social practices and normativity. *Philosophy of the Social Sciences, 37*(1), 1–11. doi:10.1177/0048393106296542

Sackmann, S., & Phillips, M. (2004). Contextual influence on culture research; Shifting assumptions for workplace realities. *International Journal of Cross Cultural Management, 4*(3), 370–390. doi:10.1177/1470595804047820

Said, E. W. (1978). *Orientalism*. London: Routledge & Kegan Paul.

Schatzki, T. (1996). *Social Practices. A Wittgensteinian Approach to Human Activity and the Social*. Cambridge, UK: Cambridge University Press. doi:10.1017/CBO9780511527470

Schatzki, T. (2002). *The Site of the Social: a philosophical account of the constitution of social life and change*. Pennsylvania State University Press.

Selmer, J., & Lauring, J. (2013). Globalization Denmark: Open-minded while ethnocentric. *International Studies of Management & Organization, 43*(1), 9–25. doi:10.2753/IMO0020-8825430101

Shove, E., & Pantzar, M. (2005). Consumers, Producers and Practices – Understanding the invention and reinvention of Nordic walking. *Journal of Consumer Culture*, *5*(1), 43–64. doi:10.1177/1469540505049846

Shove, E., Pantzar, M., & Watson, M. (2012). *The dynamics of social practice – Everyday life and how it changes*. Los Angeles, CA: Sage. doi:10.4135/9781446250655

Smith, T. L. (1999). *Decolonizing Methodologies – Research and Indigenous Peoples*. Dunedin: Zed Books, University of Otago Press.

Swidler, A. (2001). What Anchors Cultural Practices. In T. Schatzki, K. Knorr-Cetina, & E. von Savigny (Eds.), *The Practice Turn in Contemporary Theory*. London: Routledge.

Tange, H., & Jensen, I. (2012). Good teachers and deviant learners? The meeting of practices in university level international education. *Journal of Research in International Education*, *11*(2), 181–193. doi:10.1177/1475240912447849

Chapter 5

Using the Classical Rhetorical Concept of "Ethos" to Solve Online Collaboration Problems of Trust and Presence:
The Case of Slack

Benjamin J. Cline
Western New Mexico University, USA

ABSTRACT

This chapter shows how ethos continues to play an important role in contemporary online collaboration and can be applied as a methodology to those leading or even engaging in such collaborative projects. First articulated by Aristotle as a combination of the traits of phronesis (practical wisdom), aretae (virtuous character), and eunoia (goodwill), the concept of ethos is a central aspect of public communication. This chapter discusses how applying this classical concept to one's online communication in collaborative projects can solve the difficulties of presence and trust. Furthermore, the concept when applied to electronic communication goes a long way toward establishing an e-ethic. In order to accomplish this, this chapter specifically looks at the application Slack. Slack is an online collaborative tool that is built on top of the IRC framework. This chapter takes the concept of ethos and shows how it can be practically integrated into Slack communication.

INTRODUCTION

Communication is a messy phenomenon even when operating at its simplest level, one person speaking face-to-face with one other person who shares a mother-tongue, a culture, and a history. For this reason, most relationships, "even good ones, are gratifying in some ways and frustrating in others" (Huston, 2009, p. 318). The emotions that occur in such communication are "multi-layered processes in which intra-individual processes are tightly coupled and often cannot be separated from inter-individual pro-

DOI: 10.4018/978-1-5225-4094-6.ch005

cesses" (Arvid, 2013). Even within this relatively pristine communication environment communication can be "intentional or unintentional", is "irreversible", is unavoidable, is "unrepeatable," and "has both content and relational dimensions" (Adler & Proctor, 2011, pp. 14-16). All of this makes communication complicated, difficult, and messy.

Computer mediated communication (CMC) often becomes "a more messy socially-mediated translation process that occurs across shifting scales (local, national, regional, and global), and a range of cultures (online, offline, local, global, networked)" (Crosby & Notley, 2014, p. 135). Referring specifically to CMC through the popular microblogging site, Twitter, researchers have described the difficulties that arise from CMC: "The invisible college formed on Twitter is messy, consisting of overlapping social contexts (professional, personal, and public), scholars with different habits of engagement, and both formal and informal ties" (Quan-Haas, Martin, & McCay-Peet, 2015, p. 10). According to a UCLA study most CMC users surveyed felt that the technology "allows users to communicate with people they normally could not, more than three-quarters (77 percent) agreed or strongly agreed" (Cole et al., 2013, p. 59). Nonetheless that communication carries "an increased risk. As users might be placed in different contexts or cultures, misunderstandings become more likely, and enforcement of agreements and regulations becomes more difficult. As risks increase and become more difficult to evaluate, users of collaborative technologies face more complex decisions" (Riegelsbereger & Sasse, 2003, p. 785).

That risk, or at least the perception of it, is exacerbated to the point of incivility by the alienation from the communication created by electronic media; "[m]edia today, particularly digital media, have both ecological and epistemological ramifications" (Friesen, 2011, p. 175). These epistemological ramifications inherently proceed to a lack of trust: "it is important for educators and academics to clearly understand the epistemological differences that will emerge in their students. We can expect an increase in skepticism that we have never before known. Inartistic proofs will be expected to be reproducible on demand with little or no preparatory warning" (Cline B., 2012, p. 10). Simply believing somebody becomes more difficult when one is physically separate from another person.

Some of the grounds for this apparent lack of trust comes from the inadequate sense of presence that arises from a nuanced communication that can come from face to face interactions: "Because of its lack of social and contextual cues, negative online interactions are likely to generate a stronger adverse effect on victims compared to traditional face-to-face or telephone encounters as individuals lack the opportunity to seek immediate clarification or obtain dynamic feedback" (Lim, Thompson, & Chin, 2008, p. 155). The lack of physical presence means that electronic "comments in an emotionally charged situation are often misinterpreted" (Denning, 1993, p. 170). That misinterpretation of emotional information can create chain reactions of incivility which have the potential to reverberate throughout an organization and ultimately to the whole of society: "Although the intent to harm may be ambiguous, the spillover effect of an uncivil interpersonal workplace encounter on others as well as the organization should not be underestimated" (Lim, Thompson, & Chin, 2008).

This chapter will show that CMC is a symbol system that is rhetorically constructed on top of language and using language. It will argue that the problems of trust and presence in CMC are the result of a divisive linguistic phenomenon that has been well identified in the rhetorical and philosophical literature. Because of this, CMC can be rhetorically manipulated to serve ends solving trust and presence (identification) despite being a cause of those problems (division). The paper will show that the basic rhetorical nature of CMC means that the various tools of rhetoric, many of which have been in use for millennia, can be used to solve certain problems that exist in CMC. The chapter will focus on Aristotelian concept of *ethos* in order to rhetorically construct trust and presence within the framework of CMC. The chapter will show that in so doing, an e-ethic is constructed that allows CMC users to ethically develop trust with partners and presence in their online communication. In order to accomplish this, this chapter will use as its case study the software system Slack, a system developed to improve communication in project teams working at a distance.

BACKGROUND: THE CREATION AND ASCENTION OF SLACK

Slack is a company whose application is "increasingly nibbling at a significant portion of [the internet]" (LaFrance, 2016). The name of the company is an acronym meaning "Searchable Log of All Conversations and Knowledge" (Kim, 2016). The application was created as a side effect of a failed project; "Slack sprouted to life from a less successful project. While building the now-defunct MMO *Glitch*, the development team found that none of the existing communication tools fit their needs. They built their own… and as *Glitch* closed its doors (citing 'limited audience appeal'), Slack sprang forth. A little over a year later, Slack was raising $120 million at a $1.12 billion valuation" (Kumparak, 2015).

Creating a successful product from a failed game was hardly a new concept to at least one of Slack's cofounders, Stewart Butterfield. He had failed before by creating the *Game Neverending* which was "a financial failure" that spawned Flickr which Butterfield sold to Yahoo! for "somewhere between $22 million and $25 million" (Honan, 2014). Perhaps such experience turning applications created to build other projects is part of what enabled Slack to develop to the extent it has. On one hand, nothing in Slack is entirely new, being built, as it was, on the back of IRC, an internet protocol developed in the very early days of the internet back in 1988 for communicating in text;

IRC, or internet relay chat, is a 28-year-old protocol for text-based communication that's open and incredibly versatile. [Cal] Henderson [one of the cofounders of Slack] and his colleagues built their entire workflow on top of it. They shared game assets, migrated a game server with it, committed code—and, of course, maintained contact with the whole team. When they ceased work on Glitch in 2012, Henderson says, they agreed that they would never work together again without a customized IRC system like what they had (Newitz, 2016).

In addition to IRC, Slack also integrates direct messaging, private groups, a slackbot that works as a personal digital personal assistant, and integration with services such as "Google Drive, Google Hangouts, Twitter, Asana, Trello, Github, and many other services [which] can be pulled into conversations and made visible in the channel, private group, or direct message" (Augustine, 2017).

Slack can also integrate your email, however, almost from the beginning Slack was known as an "email killer:"

Even if spam filters are getting pretty good at catching the unwanted message, there's a sense in which all email is basically spam, suggests Slack cofounder Stewart Butterfield. It's simply more clunky and intrusive than most communications need to be. 'It's the least efficient way to send a message, with the most cognitive overhead,' Butterfield tells Fast Company. 'If someone I worked with emailed me, I'd probably fire them' (Zax, 2013).

Email isn't the only application that Slack is threatening to kill. Even juggernauts like Facebook and Google recognize Slack as a threat:

Slack isn't built for social performance in quite the same way that Facebook is, but Slack does seem to be more a reflection of the real-time web than Facebook. Yes, Facebook has Messenger and News Feed, but there's still something about the interface that feels old-school—at its core, Facebook is more like a bulletin board than a conversation. Perhaps more importantly, for people who are seeking a place to convene with small (ideally private) groups of friends and family online, Slack could be preferable. And there's evidence that people are increasingly seeking such social environments online. (LaFrance, 2016).

Google has had to adapt its instant messaging client in order to compete in Slack's market hoping that since "Both Slack and GSuite are practically prerequisites for many modern companies, and while before they perfectly complemented each other, Google is now turning Slack into a competitor. The integration with GSuite is going to be tough to overcome for Slack—companies already have to pay for GSuite, so why not dump Slack and use the built-in chat program?" (Amadeo, 2017). However, Slack is much more than GSuite with a chat built in. Slack is a way of creating ethos, an ancient rhetorical tool, in an online environment in a more cohesive and coherent way than has ever been contemplated since the beginning of electronic communication.

ETHOS AND SLACK

Issues, Controversies, Problems: Trust and Presence in Online Communication

The problems of trust and presence in cooperative communication that plague those who attempt to communicate in the technological environment of a contemporary business did not come into being with the introduction of CMC. They likely came into being with the coming of oral language and have been exacerbated by the various symbol systems that have developed from that initial technological innovation; "it would appear that the technological inventions of writings, print and electronic verbalization, in their historical effects are connected with and have helped bring about a certain kind of alienation within the human lifeworld" (Ong, 1977, p. 17). Such alienation is inherent in electronic communication.

What Ong calls "alienation" is closely related to what McLuhan calls "autoamputation" wherein electronic media might extend certain parts of our presence, they simultaneously amputate others:

Depending on which sense or faculty is extended technologically, or "autoamputated" the "closure" or equilibrium-seeking among the other senses is fairly predictable. It is with the senses as it is with color. Sensation is always 100 per cent and color is always 100 per cent color. But the ratio among the components in the sensation of the color can differ indefinitely. Yet if sound, for example, is intensified, touch and taste and sight are affected at once. The effect of radio on literate or visual man was to re-awaken his tribal memories and the effect of sound added to motion pictures was to diminish the role of mime, tactility and kinesthesis. Similarly, when nomadic man turned to sedentary and specialist ways, the senses specialized too. The development of writing and the visual organization of life made possible the discovery of individualism, introspection and so on (McLuhan, 1964, p. 54).

Every medium, in McLuhan's view, amputates as well as extends human senses.

This extension and autoamputation includes media such as language itself which inherently separates a person from a reality. Ogden and Richards explain that words "'mean' nothing by themselves . . . It is only when a thinker makes use of them that they stand for anything or, in one sense, have 'meaning.' They are instruments." (Ogden & Richads, 1989, pp. 9-10). These instruments never describe immediately with physical realities; "there is no relevant relation other than the indirect one, which consist in its being used by someone to stand for a referent. Symbol and Referent, that is to say, are not connected directly . . . but only indirectly" through what Ogden and Richards call a "thought" or "referent." (Ogden & Richads, 1989, pp. 11-12).

Lyotard points out that this separation is made more evident "In contemporary society and culture—postindustrial society, postmodern culture—the question of legitimate knowledge is formulated in different terms. The grand narrative has lost its credibility, regardless of what mode of unification it uses, regardless of whether it is speculative narrative or a narrative of emancipation" (Lyotard, 1997, p. 37). He recognizes the loss of grand narrative is located in the language game disrupted by the "capitalist renewal and prosperity and the disorienting upsurge of technology" and refers to the separation that causes one to distrust a "delegitimation" of knowledge (Lyotard, 1997, p. 38).

Kenneth Burke also dealt with the same problem calling it a "division." According to Burke, human beings are separated from their environments by instruments of their own design (1966). That separation from our natural environment inevitably involves a separation from each other. It is essential to note that it is the instruments that separate us from each other, namely our definitive abilities as "symbol using, symbol-making, and symbol misusing" animals (Burke, Language As Symbolic Action, 1966, p. 6), are also the same abilities that we begin to use to build back to that space of connection, or identification; "identification is affirmed with earnestness precisely because there is a division. If men were not apart from one another, there would be no need for the rhetorician to proclaim their unity" (Burke, 1962, p. 546). Therein lies the hope for those of us who would tackle this alienation. The symbolic instruments which alienate us from one another are the same basic tools that can bring us back together.

Empirical research on CMC has confirmed what the philosophical and media ecological critiques would have predicted. Even from the earliest empirical studies in CMC the autoamputation and division described above manifest themselves empirically in terms of time and information processing pressures, absence of regulating feedback, dramaturgical weakness, few status and position cues, and problems with social anonymity (Kiesler, Siegel, & McGuire, 1984). Even as technology advanced, these problems persisted in terms of status and cue positions resulting in fundamental attribution errors in many CMC situations, a mistake that could be exacerbated in cross-cultural CMC environments (Vignovic & Thompson, 2010). Furthermore, the lack of identification and the basic rhetorical problems of division have empirically been shown to decrease the social pleasure of CMC interaction in comparison with face-to face interactions (Ranney & Troop-Gordon, 2012).

That hope is an important one that the Slack company and attendant application can mediate. Whether or not Slack or a similar application can actually solve what might be called the central concern of postmodernism and, in a much older way, the basic flaw of human symbol usage, is doubtful. However, the partial mitigation of the problem, whether one calls it "alienation," "autoamputation," the gap between the "symbol" and the "referent," "delegitimation" or "division," is theoretically possible through an app through which a "rhetorician" can "proclaim unity," to adjust the Burkean notion.

While different philosophers of communication used different terminology, listed above, to describe this basic problem, this paper will use the Burkean term of "division" in order to tackle the problem from this point forward. Certainly, there are nuances between the different thinkers' terminologies that need to be taken into account. However, division seems to be the most comprehensive of the terms. Furthermore, it seems to be the most hopeful in mediating a solution in that "In pure identification there would be no strife. Likewise, there would be no strife in absolute separateness, since opponents can join battle only through a mediatory ground that makes the communication possible . . . But put identification and division ambiguously together, so that you cannot know for certain where one ends and the other begins, and you have the characteristic invitation to rhetoric" (Burke, Language As Symbolic Action, 1966, p. 549). Rhetoric contains a large body of theory and allows careful examination of symbolic problems and their remedies.

The rhetorical theory through which we can see the problem mitigated using an application such as Slack is a decidedly old one. While the theoretical solution might be as old as symbol systems themselves, it was primarily developed through Aristotle's rhetorical theory during the 5th century BC. Specifically, one can examine Aristotle's notion of invention, in general, and ethos, in particular, in order to understand how Slack can help partially mitigate the problem of division.

The concept of ethos is the means by which Aristotle said that one could examine the problems of trust, immediacy, and presence for a public speaker; "since rhetoric exists to affect the giving of decisions - the hearers decide between one political speaker and another, and a legal verdict is a decision - the orator must not only try to make the argument of his speech demonstrative and worthy of belief; he must also make his own character look right and put his hearers, who are to decide, into the right frame of mind" (Aristotle, *On Rhetoric*, 2.1.2). Thus, a rhetor must be able to create trust in his or her hearers, a decidedly difficult proposition given the difficulties described in the first section of this article.

Aristotle posits that in order to be a functional communicator, one must be perceived by her or his audience a particular way: "when they feel friendly to the man who comes before them for judgment, they regard him as having done little wrong, if any; when they feel hostile, they take the opposite view" (Aristotle, 2011, 2.1.4). Aristotle finally lays down the parts to his theory of ethos: "There are three things which inspire confidence in the orator's own character - the three, namely, that induce us to believe a thing apart from any proof of it: good sense [*phronesis*], good moral character [*aretae*], and good will [*eunoia*]" (Aristotle, 2011, 2.1.5).

While each of these can be developed in a face-to-face interpersonal or public speaking environment, they are a bit more difficult to create in mediated communication. Print media have developed standards over the centuries to achieve ethos without physical presence, many of which can be seen in this chapter. For instance, in academic writing a system of editing and peer-review have developed which are supposed to indicate that the speaker has been judged by peers to have *phronesis*. A system of citation and quotation exists to indicate humility, generosity, decorum, and other aspects of *aretae*. A tradition of suggesting further applications and uses of one's writing are used to signal that the writer has *eunoia* and sincerely wants the audience to find the communication personally and professionally useful. While these mechanisms are imperfect and perhaps even formulaic to the point of being problematic, they are recognizable means that print media have developed for constructing ethos in academic writing.

Admittedly, CMC has attempted to adopt these same tools as CMC largely developed out of academic print media. Early uses of the internet were for academics to share research; "[T]here are large data files available at individual locations which are not valuable enough to warrant duplication at every computer center, but from which segments could be obtained at any network location" (Roberts, 1968/2001). While the military played a part in what would eventually become modern day CMC because of the research done at the Rand Corporation, a largely government-funded corporation, at which the early the concepts of packet switching and of protocols, which allowed networked computing to really come into existence were developed (Kristula, 2009), CMC hardly began with military applications. The role of the military is a "myth" persists and seems to have "gone unchallenged long enough to become widely accepted as fact" (Hafner & Lyon, 1996, p. 10). The academy developed CMC and "the first big surge in the internet was mostly accomplished within universities in the western part of the United States. Stanford, UCLA, The UC Santa Barbara and the University of Utah worked together to create the internet's earliest backbone, ARPnet, in 1969" (Cline B. J., 2016, p. 9). However, the internet quickly came to be home to business.

The academic protocols mentioned above were never seen as widely functional in business arrangements. Ethos was still built face-to-face, interpersonally and in public presentations. Written communication was the legal, formal, or intercessory aspect of what was largely a series of face-to-face relationships. When a relationship was entirely textual, it was also generally hierarchical and ethos, or proof in general is not necessary when there is no equality: if one's boss says that from here on out one will refer to the sky's color as "pink" instead of "blue," a person will do that regardless of whether or not she or he is convinced because a job is at stake. When one is working among equals, in teams, however, rhetoric is a central aspect of existence. Thus the academic protocols did not work to establish ethos in a business relationship.

Secondly, the academic protocols to establish ethos were designed for a textual medium. While hypertextuality was different in many ways, allowing a more fluid interrelationship between texts (Landow, 1992), it was still tied in many ways to a phonetic alphabet. The emerging CMC is not a purely or even primarily textual medium. Video, audio, images of all kind, chunks of code that are meaningless without other programs to read them, encrypted data, spreadsheets, and databases sit alongside textual documents in the communication of today and Slack is able to not only manage all of them, it is able to manage a communicator's ethos in this radically changed environment as well.

Application of Ethos to Slack

One can see the ways in which Slack manages not only the information, but allows a rhetor within a business to manage ethos as well. When considered in terms of each part of ethos outlined by Aristotle, Slack provides a means where the business rhetor can make use of the app to improve his or her credibility and presence. Doing this will allow a person to be more believable and create a means by which trust, which is necessary for business to achieve its ends, can be elicited without the previously requisite face-to-face interactions.

Slack and Phronesis

Phronesis, translated earlier as "good sense" is also sometimes translated as "practical wisdom" or occasionally just as as "wisdom" (although the Greek word "sophia" is more often understood as wisdom). Phronesis is a communicator's ability to communicate that she or he actually knows what she or he is talking about. The main tool used to rhetorically construct phronesis in face-to-face settings has always been one's history with one's audience:

While it is possible to fake an honest manner, it is exceedingly difficult to maintain a reputation for phronēsis without actually giving a fair amount of sound advice. A vulgar rhetorician might for a time succeed in dazzling an audience by playing to their preconceptions, but a few persuasive speeches do not make a successful rhetorician. Since a rhetoricians reputation is, over the course of time, at least in part measured by the consequences of the policies he [sic] recommends, it would be difficult for a vulgar rhetorician—one who has pandered to immediate desires, without considering the real interests of his audience—to sustain a reputation for either good sense or virtue. (Rorty, 1980, p. 9)

That means that an employee can build phronesis with her or his team members over time in a regular face-to-face business setting, but that ability to have given good advice is severely compromised in an online setting where teams are often fluid and who gave what advice to the final product is often questionable.

Slack can solve some of these problems through the use of customized profiles. With custom profiles, teams can create new profile fields for any purpose. Now you can share even more details: location, start date, social profiles, and more" (Slack Help Center, 2017). The addition of social profiles, including Twitter, LinkedIn, and professional profiles such as Soundcloud or Sourceforge, where completed projects might be stored, allows other team members to evaluate the extent to which the team member has shown practical wisdom in the past. Despite the fluidity of a CMC team, phronesis can be garnered through applications of social and professional profiles which allow people to evaluate the trustworthiness of another person.

Slack and Aretae

The second Aristotelian aspect of *ethos*, translated above as "good moral character," *Aretae* is an important aspect of building trust and presence within CMC as it was in Aristotle's notion of public speaking. However it is important to note that the use of the Greek term *Aretae* can be differentiated from another term *ethika*, or *ethikon,* a closely related term to *ethos*, which could also be translated as "good moral character."

Aretae is a part of *ethikon* just as it is a part of *ethos*. *Ethikon* is a "subdivision of philosophy at least from the time of Aristotle, literally the study of character (*ethos*)" (Sedley, 2003, p. 377). Thus if we define "ethos" as "character" and aretae as "character" we are confusing the whole and the part. However, the reason for this possible confusion is because *aretae* is the most essential aspect of *ethos.* Many scholars use the term "virtue" to understand aretae "as it preserves the integrity of Aristotle's choice for the word [*aretae*] instead of ethics (ethika)" (Ward, 2001). However, virtue, with its Latin etymology meaning little more than "an aspect" is equally problematic. For the Greeks, aretae meant "goodness or excellence of any kind" (Ledell & Scott, 1925). When understood from Aristotle's point of view, however, *aretae* was "moral behavior [as] an important part of a public persona" (Cline B., The Concept of Ethos as a Tool for Assessing Civility, 2014, p. 195). A person might or might not be an excellent person, but unless the people with whom one is working perceive that person's excellence, they cannot know whether or not the person is trustworthy.

The decision as to whether or not to believe someone depends in part, then on whether or not one perceives the other as being worthy of trust. A person might know what she or he is talking about, (*phronesis*) and might want the best for people in the group (*eunoia*), but if she or he is generally a bad person, their coworkers will still not trust them. To build that trust, coworkers need to perceive the virtue within their coworker. McTavish describes these "various virtues listed by Aristotle include: justice, manly [sic] courage, self-control, magnificence, magnanimity, liberality, gentleness, prudence and wisdom" (McTavish, 2010, p. 69). These are the particular virtues that constitute *aretae* and they are often difficult to know about another person in CMC

Slack allows for the rhetorical construction of *aretae* within a rhetrorical construction generally referred to as a "project" by Slack users by allowing a more complete interaction with the combination of multiple tools. The individual project that is created over a particular channel or series of channels within IRC might demonstrate competence, but one does not decide one's ability to trust a person based solely on competence. We decide whether a person is acting from courage or self interest, is being generous or lazy, is being prudent or cowardly by the body language and tone one is using. According to Mehrabian's germinal work, 93% of our communication with others comes from these more non-verbal aspects and only about 7% from the actual words we use (Mehrabian, 1971).

While project managers might be able to cobble together a series of tools that allow users to interact in such a way as to make informed decisions about the *aretae* of their coworkers, Slack integrates features such as video conferencing with IRC chats and combined projects in order to make a seamless integration of the workflow. *Aretae* seems cobbled together otherwise and that inherently creates a lack of integrity in multiple senses of the word.

Slack and Eunoia

"*Eunoia*," as a term, does not suffer from quite the philological difficulties encountered in understanding *aretae*. Nonetheless, as with any translated technical term it is possible to understand it in different ways. Fortenbaugh preferred to translate the term as "friendship" in that it seeks to make the audience friendly to the communicator (Fortenbaugh, 1992). Certainly "goodwill" captures a literal translation and "friendship" captures much of the attitude toward the communicator that Aristotle wanted to invoke. However, "goodwill" has epistemological problems in that the audience cannot really know the "will" of the rhetor, only his or her perception of it. And "friendship" is something quite outside the rhetorical act of *ethos* creation. One should certainly feel "friendliness" from the rhetor, but that is more of an openness to friendship than friendship itself and does not necessarily engender trust.

For contemporary scholars, the Burkean term of "identification" may "come closest to Aristotle's concept of *eunoia*" (Cline B., 2014, p. 196). Such a conflation is still imperfect but: "Burke is completely aware that is not introducing a totally new concept [of identification], observing that Aristotle had long ago commented that 'it is not hard . . . to praise Athenians among Athenians' and that one persuades by 'identifying' one's ways with those of his audience" (Golden, Berquist, Coleman, & Sproule, 2011, p. 276). This concept of eunoia as identification can be summed up in the adage: "a rising tide raises all ships."

Burke described identification by saying:

A is not identical with his colleague, B. But insofar as their interests are joined, A is identified with B. Or he may identify himself with B even when their interests are not joined, if he assumes they are, or is persuaded to believe so...In being identified with B, A is 'substantially one' with a person other than himself. Yet at the same time, he remains unique, an individual locus of motives. Thus he is both joined and separate, at once a distinct substance and consubstantial with another (Burke, 1962, pp. 544-545)

This adaptation of the concept of *eunoia* as Burkean identification is useful for a number of reasons. One reason is that it is automatically paired with the concept of division that we have already shown is both created by CMC and that we must also find a way that CMC solves this problem. Secondly, it more realistically subsumes other definitions such as "friendship" or "goodwill."

Slack allows the creation of eunoia through a visual representation of the ways in which one colleague's "interests are joined" with another colleague. More emotive ties, such as "friendship" may also be facilitated, but that's not really the purpose of Slack nor or *eunoia*. Slack has "decided not to list [a dating application designed for Slack] in it's app directory" because they consider it to be "inappropriate for use in the workplace or that [it will] negatively impact the user experience" (Beck, 2017). While in Aristotle's time, the notions of the public and private sphere as developed in the 18th century were unknown (Habermas, 1991), they certainly developed out of the Greek *agora*, a place of business and politics, and oikos, a place of home-life. *Eunoia*'s role in *ethos* creation was certainly more a function of the former than latter.

The visual interface of Slack forces the user to see the roles and interactions within a project or projects with other people. "Joined interests" are immediately seen without any artistry on the part of the employees using the client. Thus, it is almost as though the role of *eunoia* in ethos creation is outsourced to Slack and hardly even needs to be justified even as much as it was through traditional face-to-face communication.

Applications and Division

Thus, Slack works as a rhetorical tool to solve the problem of division identified earlier in this chapter as the basic problem inherent in human communication. Providing more than a virtual ground on which to stand, so to speak, Slack becomes a substance in the Burkean sense

Hypostasis, literally standing under: hence anything set under such as stand, base, bottom prop support, stay, hencemetaphorically that which lies at the bottom of a thing, as the groundwork, subject-matter, argument of a narrative, speech, poem; a starting point, a beginning and then come the methaphysical meanings . . . subsistence, reality, real being. . . nature essence (Burke, 1962, p. 23).

Slack, then becomes that "thing" on which the users can "stand," in a metaphorical and metaphysical space which divided beings can "come together" more than just as a means of shared data. Data certainly can be shared but more than that, what the Slack interface calls a "workspace," a url in which workers on a particular project meet, becomes a spot of overcoming divisions. The problems of trust instituted by the philosophical problem of division which is exacerbated by mediation are also simultaneously solved through mediation. Everything is right there. The history is present for the construction of eunoia, phronesis, and arete, thus mitigating the problem of trust as will be shown in the next section.

SOLUTIONS AND RECOMMENDATIONS

Slack facilitates the construction of *ethos*, but that does not mean that there is not still rhetorical work to do for organizations and leaders using the application. The application only provides the tools. That does not necessarily mean that an organization or individual using the tools will use them correctly or effectively in order to construct ethos.

First of all, while *phronesis* can be constructed through the integration of social and professional profiles, this only happens if the organizations using Slack wish to have them integrated. Organizations using Slack must add these applications if they hope for their employees to effectively use them to create ethos and build trust.

There is also the necessity that the users effectively use their profiles. Most contemporary computer users have a number of profiles that they use socially for a number of different uses. A profile aiming at divisive politics, identity issues, private family and sexual matters, or even musical tastes are probably not relevant for most workplace conversations and can actually point to a lack of *phronesis* as it will appear that the user does not have the "good sense" to separate work-life from other aspects of life. While Slack can help facilitate *phronesis* development, it could also, if misused, reduce that facilitation.

Similarly, *aretae* can be facilitated or inhibited through the use of Slack. Slack can facilitate the non-verbal aspects of communication that allow us to assess moral excellence, but it cannot retrieve moral excellence once spoiled. In other words, Slack can be used to communicate moral excellence, but can only do so to the extent that a person is indeed morally excellent. Certainly, in the short term, a person might be able to bamboozle her or his coworkers by putting on an "honest face" that can be facilitated through Slack, but that is not a good long-term strategy. In an adage often attributed to Abraham Lincoln, "You can fool some of the people all the time and all the people some of the time, but you can't fool all the people all the time." Once defective character has been shown, it becomes increasingly difficult to rebuild one's personal brand, or as Quinitillian wrote:

For I do not merely assert that the ideal orator should be a good man, but I affirm that no man can be an orator unless he is a good man. For it is impossible to regard those men as gifted with intelligence who on being offered the choice between the two paths of virtue and of vice choose the latter, nor can we allow them prudence, when by the unforeseen issue of their own actions they render themselves liable not merely to the heaviest penalties of the laws, but to the inevitable torment of an evil conscience (Quintillian, 1922, p. 2.1.3)

Yet, we all have faults, personal failings, and other aspects of our character that we would prefer remain hidden from our colleagues. While Slack can indeed allow the nuanced communication that allows us to decide to trust someone, it also provides means to accidentally expose our flaws and faults in a professional setting by linking up aspects of our lives that we would prefer were not made public.

Finally, Slack allows for the rhetorical construction of *eunoia*, but once again, only if it is used properly. If an organization only uses Slack in its simplest and cheapest forms, the interplay between projects that show the interconnectedness of the organization do not immediately become apparent. In such cases the employees themselves will have a difficult time providing adequate proof that by meeting each other's needs, their own needs are met. A decision has to be made to provide information about who reports to whom and how the organizational structure works with regard to a project and an individual employee to allow Slack to be used to its fullest potential.

FUTURE RESEARCH DIRECTIONS

Slack is likely not the only tool that is being created to facilitate ethos creation in a business setting, however it does appear to be the most well-known and commonly used. However, the adherences of the masses do not necessarily make it the best. Future research might examine other companies that compete with Slack that perhaps do a better or worse job in the facilitation of ethos construction.

Perhaps more importantly, this chapter has shown a relationship between Slack and ethos creation that points to something more. Proper use of Slack as indicated in the previous section results in good ethos creation. This indicates that proper use of the medium likely involves good ethics, since the *Ethikon* was described earlier in the chapter as a "subdivision of philosophy at least from the time of Aristotle, literally the study of character (*ethos*)" (Sedley, 2003, p. 377). Thus, we see an electronic ethics emerging throughout this chapter that might provide some ways of deciding the ethical use of other electronic media, even media that had yet to be developed. If a medium facilitates *arete*, *phronesis*, and *eunoia*, perhaps it can be seen as a "good medium" whereas if a medium inhibits any of these it might be seen as a "bad medium." By the same token even if a medium is seen to be more or less ethically neutral, or even good, if a particular user makes use of the medium in such a way that lessens *arete*, *phronesis*, and *eunoia*, then we could say more than that it is a poor use of the medium, but even to say that it is an immoral or unethical use of the medium. The implications from a media ecology perspective, then, are profound. Future scholarship should investigate this further.

CONCLUSION

This chapter has taken an ethical look at Slack, a company that provides a tool for business communication and, we have shown, specifically facilitates the *ethos* construction of its users. This was done by exploring the Aristotelian concept of *ethos* and explaining how the application allows the users can use the tools provided by Slack to facilitating each of the constituent parts: *phronesis*, *aretae*, and *eunoia*. The chapter has shown that Slack is a rhetorical tool whereby which a rhetorician, or user, can "proclaim unity," in order to mitigate a problem referred to by communication philosophers as "alienation," "autoamputation," the gap between the "symbol" and the "referent," "delegitimation" or "division." However, this chapter has shown that Slack only works as a rhetorical tool if it is used correctly, as when it is used incorrectly, it can actually inhibit the construction of *ethos*. All of this has been shown to point to a larger picture of an electronic ethic in which both media and their uses can be seen as good to the extent that they facilitate ethos.

REFERENCES

Adler, R. B., & Proctor, R. F. (2011). *Looking out, Looking* (13th ed.). Boston: Wadsworth Cengage Learning.

Amadeo, R. (2017, March 9). *Google is Turning Hangouts Into a Slack Competitor.* Retrieved from Ars Technica: https://arstechnica.com/gadgets/2017/03/google-is-turning-hangouts-into-a-slack-competitor/

Arvid, K. (2013). The Social Regulation of Emotion. Messy Layers. *Frontiers in Psychology*, *4*, 51. Retrieved from http://journal.frontiersin.org/article/10.3389/fpsyg.2013.00051 PMID:23424049

Augustine, A. (2017, May 10). *A review of the Slack Communication Service: Slack lets you do without email.* Retrieved from Lifewire: https://www.lifewire.com/slack-sets-standard-team-communication-online-771603

Beck, J. (2017, May 9). *Slack Doesn't Want People Using it For Romance.* Retrieved from The Atlantic: https://www.theatlantic.com/technology/archive/2017/05/slack-doesnt-want-people-using-it-for-romance/525975/

Burke, K. (1962). *A Grammer of Motives and A Rhetoric of Motives*. New York: The World Publishing Company.

Burke, K. (1966). *Language As Symbolic Action*. Berkeley, CA: Univeristy of California Press.

Cline, B. (2012). The Future of Reading/Thinking: Epistemological Construction in the Age of the Kindle. *Communication +1, 1*(1), Article 2.

Cline, B. (2014). The Concept of Ethos as a Tool for Assessing Civility. In J. Bauer & C. Stevenson (Eds.), *Building Online Communities in Higher Education Institutions* (pp. 191–200). Hershey, PA: Information Science Reference (An Imprint of IGI Global). doi:10.4018/978-1-4666-5178-4.ch010

Cline, B. J. (2016). *E-thinking: How Thinking Changes in the Age of the Kindle*. Saarbrücken, Germany: Lambert Academic Publishing.

Cole, J. I., Suman, M., Lunn, R., Aquino, J.-S., Firth, D., Fortier, D., . . . Lebo, H. (2013). *The UCLA Internet Report: Surveying the Digital Future-- Year Three.* Retrieved February 14, 2017, from Center for the Digital Future: http://www.digitalcenter.org/wp-content/uploads/2013/02/2003_digital_future_report-year3.pdf

Crosby, A., & Notley, T. (2014). Using Video and Online Subtitling to Communicate across Languages from West Papua. *The Australian Journal of Anthropology*, *25*(2), 138–154. doi:10.1111/taja.12085

Fortenbaugh, W. W. (1992). Aristotle on Persuasion through Character. *Rhetorica*, *10*(3), 220. doi:10.1525/rh.1992.10.3.207

Friesen, N. (2011). Media: Digital, ecological and epistemological. *E-Learning and Digital Media*, *8*(3), 175–180. doi:10.2304/elea.2011.8.3.175

Golden, J. L., Berquist, G. F., Coleman, W. E., & Sproule, J. M. (2011). *The Rhetoric of Western Thought*. Dubuque, IA: Kendall Hunt Publishing Company.

Habermas, J. (1991). *The Structural Transformation of the Public Sphere: An inquiry into a category of Bourgeois Society* (T. Burger, Trans.). Cambridge, MA: MIT University Press.

Hafner, K., & Lyon, M. (1996). *Where wizards stay up late: The origins of the Internet*. New York: Simon and Schuster.

Honan, M. (2014, August 8). *The Most Fascinating Profile You'll ever Read About a Guy and His Boring Startup.* Retrieved from Wired.com: https://www.wired.com/2014/08/the-most-fascinating-profile-youll-ever-read-about-a-guy-and-his-boring-startup/

Huston, T. L. (2009, September 30). What's love got to do with it? Why some marriages succeed and others fail. *Personal Relationships*, *16*(3), 301–327. doi:10.1111/j.1475-6811.2009.01225.x

Kiesler, S., Siegel, J., & McGuire, W. T. (1984). Social psychological aspects of computer-mediated communication. *The American Psychologist*, *39*(10), 1123–1134. doi:10.1037/0003-066X.39.10.1123

Kim, E. (2016, September 28). *Slack, the red hot $3.8 billion startup, has a hidden meaning behind its name.* Retrieved from Business Insider UK: http://uk.businessinsider.com/where-did-slack-get-its-name-2016-9?_ga=1.138509299.1175501442.1475074973?r=US&IR=T

Kristula, D. (2009). *History of the Internet.* Retrieved 1 2014, 10, from History of the Internet: http://www.davesite.com/webstation/net-history1.shtml

Kumparak, G. (2015, February 5). *Slack's Co-Founders Take Home The Crunchie For Founder Of The Year.* Retrieved from Tech Crunch: https://techcrunch.com/2015/02/05/slacks-co-founders-take-home-the-crunchie-for-founder-of-the-year/

LaFrance, A. (2016, June 21). *Slack, The Facebook Slayer.* Retrieved from The Atlantic: https://www.theatlantic.com/technology/archive/2016/06/slack-eats-internet/488033/

Landow, G. P. (1992). *Hypertext: The Convergence of Contemporary Critical Theory and Technology.* Baltimore, MD: Johns Hopkins University Press.

Ledell, G. H., & Scott, R. (1925). *Arete.* Retrieved May 15, 2017, from A Greek-English Lexicon: http://www.perseus.tufts.edu/hopper/text?doc=LSJ%20a)reth/&lang=original

Lyotard, J.-F. (1997). *The Postmodern Condition: A report on Knowledge* (G. Bennington & B. Massumi, Trans.). Minneapolis, MN: University of Minnesota Press.

McLuhan, M. (1964). *Understanding Media: The Extensions of Man.* New York: McGraw Hill.

Mehrabian, A. (1971). *Silent Messages: Implicit Communication of Emotions and Attitude.* Belmont, CA: Wadsworth Publishing Compan.

Newitz, A. (2016, March 9). *What Slack is doing to our offices—and our minds.* Retrieved from Ars Technica: https://arstechnica.com/information-technology/2016/03/what-slack-is-doing-to-our-offices-and-our-minds/

Ogden, C. K., & Richads, I. A. (1989). *The Meaning of Meaning: A Study of the Influence of Language Upon Thought and Of the Science of Symbolism.* New York: Harcourt Brace Javonovich Publishers.

Ong, W. J. (1977). *Interfaces of the Word: Studies in the evolution of consiousness and culture.* Cornell University Press.

Quan-Haas, A., Martin, K., & McCay-Peet, L. (2015). Networks of Digital Humanities Scholars: The informational and social uses and gratifications of Twitter. *Big Data and Society*, *2*(1), 1–12.

Quintillian. (1922). *Quintilian, Institutio Oratoria.* Retrieved from http://www.perseus.tufts.edu/hopper/text?doc=Perseus:text:2007.01.0069:book=12

Ranney, J. D., & Troop-Gordon, W. (2012). Computer-Mediated Communication With Distant Friends: Relations With Adjustment During Students' First Semester in College. *Journal of Educational Psychology, 104*(3), 848–861. doi:10.1037/a0027698

Riegelsbereger, J., & Sasse, J. M. (2003). The Researcher's Dillema: Evaluating Trust in Computer Mediated Communication. *International Journal of Human-Computer Studies, 58*(6), 759–781. doi:10.1016/S1071-5819(03)00042-9

Roberts, L. G. (1968/2001). *Resource Sharing Computer Networks.* Retrieved October 1, 2015, from Packet.cc: http://www.packet.cc/files/res-share-comp-net.html

Rorty, A. O. (1980). Structuring Rhetoric. In A. O. Rorty (Ed.), *Essays on Aristotle's Rhetoric* (p. 9). University Of California Press.

Sedley, D. (2003). *The Cambridge Companion to Greek and Roman Philosophy.* Cambridge, UK: Cabridge University Press. doi:10.1017/CCOL0521772850

Slack Help Center. (2017). *Customize your team's profiles.* Retrieved from Slack.com: https://get.slack.help/hc/en-us/articles/212281478-Customize-your-team-s-profiles

Vignovic, J. A., & Thompson, L. F. (2010). Computer-Mediated Cross-Cultural Collaboration: Attributing Communication Errors to the Person Versus the Situation. *The Journal of Applied Psychology, 95*(2), 265–276. doi:10.1037/a0018628 PMID:20230068

Ward, L. (2001). Nobility and Necessity: The problem of Courage in Aristotle's Nicomachean Ethics. *Political Science Review, 95*(1), 71. Retrieved from http://go.galegroup.com/ps/i.do?p=AONE&sw=w&u=nm_a_wnmu&v=2.1&it=r&id=GALE%7CA73021417&asid=b295edb474117734b0f7643af4e2ac8a

Zax, D. (2013, August 14). *Flickr Cofounders Launch Slack, An Email Killer.* Retrieved from Fast Company: https://www.fastcompany.com/3015730/innovation-agents/flickr-cofounders-launch-slack-an-email-killer

ADDITIONAL READING

Adgani, P., Bowen, H., & Daniels, M. (2007). Aristotle's forms of proof: the keys to teaching recreation marketing. *Schole: A Journal of Leisure Studies and Recreation Education, 22*, 47-50.

Baumlin, J., & Baumlin, T. (1994). *Ethos: New Essays in Rhetorical and Cultural Theory.* Dallas, TX: Southern Methodist University Press.

Cherry, R. (1988). Ethos Versus Persona: Self Representation in Written Discourse. *Written Discourse, 5*, 251–276.

Corbett, E., & Connors, R. (1998). Classical Rhetoric for the Modern Student (4th Edition ed.). New York: Oxford University Press.

Corts, T. (1968). The Derivation of Ethos. *Speech Monographs*, *35*(2), 201–202. doi:10.1080/03637756809375583

Cummings, K., & Stanescu, J. (2007). Argumentation and democratic disagreement: On cultivating a practice of dissent. *Controversia*, *5*(2), 55–76.

Enos, T., & Borrowman, S. (2001). Authority and Credibility: Classical Rhetoric and the Internet and the Teaching of Techno-Ethos. In L. Gray-Rosendale, & S. (. Grubner, Alternative Rhetorics: Challenges to the Rhetorical Tradition (pp. 93-110). Carbondale, IL: Southern Illinois University Press.

Fleckenstein, K. (2005). Cybernetics, Ethos and Ethics: The Plight of the Bread and Butter-Fly. *Journal of Advanced Composition*, *25*(2), 323–345.

Frobish, T. (2003). An Origin of a Theory: A Comparison of Ethos in the Homeric Iliad with that Found in Aristotle's Rhetoric. *Rhetoric Review*, *22*(1), 16–30. doi:10.1207/S15327981RR2201_2

Furley, D., & Nehamas, A. (1994). *Aristotle's Rhetoric: Philosophic Essays*. Princeton, NJ: Princeton University Press.

Golden, J., Berquist, G., Coleman, W., & Sproule, J. (2011). The Scientific Approach of Aristotle. In J. Golden, G. Berquist, W. Coleman, & J. Sproule (Eds.), *The Rhetoric of Western Thought* (pp. 67–82). Dubuque, IA: Kendall-Hunt.

Grimaldi, W. (1972). *Studies in the Philosophy of Aristotle's Rhetoric*. Weisbaden, Germany: Franz Steiner Verlag.

Halloran, S. (1982). Aristotle's Concept of Ethos, or if not His, Someone Else's. *Rhetoric Review*, *1*(1), 53–63. doi:10.1080/07350198209359037

Hicks, D., & Langsdorf, L. (1999). Regulating Disagreement, Constituing Participants: A Critique of Procedural Theories of Democracy. *Argumentation*, *13*(2), 139–160. doi:10.1023/A:1007773607213

Howes, R. F. (1961). *Historical Studies of Rhetoric and Rhetoricians*. Ithica, NY: Cornell University Press.

Hyde, M. (2004). *The Ethos of Rhetoric*. Columbia, SC: South Carolina Press.

Johnson, N. (1984). Ethos and the Aims of Rhetoric. In R. Connors, L. Ede, & A. Lunsford (Eds.), *Essays on Classical Rhetoric and Modern Discourse* (pp. 98–114). Carbondale, IL: Southern Illinois University Press.

Johnstone, C. (1980). An Aristotelian Trilogy: Ethics, Rhetoric, Politics and the Search for Moral Truth. *Philosophy & Rhetoric*, *13*, 1–24.

Kallendorf, C., & Kallendorf, C. (1985). The figures of speech, Ethos, and Aristotle: Notes toward a rhetoric of business communication. *Journal of Business Communication*, *22*(1), 35–50. doi:10.1177/002194368502200102

McCroskey, J., & Young, T. (1981). Ethos and Credibility: The Construct and its Measurements after Three Decades. *Communication Studies*, *32*, 24–34.

Miller, A. (1974). Aristotle on Habit and Character: Implications for Rhetoric. *Communication Monographs, 41*(4), 309–316.

Pross, E. (1952). Practical Implications of the Aristotelian Concept of Ethos. *Southern Journal of Communication, 17*, 257–264.

Reynolds, N. (1993). Ethos as Location: New Sites for Discursive Authority. *Rhetoric Review, 11*(2), 325–338. doi:10.1080/07350199309389009

Rosenthal, P. (1966). The Concept of Ethos and the Structure of Persuasion. *Communication Monographs, 14*, 114–126.

Rosenthal, P. (1971). Specificity, Verifiability, and Message Credibility. *The Quarterly Journal of Speech, 57*(4), 393–401. doi:10.1080/00335637109383084

Sattler, W. M. (1947). Conceptions of Ethos in Ancience Rhetoric. *Communication Monographs, 14*, 55–65.

Warnick, B. (1989). Judgement, Probablitily and Aristotle's Rhetoric. *The Quarterly Journal of Speech, 75*(3), 299–311. doi:10.1080/00335638909383879

Willams, J. (2009). Aristotle And the Systemization of Rhetoric. In J. Williams (Ed.), *An Introduction to Classical Rhetoric: Essential Readings* (pp. 222–270). Malden, MA: Wiley-Blackwell.

KEY TERMS AND DEFINITIONS

Alienation: Walter Ong's term for the separation from the lifeworld that comes from the use of symbols.

***Aretae*:** Aristotle's term for virtue, moral character, or personal excellence.

Autoamputation: Marshall McLuhan's term for the reduction in particular senses because of the extension of other senses through the use of media.

Computer-Mediated Communication (CMC): Communication using computers.

Ethos: Aristotle's concept of rhetorical character consisting of the concept of *Phronesis, Aretae*, and *Eunoia* that allows a person to be believed apart from any other form of proof.

***Eunoia*:** Aristotle's term for goodwill or friendship. It is a sense that a communicator wants what is best for her or his audience.

App (Application): In CMC or computer terminology, an end-user-focused computer program.

Internet Relay Chat (IR): An early text based protocol for internet communication.

Mediated Communication: Communication that is achieved through a medium rather than face-to-face.

***Phronesis*:** Aristotle's term for good sense. Often translated practical wisdom it is an audience's belief that a communicator knows about what he or she is communicating.

Referent: Ogden and Richards' term for the physical thing referred to through a thought by a symbol.

Symbol: Something that stands in for something else, in rhetorical studies often a word.

Thought: In Ogden and Rirchards' terminology a thought is an internal mental process that relates a referent to a symbol.

Chapter 6
Online Leaders Increase Three Types of Capital

Ditte Kolbaek
Aalborg University, Denmark

ABSTRACT

This chapter explores 10 years of development in online leadership by asking, How may information and communications technology (ICT) increase the economic, cultural, or social capital of online leaders in a global information-technology company classified as big business? Drawing on practice theory, this chapter is aimed at investigating online leaders' approach to their role seen "from inside," particularly in regards to three types of capital (social, cultural, and economic capital). This qualitative case study employs the methods of memory work and document analysis covering the development of ICT by Oracle, a global IT company from 2002 to 2012. The analysis provides insights into four themes: first, establishment of common ground for cooperation; second, working practices; third, critical competencies; and fourth, stock valuation trends. The findings indicate that ICT supports the enhancement of all three types of capital.

INTRODUCTION

This study takes as its point of departure the author's many years of experience as an online leader. This sparked her curiosity about how other online leaders approach the online work environment and how current theories and methodologies might support the understanding and development of work mediated by information and communications technology (ICT). Applying the practice-based approach (Gherardi, 2012) from organizational studies, this study explores how leaders in a global setting use the available ICT to accomplish tasks and how they create a sense of meaning concerning their actions in this context.

ICT has enabled organizations to transform from local, physical entities into global entities in which employees at all levels cooperate online rather than face to face. The advantages of globalization for such companies are well known: the work follows the sun, and tasks are solved 24 hours a day, 7 days a week, 365 days a year. A company can hire the best and least-costly employees regardless of geographical location (Daim, Reutiman, Hughes, Pathak, & Bynum, 2012). Online leaders have one-on-one ac-

DOI: 10.4018/978-1-5225-4094-6.ch006

cess to their team members and can communicate with their teams at any time (Hunsaker & Hunsaker, 2008). This online work environment raises various questions about leadership and teamwork because the development of ICT has changed traditional ways of communication and collaboration (Hambley, O'Neill, & Kline, 2007). Online leadership entails leading geographically dispersed teams that rarely or never meet face to face but instead cooperate and collaborate online (DasGupta, 2011).

This chapter explores the use of ICT to lead geographically dispersed team members in a large, global IT company and discusses the deployment of new technology in this company from 2002 to 2012. Online leadership and the thoughts and considerations involved are hidden in the heads and hearts of online leaders. This chapter, therefore, proposes memory work (Schratz-Hadwich, 2005) as a method to uncover these hidden processes and clarify how ICT impacts online leaders to enhance the three types of capital being social, cultural and economic capital. As well, this study uses document analysis to explore the 2002–2012 ICT development in the global IT company. To investigate both memory and ITC development in the same company, data are collected from only one company. The study thus can be classified as a case study (Flyvbjerg, 2006), and the research setting and data collection should meet the requirements of a case study for rich, qualitative data that can be analyzed and interpreted. This study collects data on 2002–2012 ICT development in the global IT company, working practices and critical competences for online leaders in the global IT company, and measures of corporate valuation. The objective is two-fold: 1) to provide insights into online leaders' approaches to their role and to clarify how ICT supports enhancing cultural, social, and economic capital; and 2) to explore memory work as a new approach in organizational research.

BACKGROUND

Bourdieu (1986) conceptualizes value in three forms of capital: economic, cultural, and social capital. Economic capital is easily measurable, and in most companies, it is the most important measure. Online leaders are assigned specific economic targets, and the processes of online leadership should ensure their achievement. To do so, online leaders engage cultural capital such as English language skills, education, and hierarchical status; and social capital e.g., building relationships with team members (Bourdieu, 1986). While online social networks may be perceived as social capital (Putnam, 2001), the theoretical foundation of this framework lies in Bourdieu's (1986) work. This chapter discusses economic capital only briefly, focusing instead on the thoughts and considerations of online leaders themselves.

Pigg and Crank (2004) point out that research has rarely addressed the question of how ICT may enhance social capital. This chapter makes a small contribution to fill this gap. A literature review of EBSCO, JSTOR and IEEExplore conducted in February 2017 found no English-language books or articles proposing memory work as a method for organizational studies or leadership and online leadership studies. This study, therefore, presents a new method in this regard. The findings suggest that ICT may support online leaders in increasing social capital and according to Fraser and Michell (2014) memory work is suitable for investigating sensitive matters, hence memory work seems a suitable method for organizational studies.

THEORETICAL FOUNDATION

This section presents the theoretical foundation of this study, which builds on practice theory as developed by Gherardi (2012). Bourdieu's (2011) three types of capital are also briefly presented.

Practice Theory

This study focuses on online leaders and their perceptions and actions as understood from the perspective of practice theory. A working practice is a site of knowing, working, and organizing and is situated in a context shaped by factors, such as its physical surroundings, historical background, sociological aspects, and technologies (Gherardi 2012; 2015; Engeström, 2001; Wenger, 1998). Technology affects working practices and frames how practitioners interact with technology and relate to each other. When technology changes, practitioners realign their working practices to the new possibilities or limitations (Gherardi, 2012). Important aspects of the social context include organizational structures, critical competencies, languages, and communication channels (Engeström, 2001; Gherardi, 2015; Kolbaek, 2014). The historical background frames the present context by providing insights into earlier events and their consequences. The historical background includes the history of technological development and its consequences (Gherardi, 2015; Nonaka & Takeuchi, 1995), while the relevant physical surroundings include the geography of the workplace and the layout of the buildings in which working practices take place (Kolbaek, 2014).

A view of leadership as a working practice does not define it in terms of norms but by how it is lived and reproduced through action, discussion, and negotiation in the workplace context (Gherardi, 2012). Interpreting online leadership as a working practice requires exploring how online leaders understand their roles, position themselves in their roles, and actually do in online leadership. In a practice-based approach, the researcher studies patterns of activities and understands how a practice is seen "from inside" (Gherardi, 2012). This requirement influences the methods chosen because the researcher needs to encourage practitioners to express themselves and to accept that the object of research is in constant transformation as the practice under observation continues to develop (Gherardi, 2012).

Online leadership is "a social influence process mediated by advanced information technologies to produce changes in attitudes, feelings, thinking, behavior, and/or performance of individuals, groups and/or organizations" (DasGupta, 2011, p. 1). It requires the ability to create common grounds for cooperation, including a shared language and access to common ICT. Online leadership encounters challenges in the social aspects, particularly in the need to establish strong relationships, coherence, and trust in the team. Most communication takes place in written form, so spontaneity and small talk are often lacking (Hunsaker & Hunsaker, 2008). In other words, online leadership seems to require social capital. Supportive online leadership behaviors include setting clear goals, giving constructive feedback, and maintaining dialogue, whereas obstructive online leadership behaviors include missing or vague goals, the use of threats or punishments, and one-way communication from the online leader to team members (Zhu, Kraut, & Kittur, 2012).

Three Types of Capital

The three forms of capital defined by Bourdieu (1986) are briefly described as follows. *Cultural capital* (Bourdieu, 1986) includes the ability to utilize artifacts, such as machines, instruments, books, and pictures, to achieve other types of capital. Other important aspects of cultural capital include educational levels, both university degrees and practical skills, and the ability to influence laws or norms. Cultural capital is accumulated from birth because it is embodied in the family. The link between cultural and economic capital can be measured by the time needed to develop cultural capital.

Social capital (Bourdieu, 1986) consists of social network and relationships, for example, belonging to certain groups. Within these groups, the members exchange acknowledgement by giving and receiving feedback. Social capital enables the production and reproduction of knowledge and recognition through, for example, sharing considerations and thoughts. If social capital is "the aggregate of the actual or potential resources which are linked to possession of a durable network of more or less institutionalized relationships of mutual acquaintance and recognition" (Bourdieu, 1986, p. 21), online leaders need to be aware of relationship building and providing recognition to team members (Zhu et al., 2012). Social capital may also be perceived as social networks in which participants "do things for each other" (Putham, 2000, p 19), or social relations through which information, influence and solidarity flow (Adler & Kwon, 2000); but this chapter sticks to Bourdieu's work.

Economic capital (Bourdieu, 1986) is directly convertible into money, and companies' financial statements or stock value indicates their economic capital. Online leadership supports the enhancement of economic capital as online leadership enables companies to hire cheap labor independently of geography and time, so that the company can run and make money 24 hours a day, 365 days a year.

Based on these insights, online leadership can be stated to involve the establishment of common ground for cooperation, effective working practices, and critical competencies. Establishment of a common ground for cooperation requires a clear organizational structure, shared language, and common ICT framework incorporating communication channels. Effective working practices include goal-setting, actions, discussion, and negotiation. The critical competencies of an online leader extend to strong communications skills.

METHODOLOGY AND METHODS

In this section case study will be presented as the overall methodology, whereas Memory Work and Document Analysis are the underlying methods. Data are presented in the next section named Research Setting and Data Collection.

Case Study

This researcher interprets working practices by investigating what leaders understand, say and do to fulfil their roles. Working practices interact with their surrounding context, and each influences the other. Investigating working practices, therefore, presupposes awareness of and inquiry into their context (Gherardi, 2012). The case study method seems suitable for investigating online leaders in a global IT company classified as big business, as case study uncovers context-dependent knowledge and situated expertise by investigating real-life situations described in narratives and provides rich details that include interpretations held by the practitioners, researchers, and others, so that the story unfolds from many differing perspectives. Case studies allow researchers to learn from concrete, context-dependent experience (Flyvbjerg, 2006). However, this may lead to bias, so researchers need to consider their role and describe their assumptions.

A case study may uncover issues or relations that quantitative research cannot because they are rare and consequently disappear in statistics. Online leadership is a human activity that cannot be described by predictive theory, so theory and practices emerge from cases based in different contexts. Case studies provide context-dependent knowledge (Flyvbjerg, 2006); consequently, the results may not be due to formal generalization, but "formal generalization, be it on the basis of large samples or single cases, is considerably overrated as the main source of scientific progress" (Flyvbjerg, 2006, p. 424). Online leadership is also a process, so quantitative data are not sufficient to look at the matter from informants' inside perspective (Valsiner, 2012). Instead, qualitative methods are suitable for investigating processes because processes are unmeasurable (Valsiner, 2012).

Qualitative research might not be repeatable because data are gathered from people by people influenced by emotions and time. Transparency then becomes important so that others can follow researchers' actions and intentions, which may be considered in relation to the ethical issues involved. Furthermore, researchers can minimize bias by sharing their intentions and pre-assumptions with the audience (Justesen & Mik-Meyer, 2012).

Case studies combine different data sources to achieve in-depth analyses. The design of this case study takes advantage of the researcher's situational knowledge by collecting data from only one global IT company. This case study employs two qualitative methods: memory work and document analysis. Qualitative methods are recommended to investigate participants' perceptions and insights (Justesen & Mik-Meyer, 2012; Valsiner, 2012). Practitioners should be encouraged to express themselves with as little interference as possible from the researcher, who should choose appropriate methods accordingly. The researcher must define the research environment for qualitative and quantitative studies, including who will participate and what they should do to produce valid data (Brinkmann &Tanggaard, 2010). This study is aims at understanding what practitioners comprehend, say, and do. The researcher wants the informants to share perspectives without interference. Memory work, therefore, seems to be a suitable method because it allows the researcher to capture the informants' thoughts and considerations. In other words, memory work permits the researcher to look at the matter "from inside" the informants' perspective.

Memory Work

This study aims to investigate online leaders' approach to their role seen "from inside", hence the methods should allow the people under study to reveal their insights preferably with as little interference from the researcher as possible. Memory work seems to fulfill this requirement, and besides it is interesting to explore "new" methods and possibly provide new approaches to the field of research.

Memory work originated from feminist research in the 1960s based on the premise that individual experience can bridge the gap between subjects and objects in research fields where they are one and the same (Schratz-Hadwich, 2005). Time, place, and culture affect memories (Fraser & Michell, 2014), so they are not true records of the past; rather, memory constitutes an identity through reflections on the past. Memory work allows participants to express experiences and insights that reflect knowledge beyond their conscious awareness. The format is especially appropriate for sensitive matters because it can create distance between the author and the content (Fraser & Michell, 2014). Memory work consists of three phases:

- **Phase One:** Individual reflections that lead to a written description of a memory.
- **Phase Two:** Collective group work with memories, resulting in a joint understanding.
- **Phase Three:** Theorization of the material from the written memories and the collective discussion of them (Onyx & Small, 2004).

Phase one of memory work consists of writing a story about a personal experience or an event from everyday life. This phase has four basic rules: 1) the text must be only one or two pages long, 2) be written in a third-person voice, 3) include as much detail as possible, 4) and not include interpretations, explanations, and biography (Onyx & Small, 2004). This format enables the author to perform the writing task within a few hours and maintain focus on what really matters in the situation. The third-person voice forces the author to present the experience as a stranger even while including many details. Doing so distances the author from self-image (Haug, 2008), and the ban on interpretations, explanations, and biographical context encourages the author to stick to factual information.

In phase two, the author and the researcher collaborate to analyze the memory work and remove clichés, add theoretical perspectives to the content, and identify gaps or incomplete information (Onyx & Small, 2004). Phase two allows the authors to create a common understanding of the issues under study, and the intention is to enable the informants to influence the situation studied. This study excludes phase 2 because the authors no longer work together and did not want to spent time on a common discussion.

Phase three theorizes the inputs from phases one and two. This is typically done by the researchers without the participation of the authors involved in phase one. Memory work allows the researcher to glimpse the emotions and reflections of the informants, who provide their individual insights without interference from the researcher. Memory work provides personal and subjective data (as opposed to objective data), and the researcher must consider the informants' motivation for sharing memories in order to maintain a critical perspective (Onyx & Small, 2004).

Document Analysis

The research question *"How may information and communications technology (ICT) enhance the economic, cultural, or social capital of online leaders in a global information-technology company classified as big business?"* demands investigation of the ICT development in this global IT company from 2002 to 2012. Document analysis is utilized for this purpose. *Documents* refer to various kinds of material produced independently of the research, unlike memory work, where the data production is integral to the research. In this study, the documents consist of writings, pictures, and sounds available online or in different physical forms that provide insight into the 2002–2012 ICT development by the global IT company. Again, the context determines what documents are relevant and accessible, and the findings depend on this specific organization (Justesen & Mik-Meyer, 2012).

Documents may be divided in three types. Primary documents are accessible only to a limited number of people, secondary documents are accessible to anyone at any time, and tertiary documents are accessible to anyone for a limited period of time (Brinkmann & Tanggaard, 2010). This study uses primary and secondary documents. Primary documents include websites, web conferences, and shared files; the category depends on the application of the document (Justesen & Mik-Meyer, 2012). Secondary documents consist of records of the stock value of the global IT company.

The selection criteria for documents are accessibility and ICT-relevant content. The researcher had access to the documents as an employee of the global IT company from 2002 to 2012. Her role included to be first mover on new technology and introduce new technology to colleagues.

RESEARCH SETTING AND DATA COLLECTION

This case study takes advantage of the researcher's situational knowledge from her employment in the global IT company, Oracle. This organization seems suitable for this study for several reasons. First, Oracle was the first global IT company to utilize the (then) new Internet as the basis for business databases, applications, and operating systems, as well as for internal communication and collaboration. Second, Oracle provides ICT for customers and uses new interventions internally to stress test products and services. These activities make the company a first mover in applying the possibilities offered by ICT. Third, in 2005, Oracle started to transform from a multinational company with local entities in various countries into a global company with lines of business that cross geographical borders and require online leaders—again making the company a first mover. Oracle senior managers, therefore, are highly experienced online leaders who can share worthwhile lessons. Fourth, the relevant documents and relevant informants were available to the researcher.

In the data collection for this study, the primary documents are accessible only to Oracle employees. These include screenshots of the Oracle Intranet from 2003 to 2012; meeting minutes from the Knowledge Management Team, Europe, Middle East, and Africa (EMEA) and the Organizational Learning Team, EMEA; besides teaching materials for the Oracle internal Facilitator Community and Web 2.0 community; and stories from internal newsletters and annual reports on the achievements of the EMEA Knowledge Management and Organizational learning teams. In addition, this study draws on two sources of secondary data: The Oracle external website (www.oracle.com) and New York Stock Exchange website (www.nyse.com).

Following the study's theoretical framework, the data collection is aimed at revealing how ICT may enhance social and economic capital. The data collection focuses on three themes: 1) establishment of common ground for cooperation, which includes a clear organizational structure, shared language, and common ICT framework incorporating communication channels; 2) working practices and critical competencies, including goal-setting, actions, discussion, negotiations, and, essential for an online leader, strong communications skills to give and receive feedback and recognition and share observations and insights; and 3) stock value trends indicating the development of Oracle's economic capital.

Common Ground

Establishing a common ground for cooperation requires an organizational structure, shared language, and common ICT framework. Oracle's organizational structure changed over time as it developed new products and services. It performed only business-to-business transactions and delivered hardware, middleware, and software to market leaders in banking, transportation, healthcare, and other industries (Rao, 2003). In 2011, Oracle company had revenue of US$36.7 billion US and approximately 108,000 employees, including 30,000 people in the EMEA region. In 2005, Oracle changed its approach, shifting from a multinational company with local organizational entities in multiple countries to a global company with global business lines whose employees from different countries worked in teams and were managed online. From 2005 to 2012, Oracle merged more than 60 companies and a vast number of new employees into itself. Online leadership and cooperation thus became increasingly important (Kolbaek, 2014). The shared language was English, and the common ICT framework developed over the years; for more details, see Table 1 and the following discussion. These technological developments are important to understand what the online leaders could actually do in their roles (Gherardi, 2012).

The development of common ICT shows that the technology available to the employees transformed over time. In Table 1, this process can be seen in the changes to the external and internal websites and intranet in 2005, followed in 2006 by the attempt to achieve quicker online cooperation through using a shared repository for files and by the use of blogs to avoid e-mail spam within teams and make it easier to gather information about other teams.

As Oracle's acquisitions increased, it needed additional online communication channels. In 2007, an Oracle version of an instant messaging service was deployed, allowing team members to converse by writing short sentences to each other over the Internet. Consultants heavily used this tool to ask their colleagues for support with difficult tasks while on site with customers. Questions could be sent to all

Table 1. New Oracle technology available for employees and leaders

Year	New Technology
2002	Mobile phones and landline phones Intranet based on the Internet, included databases, work processes, and local websites in local language E-mail Web conferencing
2003	Internet protocol phones and mobile phones
2004	Websites for communities of practice open to members
2005	Customer portals to interact with external parties, such as customers and partners
2006	New global website for all employees incorporating the intranet and external website Online repository for shared files Websites for communities of practice open to all employees Blog providing a communication channel for leaders
2007	Instant messaging service
2008	Web 2.0 tools and concepts, including New EMEA website Online idea center for innovations Internal people search including all employees Applications for capturing feedback on strategic issues from team members to leaders
2009	More social media Wikis Oracle Facebook-type groups Instant polls, online questionnaires, and chats Web and video conferencing Global project management tools
2010	Improved video conferencing
2011	Competency profile definition in the competency system, e-cards for e-mails, and recognition stamps in the internal online people search
2012	Best practices for remote meetings using social media

team members without the customer's awareness, and team members could answer without talking on the phone or scrolling through emails.

In 2008, Oracle provided new online leadership tools, including additional website functionalities. For example, the idea center allowed teams to come up with and work on ideas independently of time and space and then deliver these elaborated ideas to their online leaders. As well, the people search enabled employees to identify their colleagues and managers to identify team members, as well as potential team members and their managers. Oracle's acquisitions drove up the number of newcomers and demand for effective feedback and learning. The company, therefore, developed a learning process that included new procedures and an application to collect feedback for team leaders and their managers from the geographically dispersed teams (Kolbaek, 2012, 2014).

In 2009, Oracle added social media resources to its corporate tool box, including wikis defined and maintained by team members. Oracle's Facebook-type groups were based on the company's own technology and did not link to Facebook but offered similar functionalities, allowing employees to connect, communicate, and share in a written format as they desired (according to the company's rules and regulations). In addition, online chat, polls, and questionnaires were made accessible as additional options for online leaders and teams to create common ground or identify differing opinions among web conference participants. Furthermore, Oracle launched initial attempts at video conferencing, but differences in the available bandwidth throughout the EMEA region made using this option challenging, and the application did not work for more than four participants (a small team in Oracle). Projects developed from being based in a local workplace to including team members from several countries. Next, globally aligned tools for project management became available.

In 2010, Oracle launched an improved video-conferencing system at its offices, including most EMEA countries. In 2010 and 2011, the company developed an online system for managing performance, competencies, and skills and incorporated it as the basis for its annual personnel performance interview and talent program. These experiences with online leadership and online teams uncovered a need for different forms of recognition than employed in face-to-face cooperation. Consequently, Oracle developed processes and designs for online awards, including a recognition stamp placed on employees' profile in the people search system. In 2012, creating engagement and participation in online meetings remained challenging, so Oracle gathered, formulated, and distributed best practices for remote meetings using social media on an internal webpage.

Working Practices and Critical Competencies

According to the theoretical framework, goal-setting, actions, discussions, and negotiations are important work practices for online leaders. The critical competencies of an online leader include strong communication skills to give and receive feedback and recognition and share observations and insights.

The history of online leadership technology does not reveal their accompanying working practices, but memory work enables the researcher to look at online leadership as a working practice "from inside" the informants (Gherardi, 2012) and to get an impression of how online leaders comprehended their various roles, positioned themselves in relation to their role, and what they actually did as online leaders (e.g., setting clear goals, participating in discussion and negotiation, and giving and receiving feedback in online teams).

To participate in the memory work in this research, online leaders had to have left Oracle, so they would not need its approval to participate in this study. As well, the informants had to be representative of the gender composition of the Oracle workforce (66% male and 33% female). To avoid beginner difficulties in leadership or online leadership, the informants in the memory work had to be senior managers. Leadership roles become more complex with higher positions in the organizational hierarchy, consequently top-level managers from EMEA seem to constitute an appropriate level for gathering rich data. In the spring 2017 seven former top-level managers in Oracle EMEA (two women and five men) were asked to be informants for this study. They were sent an invitation email requesting that they "write one to two pages about a particular episode, action, or event that made an impression on you. Please write in the third person using a pseudonym, and add as much detail as possible, including even what might be considered to be trivial or inconsequential. Please describe your experience without interpretation, explanation, or biography." Six online leaders accepted the invitation within one week, while one did not answer due to an invalid email address. Four informants delivered responses on time, while two did not deliver responses. A meeting of the informants to discuss the results was not possible, so phase two of memory work was not completed. The analysis and the conclusions in this chapter are a provisional form of phase three. In the following text, the available memory works are cited, and the informants are anonymized. Their names are available in the reference (informants in the memory work).

Development of Oracle Stock Value

On January 4, 2002, Oracle stock had a value of US$15.45 US, which increased by 134.85%, reaching US$33.02 on December 28, 2012. According to the New York Stock Exchange, the value of Oracle stock was US$38.45 on December 20, 2016. From January 4, 2002 through December 30, 2016, the value of Oracle's stock increased by 175.11%, thereby also increasing the company's economic capital (New York Stock Exchange, 2017)

ANALYSIS

This analysis examines four themes of online leadership: establishment of common ground for cooperation, effective working practices, critical competencies, and economical results. This analysis includes quotations from the memory work done by four former top-level managers in Oracle EMEA.

By looking into the memory works, the researcher looked for utterances that could be coded or classified as "common ground", "effective working practices" or "critical competences". These categories were also applied for the analyses of the documents. Besides the analysis included "economical results" which were most visible in documents in opposition to the utterances in the memory work.

Common Ground

Beginning in 2006, Oracle provided new ICT for online cooperation. One-way communication included a new extended intranet with global file sharing, local websites, blogs, and people-search functionality. Online cooperation was enabled by a new online project-management system, instant messaging service, and web- and video-conferencing systems which could be made more engaging by incorporating online polls and questionnaires. Online leaders could access people-search functionality and an online

Figure 1. Trends in Oracle stock value, (New York Stock Exchange, n.d.)

idea center, an application for capturing feedback and innovative ideas. In addition, online leaders were equipped with a new ICT competency system for giving feedback and recognition and a recognition stamp to put on team members' profiles in the people search. The new ICT supported the alignment of the project management goals and the ICT framework and provided employees with additional information and common work processes on a renewed intranet. ICT enabled the online teams and online leaders to cooperate synchronously through instant messenger and video conferences and asynchronously through shared files and the idea center. This ICT enabled the online leaders to give feedback and recognition through the new competency system and recognition stamps. ICT provided online leaders with feedback and innovative ideas from their teams, enabling them to build stronger relationships with and coherence within their online teams. ICT support, therefore, enhanced social capital.

Working Practices

The interpretation of online leadership as a working practice requires exploring how online leaders comprehend their *roles*. In this section, the names of individual participants are followed by quotations from their (third-person) memory work. The participants shared their comprehension of their roles in following statements.

Hayden was a "senior manager who worked for thirty years in international roles in large multinationals. Hayden collaborated with people across many countries and five continents and managed organizations of up to a thousand people spread across a dozen locations."

Steve "had worked for a global IT company for several years and was based in the UK. The company had an offshore group in India, which was well established and had been operating for about 12 months. Steve moved into a role that included the management of this offshore group. Steve put a lot of effort into getting to know this group, understanding their work and their people."

Linda "had a team of people working in 20 different countries. She tried to visit the company's various locations at least once per year."

Theobald "was responsible for customers of XX services in Europe, and he participated in meetings from Denmark with team members from USA, Iceland, and New Zealand."

ICT was necessary for the online leaders. They understood that they had to be familiar with technology and adapt when new ICT tools became available. Furthermore, they needed to share a language with their online team members, and the common language of company was English, which was not the first language of Hayden, Theobald, or Linda. The role of an online leader required continuous learning, and the online leaders studied were aware that they needed to utilize their language skills and improve their competences along with the development of ICT. ICT thus forced the online leaders to enhance their cultural capital.

The memory work showed how the online leaders participated in *discussions and negotiations*.

Hayden "often wrote long messages. But when you read his prose, it often felt like he was *talking* to you, not writing a formal instruction or an official statement. ... He was creating a sense of proximity with the reader."

Steve gave an example: "After another week, he still had not heard anything on the prototype. He called the Indian team member. ... Deepak was a little vague to start with, saying that the report was still not ready. It was an awkward conversation."

Linda explained that "facial expressions ... tell us something about someone's mood and state of mind. When communicating remotely in web conferences, we must express how we feel, so we start our calls with telling each other how we feel."

ICT enabled online leaders to communicate across long distances. Hayden preferred to write as he talked to his team. Steve grabbed the phone and made a call when the online team did not meet his expectations. Linda missed seeing the online team and added emotional cues to web conferences without video connections. This memory work showed that the online leaders appreciated oral communication to connect team members, probably because they wanted to build their relationships with the online team members. ICT seems to have enhanced users' communication skills, thereby building relationships between the online leaders and their teams. ICT thus helped enhance the online leaders' social capital.

The memory work showed that *emotions* needed to be explicitly included in discussions. Hayden tried to create a sense of proximity, Steve found situations to be "awkward," and Linda insisted making emotions visible. These online leaders stressed the importance of emotions in communication and negotiations to create proximity, minimize awkward conversations, and share states of mind, probably with the aims to establish coherence and to build relationships within the online team. The online leaders all turned to more oral formats—Hayden by writing as he talked, Steve by making phone calls, and Linda by connecting via web conferences. Although it seemed as if ICT should allow for oral communication to mediate the emotional expressions needed to establish strong relations, the capacity of ICT to do so seemed a bit weak. Indeed, ICT might have hindered the relationship building and thus limited the online leaders and teams' social capital.

The memory work uncovered different forms of *trust* in the various online teams.

Steve, for example, "was very busy over the next month, trying to do his regular duties whilst travelling to parts of Europe and the U.S. He had no contact with the India group during this time. On his return to the UK, he realized that there was nothing from the India team on the new report."

Linda was complimented by a team member with whom she used to have a rather difficult relationship: "I am so glad that you stopped looking over my shoulder all the time and started treating me as a responsible professional. I take much more pleasure in my work now."

Theobald recalled that a "good introduction of the clear roles and responsibilities of the online team members and a well-defined agenda enable[d] the online team members to achieve very high goals."

In these scenarios, Steve realized that the Indian team had not responded to him even though the deadline had passed—they did not have the trust or the courage to address their failure. Linda's team member gave her face-to-face feedback, indicating that Linda was able to build up enough trust online to enable a team member to give very positive, direct feedback. Theobald stuck to the formalities by introducing roles and responsibilities, which seemed sufficient to achieve his goals.

It seems that ICT enabled the online leaders to build up mutual acquaintanceships and trust with online team members. ICT thus supported the growth of the social capital of the online teams, including their leaders.

Critical Competencies of Online Leaders

The memory work showed that ICT forced the online leaders to change their *communication styles*. Hayden recalled that "he often punctuated his email messages with little jokes, sometimes smilies." Steve stated that the "main communication with the team became instant messaging, having established a close relationship at the outset with the regular conference calls." Linda "decided to not only verbally inform people about what she expected from them and what their deliverables and deadlines were, but to put that in writing. In that way, both the team member and Linda herself had something to fall back on and validate if their thinking was still in line with the request."

Hayden turned written language almost into spoken communication and improved his written language by utilizing pictures. Steve learned to use instant messaging when it became available, so he could maintain close contact with his online team in India. Linda utilized ICT to ensure a common understanding of deliverables and deadlines. These online leaders adapted their communication styles to the ICT available to share their considerations and thoughts. ICT allowed the online leaders to stay in contact with their online team members and clearly express their thoughts and expectations, which, in turn enhanced social capital. In this way, ICT enhanced the social capital of the online team and online leader.

The memory work uncovered how ICT encouraged the online leaders to influence *the norms*. Hayden explained that "by informing people and treating them as the intelligent human beings that they are, his team was more autonomous, better able to make decisions on its own, more effective." Steve recalled one "issue [that] was easily resolved when Steve explained from where that data could be obtained." Linda "explicitly told the team members that she would not be micro-managing. There would be no ad-hoc calls to check on people and see if the progress was in line with the plan. Instead, she told people that she would be there to answer their questions and help them resolve any issues they were encountering.

Delivering the right thing at the right time would be everyone's responsibility, with Linda supporting whenever they would ask for it. It felt like a big gamble."

Hayden seemed to influence the norms as determined to "inform people and treat them as the intelligent human beings that they are." Meanwhile, the telephone mediated Steve's considerations and thoughts regarding a solution for the Indian team, and he used ICT to enhance his influence on his team. Linda felt that she was taking a big risk by stopped micro-managing online. She deliberately left her comfort zone to see what would happen if she handed over some responsibility to the members of her online team. This action can be perceived as a form of real-life education.

ICT required the online leaders to develop new competences. Hayden influenced the norms when he improved the autonomy of his online team members, Steve strengthened his influence by talking on the phone to his online team members, and Linda changed the norms when she handed over responsibilities to her online team members. These three online leaders improved their critical competences because their distance from their online team members demanded it. ICT thus encouraged the online leaders to improve their cultural capital.

Economic Results

The memory work revealed how the online leaders set goals and the impact of ICT on how they conveyed the goals to team members.

Hayden explained that "rather than just telling them what to do and expecting his instructions to be obediently executed, Hayden often wrote or spoke messages to explain or repeat the background, the context, the reasons for a certain course of action."

Steve recalled a case in which "a major requirement came up for the India group. … Steve sent the specifications of the report to the group and had a conference call with the relevant managers in India. … The Indian team [said] that yes, they understood the requirement."

Linda "decided to not only verbally inform people about what she expected from them and what their deliverables and deadlines were, but to put that in writing. In that way, both the team members and Linda herself had something to fall back on and validate if their thinking was still in line with the request."

Theobald "experienced … failures in the IT delivery which Theobald was responsible for. The failures needed to be fixed within 24 hours."

The online leaders set clear goals to achieve fiscal and other targets. Hayden eagerly sought to make his online team members understand the background and reasons for the goals, whereas Steve insisted on the goal itself, sharing the specification followed up by a conference call. Linda initiated a pincer movement, starting with oral communication and following with written instruction. Theobald recalled a deadline but did not say how he informed his online team about it.

ICT created an arena in which goals could be made explicit. Steve and Linda were very specific and repeated their goals, whereas Hayden expected his online team members to understand the importance of the goal when he explained its background. The memory work did not reveal how Theobald communicated the goal to his online team. ICT supported goal-setting by enabling online leaders to communicate orally and in writing. Oral communication ensured a common understanding of the goal, while written communication aligned expectations over time. The goals were connected to economic results; therefore, it seems that ICT supported the formation of economic capital.

Figure 1 shows the increase in Oracle's stock value from 2005, when the corporate growth strategy was changed, and employees began to work primarily in online teams. This ICT framework included worldwide intranet access. As well, more systems and communication channels for online leadership and cooperation were developed over the next seven years, leading to successful globalization from an economic perspective. The stock value grew partly due to the strategy shift from organic growth to acquisitions and support for globalization of the work process from technological development and new approaches to leadership and cooperation. ICT seems to have provided the foundation for increasing the economic capital of this globalized IT company.

DISCUSSION AND RECOMMENDATIONS

The use of qualitative data has dominated in this study because the aim is to investigate online leaders' approaches to their role. The case study method is useful for exploring how phenomena unfold in a given context (Justesen & Mik-Meyer, 2012), in this case, the context of Oracle EMEA. The researcher knew about Oracle and the informants, which may be perceived as an advantage in obtaining access to the informants and supporting her understanding of the context and the challenges the informants faced in their role as online leaders (Flyvbjerg, 2006; Brinkmann &Tanggaard, 2010). The four years elapsed since the time of collegiality may decrease researcher bias.

Memory work is a widely used method in studies of social phenomena in other fields but is relatively new in organizational studies. The methodology of this study has several weaknesses: the researcher cannot draw on earlier experience, the number of informants is small, and memories are not true but biographical reflections on the past. In addition, phase two of memory work was not performed, which could have led to a lack of learning among the online leaders and weakened the results of the analysis. However, the study shows that memory work may be suitable for exploring sensitive matters in organizational studies, and it has proven to be useful for investigating the practice of online leadership "from inside" the informants.

The informants did not refer often to technology, so their experiences could not be directly connected to the development of ICT. Nevertheless, the data provide insights into the online leaders' relationship to their common ground and working practices. Further research should include interviews with online team members to add information about the value and role of online cooperation.

FUTURE RESEARCH DIRECTIONS

According to the literature review it seems as Memory Work is still a "new" methodology for organizational studies. Hence future research could benefit from studies based on this method, and future research should provide larger numbers of memories and include the second phase in which the memory authors discuss their memories and negotiate a common understanding of shared issues.

Future research in online leadership could include theory of online social networks for investigating social capital. Actor–Network–Theory (ANT) may serve as a methodological approach for investigating cultural capital, which includes the ability to utilize ICT. ANT accords humans, relationships, and technology the same level of importance in the creation of social situations, such as online leadership. Future research could focus more intensively on online leadership's impact on economic capital, which seems to require more quantitative research than possible in this study.

CONCLUSION

This chapter has explored the research question *"how may information and communications technology (ICT) enhance the economic, cultural, or social capital of online leaders in a global information-technology company classified as big business?"* This study takes its theoretical framework from Gherardi's (2015) practice theory and Bourdieu's (1986) three capital types to capture how ICT may enhance the three different kinds of capital in a global IT company. The study is aimed at understanding the participants' inside view of the practice of online leadership. The methods of document analysis and memory work generate important insights into how online leaders understood their roles, how they positioned themselves in their roles, and what they actually did as online leaders. The data collection cover ITC development at Oracle from 2002 to 2012, changes in the company's stock value from 2002 to 2016, and top managers' memories of their careers as online leaders. The use of memory work as a method in organizational studies is an original contribution to this field of research.

ICT seems to support the enhancement of economic capital. From 2002 to 2016, the value of Oracle shares increased by 175.11%. ICT enabled online leaders to communicate clear, explicit goals and develop mutual acquaintanceship and recognition with their online teams, which seem to contribute to increased economic capital. ICT changed the ways available to lead online teams—from "micromanagement" to "treating team members as the intelligent human beings they are" and giving them more responsibility. ICT enabled the online leader to follow up through different communication channels and styles, likely increasing efficiency and economic capital.

ICT enhanced cultural capital, in particular, the critical competences for using foreign languages to work online (technical skills). ICT supported the online leaders and team members to improve their technological skills to communicate synchronously, for example, through instant messaging and video- or web conferences and to communicate asynchronously using a shared file repository, the idea center, internal websites, and the Facebook-type groups. Furthermore, ICT required improved foreign language skills of the online leaders and team members. Most did not speak the common language (English) as their first language, but using the English language was a prerequisite for utilizing ICT and cooperating online.

ICT may have enhanced social capital because it enabled team members to give feedback and provide ideas to their online leaders. As ICT enhanced communication skills, it helped build relationships between the online leaders and their teams. ICT provided a foundation for the online teams to solve tasks and build relationships among members and their leaders. ICT enabled online leaders to build coherence within online teams and strong relationships with other team members. By supporting strong relations, ITC enhanced social capital.

However, ITC seems to have a weak capacity to express emotions, which can hinder online relationship building. ICT thus may have limited the social capital of online teams, including their leaders. ICT enhanced the cultural and economic capital of online leaders

This study contributes to the research field of online leadership by proposing memory work as a new method for investigating practice of online leadership "from inside" of the practitioners.

ACKNOWLEDGMENT

I would like to thank the managers who accepted the invitation to do this memory work. It still impresses me that managers at this level take their time to do this kind of voluntary work.

The informants in the memory work are Heidy van der Linden, Stéphane Rousset, Stuart Storr, and Teddy Thorup.

REFERENCES

Adler, P. S., & Kwon, S. (2002). Social Capital: Prospects for a new concept. *Academy of Management Review*, *27*(1), 17–40.

Bourdieu, P. (1986). The forms of capital. In J. Richardson (Ed.), *Handbook of theory and research for the sociology of education* (pp. 241–258). Westport, CT: Greenwood.

Bourdieu, P. (2011). The forms of capital. In Cultural theory: An anthology (pp. 81–93). Academic Press.

Brinkmann, S., & Tanggaard, L. (2010). *Kvalitative metoder: En grundbog. "Qualitative Methods: The foundations*. Copenhagen, Denmark: Hans Reitzels Forlag.

Daim, T., Ha, A., Reutiman, S., Hughes, B., Pathak, U., Bynum, W., & Bhatla, A. (2012). Exploring the communication breakdown in global virtual teams. *International Journal of Project Management*, *30*(2), 199–221. doi:10.1016/j.ijproman.2011.06.004

DasGupta, P. (2011). Literature review: Online leadership. *Emerging Leadership Journeys*, *4*(1), 1–36.

Engeström, Y. (2001). Expansive learning at work: Toward an activity theoretical reconceptualization. *Journal of Education and Work*, *14*(1), 133–156. doi:10.1080/13639080020028747

Flyvbjerg, B. (2006). Five misunderstandings about case-study research. *Qualitative Inquiry*, *12*(2), 219–245. doi:10.1177/1077800405284363

Fraser, H., & Michell, D. (2014). Feminist memory work in action: Method and practicalities. *Qualitative Social Work: Research and Practice*, *0*(00), 1–17.

Gherardi, S. (2012). *How to conduct a practice-based study: Problems and methods*. Northampton, MA: Edward Elgar. doi:10.4337/9780857933386

Gherardi, S. (2015). How the turn to practice may contribute to working life studies. *Nordic Journal of Working Life Studies*, *5*(3a), 13–25. doi:10.19154/njwls.v5i3a.4831

Hambley, L. A., O'Neill, T. A., & Kline, T. J. (2007). Virtual team leadership: The effects of leadership style and communication medium on team interaction styles and outcomes. *Organizational Behavior and Human Decision Processes*, *103*(1), 1–20. doi:10.1016/j.obhdp.2006.09.004

Haug, F. (2008). Memory work. *Australian Feminist Studies*, *23*(58), 537–541. doi:10.1080/08164640802433498

Hunsaker, P. L., & Hunsaker, J. S. (2008). Virtual teams: A leader's guide. *Team Performance Management*, *14*(1/2), 86–101. doi:10.1108/13527590810860221

Justesen, L., & Mik-Meyer, N. (2012). *Qualitative research methods in organisation studies*. Copenhagen, Denmark: Hans Reitzels Forlag.

Kolbaek, D. (2012). Proactive review – How to make your organisation learn from experience. BoD.

Kolbaek, D. (2014). *Learning from experience in the context of work*. Aalborg, Denmark: Aalborg University Press.

New York Stock Exchange. (n.d.). *Oracle stock*. Retrieved 20 February 2017 from https://www.nyse.com/quote/XNYS:ORCL

Nonaka, I., & Takeuchi, H. (1995). *The knowledge-creating company*. New York, NY: Oxford University Press.

Onyx, J., & Small, J. (2004). Memory-work: The method. In J. Small (Ed.), Memory-work. Qualitative research in tourism: Ontologies, epistemologies and methodologies. Routledge.

Putnam, R. D. (2001). *Bowling alone: The collapse and revival of American community*. New York, NY: Simon and Schuster.

Rao, M. (2003). *Leading with knowledge*. New Delhi, India: TATA McGraw-Hill.

Schratz-Hadwich, B. (2005). Collective memory-work: The self as a re/source for re/search. In M. Schratz & R. Walker (Eds.), Research as social change: New opportunities for qualitative research. New York, NY: Routledge.

Valsiner, J., & Rudolph, L. (2012). Who shall survive? Psychology that replaces quantification with qualitative mathematics. In E. Abbey & S. Surgan (Eds.), *Emerging methods in psychology* (pp. 121–140). New Brunswick, NJ: Transaction Publishers.

Wenger, E. (1998). *Communities of practice: Learning, meaning, and identity*. New York, NY: Cambridge University Press. doi:10.1017/CBO9780511803932

Zhu, H., Kraut, R., & Kittur, A. (2012) Effectiveness of shared leadership in online communities. In *Proceedings of the ACM 2012 Conference on Computer Supported Cooperative Work, ACM*. (pp. 407-416). New York, NY: ACM. doi:10.1145/2145204.2145269

KEY TERMS AND DEFINITIONS

Cultural Capital: The accumulated labor a person has put into skills and capabilities that enable the person to influence other people and institutions.

Document Analysis: A qualitative research method based on analyzing texts, understood in a broad sense to include written material, pictures, graphs, sound, websites, and film produced independently of the research.

Economic Capital: Accumulated labor measured in money.

E-Leadership: See *online leadership.*

Memory Work: A qualitative research method that enables informants to share memories they find important and use them to consciously change practices.

Online Leadership: The role of leadership mediated by ICT.

Online Team: Team members whose cooperation and collaboration are mediated by ICT.

Practice Theory: A theory that analyzes working life by looking at what people do (their practices) and their interactions, physical surroundings, historical background, and sociological and technological characteristics.

Social Capital: The accumulated labor a person possess that enables the person to participate in a durable network and to give and receive recognition and feedback.

Virtual Teams: See *online team.*

Chapter 7
Challenges in Online Collaboration:
The Role of Shared Vision, Trust and Leadership Style

Kimiz Dalkir
McGill University, Canada

ABSTRACT

This chapter describes a multi-year study on how online collaboration could be effective in the design of a research equipment sharing lab with university and industry partners. More specifically, this chapter focuses on the "soft" challenges of how to establish a shared vision of the project goals, trust among diverse partners, most of whom had never collaborated before this project, and the different leadership model that was required for effective online interactions. Internal obstacles included differences between academic and industrial organizational cultures (publication/patent tension) and the difficulty in establishing trust through online interactions. External obstacles included differences in leadership style (commanding vs. collegial agreement). While there were also logistical issues in the actual sharing of resources, the focus was on online interactions that crossed disciplinary boundaries (amongst university partners) and competition (amongst industrial partners). Collaboration models were useful in assessing online collaboration effectiveness.

INTRODUCTION

Collaboration is a challenging undertaking. The original example of collective pooling of resources that are then shared by a community was the shared commons (Hardin, 1968). The tragedy of the commons refers to the fact that a shared resource can only be effectively shared if everyone "behaves" in a non-selfish manner. Should any one individual over-consume, then the resource may be destroyed, thus depriving everyone. Also referred to as the shared economy or collaborative consumption, the only way effective resource sharing can occur is if everyone shares the value that the common good is greater than the individual gain of any one participant. This principle of synergy, that more value is produced

DOI: 10.4018/978-1-5225-4094-6.ch007

by the whole than the simple addition of all its parts, extends to any collaboration, including online collaboration. The overall research objective in this study was to design and implement a way of sharing expensive research equipment across different universities and companies. What proved to be the "scarce resource" with respect to online collaborations between the different partners was the information resources that the university researchers discovered. These included scholarly works on best practices as well as benchmarking studies on how other universities worldwide had designed their resource sharing systems. The "Information Commons" has become the new shared pasture (Rao et al., 2004).

Collaboration that is co-located in the same physical location and that occurs in a face-to-face simultaneous mode has become somewhat of a luxury in our global and fast-paced workplace today. As a result, increasing use is being made of technology-mediated interactions in order to collaborate online. While this is not necessarily novel, there are contexts in which online collaboration may prove to be particularly challenging. One factor is when the participants have not worked together extensively or not at all in the past. Others include not being able to realize that everyone does not have the same shared vision, understanding of their roles in addition to having different personality, communication and leadership styles. To this end, an opportunity presented itself when a research group was formed with six universities and five private sector companies. Each was in the same city but not in close proximity. The majority of interactions occurred online with one or two face-to-face meetings in each year of the project. While the companies were all in the same vertical industrial sector, the universities spanned different disciplines and languages.

This chapter addresses the research question of how online collaboration could be effective research equipment and expertise sharing with university and industry partners. Historically each participant purchased their own research equipment and, often, the exact same equipment. The project was conducted over a four-year period and included face-to-face and online interactions. More specifically, this chapter focuses on the "soft" challenges of how to establish a shared vision of the project goals, trust among diverse partners, most of whom had never collaborated before this project, and the different leadership model that was required for effective online interactions.

THEORETICAL FOUNDATION

This section presents a literature review on collaboration and online collaboration, as well as insights in potential barriers and success factors for online collaboration.

Collaboration

Bardach (1998) defines collaboration as

any joint activity by two or more agencies that is intended to increase public value by their working together rather than separately (p. 8)

whereas Lawrence et al.(1999) see collaboration as a

cooperative, interorganisational relationship that relies on neither market nor hierarchical mechanisms of control but is instead negotiated in an on ongoing communicative process (p.282).

Nahapiet and Ghoshal (1998) define collaboration as the creation of a network of people who share a common understanding and have high levels of trust, mutual respect and reciprocity. Mattessich et al.(2001) propose the following working definition:

Collaboration is a mutually beneficial and well-defined relationship entered into by two or more organizations to achieve common goals. The relationship includes a commitment to mutual relationships and goals; a jointly developed structure and shared responsibility; mutual authority and accountability for success; and sharing of resources and rewards (p. 47).

Schrage (1990) provides perhaps the most comprehensive definition of collaboration as

...the process of shared creation: two or more individuals with complementary skills interacting to create a shared understanding that none had previously possessed or could have come to on their own. Collaboration creates a shared meaning about a process, a product, or an event....Something is there that wasn't there before (p.140).

Parkinson et al.(2006) outline a number of advantages of effective collaboration between universities and industry including connecting organizations that would otherwise not meet, sharing of physical and intellectual resources such as ideas and information, and complementary expertise that can be applied to solve a common problem or complete a common project. McLaughlin (2004) notes that collaborating partners share ideas, knowledge and resources in order to increase the chances of attaining goals. Payne (2000) emphasizes that the fact that participants are collaborating with one another will lead to a better use of shared resources and uncover any duplication of effort that was occurring.

As organizations become more global and projects become more complex, it becomes increasingly harder for any one individual or organization to possess all the necessary expertise and requirements to successfully meet their objectives. Collaboration has and will continue to be a preferred and extensively used way of working for all organizations, regardless of size, maturity or sector. Universities have followed a similar evolution, with research funding being provided to multidisciplinary research teams rather than individual researchers. Organizations are thus in great need of conceptual models and practical guidelines to help them collaborate effectively.

Online Collaboration

Given that collaboration in general is a challenging task, online collaboration is even more difficult. This is because the interactions now have to depend on one or more technological tools in order to connect, communicate and work with others, who are typically not located in the same place (and sometimes, not in the same time zone). Fichter (2005) notes that:

Online collaboration tools range from the simple to the complex, inexpensive to expensive, locally installed to remotely hosted, commercial or open source, large versus small (p. 49).

All collaboration tools offer a way to communicate, to share documents and to discover who the other members are. They may also have additional features such as an online calendar, shared whiteboard spaces, chat rooms, web conferencing and member profiles. The tools themselves are sometimes additional obstacles if participants are unfamiliar with how to use them or they are not reliable (e.g. Internet connection fails during the interaction). However in general,

failure in online collaboration is usually due to overly complex routines, lack of organizational readiness, governance, and communication value to the individuals (Fichter, p. 48).

Online collaboration has been studied in a variety of contexts. Three are highlighted here in order to illustrate typical obstacles and critical success factors: online learning, online scientific research collaboration and collaborative problem solving. These areas were chosen as they had counterparts in the research project described here. Both industry and university partners engaged in significant online learning, in online problem solving and the project was one of collaborative scientific research.

Tseng et al.(2009) found that online instructors need to work harder to engage students in order to ensure effective learning takes place as it is more difficult to form interpersonal relationships, feedback is less interactive and immediate. (Tseng et al., 2009). The Pathfinder project made use of citizen scientists who become informal research assistants who collected data, participated in making sense of the data, shared information, asked questions and proposed hypothesis online with the researchers and with other citizen scientists (Luther et al., 2009). An example of collaborative problem solving is the Polymath project consisted of mathematicians who worked together online to solve open mathematics problems that are not easily broken up into independent sub-problems or sub-task (Cranshaw & Kittur, 2011). Allan and Lawless (2005) found that participants in online collaborative problem solving projects found the experience stressful when roles were not clearly defined. There was even greater stress when participants were not all from the same organization. Allan and Lawless (2005) also found that leadership, teaching and facilitating roles were critical success factors. The leader must actively and frequently encourage everyone to participate, to ask questions, to challenge assumptions, ask for clarification and engage in a highly participatory form of project management.

In this study, partners had to first collaborate online in order to design how the resource sharing would take place for both research equipment and the knowledge and experience in using this equipment. The research equipment being shared is highly complex and requires expertise not so much to operate but to properly design the research that would make use of the equipment. This article focuses on the preliminary stages involved in designing the sharing of both physical and intellectual resources.

Most collaboration models describe differing levels of collaboration on an evolutionary or developmental scale where it is assumed that the groups will pass from the lower to the higher stages when collaboration effectiveness increases (Frey et al., 2006; Peterson, 1991; Hogue, 1993; Bailey & Koney, 2000; Gajda, 2004). All these models have at their core major stages consisting of communicating, co-operating and collaborating. While these models applied well to the university partners, as our project involved both industrial and university partners, they were not inclusive enough for our purposes. The Shah (2012) and Crescendo (2009) models proved to be better matches.

Shah (2009) defines collaboration as "putting two or more heads together to in pursuit of a common goal" *(p.29).*

The collaboration model presented by Shah (2012) consists of five layers as shown in Table 1. Communication is the process of exchanging information. Contribution is an informal relationship between entities help one another to attain their own individual goals, Coordination implies a harmonious sharing of resources, responsibilities and goals, Cooperation is the stage where similar interests and negotiated roles of the partners move them towards common goals while they share resources and follow collectively established rules of interaction. The ultimate level of the model is Collaboration seen as a process in which partners observe different facets of the problems.

They engage in a process through which they can go be-yond their own individual expertise and vision by constructively exploring their differences and searching for common solutions (Shah, 2009).

Collaboration brings together diverse but complementary perspectives, types of expertise and vision to search for a common solution to a common challenge or problem. Synergy is encountered at this level and the solutions found represent more than the sum of partners' contributions.

The Crescendo Report (2009) has four levels of interaction as shown in Table 2. This last stage is the one where partners have common benefits, act collectively and aspire to reach a shared goal. Entities are still independent and autonomous, but connected through a communication network. They exchange services while they continue to maintain their own logic of action.

A comparison between the Shah and Crescendo models show a fair degree of overlap. Both of these models were found to be a good fit for the project.

Potential Barriers

Collaboration poses a number of challenges such as a technology, trust, power and competition.

As always, we get a little dicey because we are engaging the right research at the universities, usually getting into territory that is competitive, so if it is competitive we don't want [ACME] to know that I am working with University X on a brand new product, so we won't tell anybody (Participant N1).

Technology

Online collaboration is even more challenging as the lack of co-location and face-to-face interactions make it much more difficult to establish trust. Non-verbal cues such as facial expressions, eye contact and other gestures are missing which makes communication more prone to misunderstandings. Online collaboration requires higher media richness (Daft & Lengel, 1984; Kahai & Cooper, 2003) and social

Table 1. The Shah model of collaboration

Stage	Description	Example
Communication	Sending/receiving information	Post to Facebook
Contribution	Help others achieve their goals	Contribute to a fund raiser
Coordination	Multiple actors share goals, resources, responsibilities	An election campaign team that shares an office
Cooperation	Plan activities, negotiate roles and share resources to attain a common goal	Two association presidential candidates agree that one runs for President while the other runs for VP
Collaboration	Complementary perspectives on a problem; agree to search for common solution	Two political parties with differing platforms decide to merge

(Shah, 2009)

Table 2. The Crescendo (2009) model of collaboration

Level	Description	Example
0 - Independence	No inter-organization interaction	Each organization pursues its own goals on its own
1 – Communication	Inter-organizational transfer of information; low degree of interdependency; no common goal.	A teleconference to discuss a potential project.
2 – Contribution	Entities help each other to achieve individual goals; informal relationship (no contracts).	Each participant from a different organization agrees to participate in writing a proposal for the potential project.
3 – Coordination	Entities contribute to a collective set of tasks; their contributions are then integrated; mutual interdependencies are managed to yield reciprocal benefits.	Each participant contributes successful project proposals from their organization so that everyone can make use of to enhance their proposal's chances of being accepted.
4 – Cooperation	Collective actions lead to a common benefit to meet common goals; use of a common framework and set of rules; beneficial when involved in a reciprocal, collaborative problem solving activity.	Participants brainstorm how to get stakeholder engagement in their proposed project. They plan and implement an event to publicize their project to potential collaborators, potential funders and to those who stand to benefit from the project.
5 – Collaboration	Entities seeing different aspects of a problem find a common solution to a problem; more intense than cooperation level (processes are included); individual aims of the participants are more closely aligned; clear distinction between the actions of the partners not possible anymore Authority is in the collaborative entity instead of the individual parties; new entity is formed. More than the sum of the parts: sharing responsibility, authority and accountability for achieving results.	A new non-profit organization is incorporated. All participants are members and contribute time of their staff as well as physical resources that can be shared by all the other members. There is a formal constitution and rules governing intellectual property, insurance, financial management, and conflict resolution. All members formally agree that they will comply.
6 – Integration	Entities are interdependent (cannot always be separated); great degree of coherence and standardization.	International Space Station CERN

presence (Short et al., 1976; Biocca et al., 2003). The former refers to the ability of the medium to accommodate images, videos in addition to text, synchronously as well as synchronously. An example would be Skype which offers more media richness than email. Social presence refers to the degree to which people feel they are interacting with other people rather than with a technology. Face-to-face interactions have the highest degree of social presence. Texting has greater social presence than email due to its synchronous nature (Sun & Cheng, 2007).

Time

Herbsleb and Mockus (2003) point out another factor in online collaboration, the fact that resolving any issues let alone outright conflicts requires much more time. There is an order of magnitude difference. Something that could have been resolved with a few minutes of face-to-face conversation may take several days to tackle due to the time needed to explain the problem, send it to the right person or persons, then wait for them to read it and then reply. Just writing out thoughts on a complicated issue requires much more time than simply thinking about it and discussing it live with the person in front of you.

Content

Online collaboration produces vast amounts of content that needs to be managed: correctly captured, classified, stored, shared and used. As collaboration involves participants with different background, it becomes paramount to correctly label content to describe it so that others will understand and be able to find and use this content when needed. Traceability is another important factor to consider. As content is being created simultaneously (and revised continuously) it becomes more difficult to know which version was produced by whom when and for what reason. A good tracking system is needed such as a document revision history and a decision audit trail. Solid information and knowledge management practices are crucial (Dalkir, 2011).

Trust

Last but not least, the issues of credibility and trust have some bearing on deciding whether or not to use the content created (Larsson, 2005; Evans 2012). This decision will depend on who the author or content contributor is, what others know and think about their expertise, their abilities and advice or recommendations they make. Trust can also be very contextual and not necessarily a fixed characteristic of a given individual. For example, someone may engender a great deal of trust when he or she makes recommendations on which collaborative work platform to use but less credible when it comes to recommendations about how to resolve an interpersonal conflict that has arisen. Abrams et al.(2003) found that employees tend to use informal networks to find information. The authors note that effective knowledge sharing, which is a prerequisite for collaboration, involves two forms of interpersonal trust: role-based trust (e.g. credibility, reputation, and role) and personal trust (that the person is benevolent, altruistic and will reciprocate). Both types of trust appear to be crucial in effective online collaboration efforts. Benevolence-based trust is based on mutual caring about goals and general well-being while role-based or competence-based trust relies on the fact that collaborating individuals possess specific expertise and can be depended upon for that reason (Gambetta, 1988; Mayer et al., 1995; Kramer & Tyler, 1996; Dirks & Ferrin, 2001; Evans & Wencsley, 2009; Evans et al., 2015).

Power

Most research, such as the Gartner Group surveys (2002, 2007), find that people still prefer to ask their peers for any knowledge they need. Borgatti and Cross (2003) were among the first to note the increasingly important role played by social networks in work organizations both in creating new knowledge and sharing existing knowledge. Levesque et al.(2017) found that when multiple organizations are trying to control the outcome of collaboration, distrust is more likely to form than trust. Power, trust and learning will influence how successful the collaboration turns out to be. If power is roughly equal, then decision making will be more transparent. If trust is established, then information will be shared. If learning can occur, then there will be a common understanding of goals and issues. Unequal power means that some partners will have a greater ability to gain outcomes favorable to them. This may cause less powerful groups to have less commitment, to disagree and to feel their identity threatened. Participatory decision making, access to information, clear definition of roles and information rules can help address power issues. Reciprocity is a big factor to help develop trust. When someone has established a good track record, they are more likely to trust them and in return to share more information with them. In parallel, collaboration also requires a different form of learning as well as leadership: social learning rather than individual learning, where expertise from different disciplines must be aggregated or synthesized. (Wenger, 1988; Lave & Wenger, 1991). Opportunities to exchange information are essential – a time and space for collaborating whether it be face-to-face or virtual. Facilitated interactions can help ensure that everyone has a voice regardless of status, authority, or level of power.

Success Factors

Mattessich et al. (2001) and Parkinson et al. (2006) identified a number of critical success factors specifically for collaborations between for-profit businesses and non-profit organizations. Table 3 below shows a high-level regrouping of the original 20 critical success factors identified by Parkinson et al.to make collaboration a success.

Table 3. Ten critical success factors for collaboration

Criterion	Description
A history of collaboration or cooperation in the community	Some of the partners have already worked together in the past
Collaborative group is seen as a legitimate leader in the community	Reputational capital of partners is strong, perceived as credible, reliable, competent
Favorable social and political climate	Opinion-leaders, political interests, funding agencies and the general public support (or at least do not oppose) the mission
Mutual goals, respect, understanding and trust	Members share an understanding of other member' organizations (culture, processes, values, constraints, expectations); they respect one another.
Appropriate cross-section of members	Each segment of the community is represented in the group.
Members have a stake in both the process and the outcome	All participants have "ownership" of how the group works and what the group produces; all levels (operational, management, senior management) of each organization are represented and involved; participants have the same vision, agreed-upon mission, objectives and strategy.
Flexibility and adaptability	All participants are open to working in a variety of different ways in order to accomplish goals; open to making changes as needed to deal with changing conditions.
Develop clear roles and policy guidelines	Everyone understands their role, rights and responsibilities; how to carry out their responsibilities; make sure goals are concrete, attainable and clear to all participants.
Open and frequent communications	Interact often, update everyone, make sure everyone receives information, discuss issues openly.
Ensure sufficient resources (funds, staff, materials and time) and effective leadership	There must be sufficient resources – the goal must be realistic. The group leader must have strong organizing and interpersonal skills and be able to command respect and legitimacy.

(adapted from Mattessich et al.,2001; Parkinson et al.,2006)

ACTION RESEARCH

An action research model was used to study the effectiveness of collaboration among industry and academic partners. Altrichter et al.(2002) define action research as a form of inquiry where researchers share research findings with study participants as they are produced in order to improve their general situation and specific practices. Action research can be thought of as being at the opposite end of the spectrum from experimental research, in which great care is taken to not have any researcher bias and to ensure participants are isolated from the actual research data (Greenwood & Levin, 2006). The authors conclude that "action research is enquiry with people, rather than research on people" (p. 13).

The people under study are part of the research endeavor – they actively participate in research decisions and apply research results as they become available. Bradbury Huang (2010) notes that action research represents "a transformative orientation to knowledge creation in that action researchers seek to take knowledge production beyond the gate-keeping of professional knowledge" (p.1).

Broydon-Miller et al. (2003) concur that action research is a very participatory process where there are scheduled reflective meetings to go over results as they are obtained and to implement decisions and any necessary changes as the project progresses. In other words, unlike experimental research, everyone knows who the researchers are and what they are doing. There is no need to wait until the study is completed before sharing the data and the results with all the participants. The authors also note that "action research is a work in progress" (p. 11)

as there is no single recipe for carrying it out, action research promotes broad participation in the research process and supports action leading to a more just, sustainable, or satisfying situation for the stakeholders.

Together, the professional researcher and the stakeholders define the problems to be examined, cogenerate relevant knowledge about them, learn and....take actions (McNiff, 2013, pp. 3-4).

As the researchers were also participants, the study was thoroughly reviewed and approved by the university ethical boards in order to avoid any potential observer bias. The goal of action research is to purposefully intervene throughout the study in order to implement incremental improvements during the study so it was important to communicate interim results with all participants as they were obtained.

The Research Setting

As university researchers, our team was involved in helping to set up a shared collective pool of resources among three universities and four industrial partners. As a subset of the larger project, our team elected to use action research in order to study how well the participants collaborated on the initial stages of the project, which were to conduct background research to identify a good model of collaboration as well as some key enablers and obstacles to effective collaboration. The extent to which a shared vision existed among all participants, the extent to which participants felt they trusted other participants and the leadership style exercised by the project head were studied and the results are presented in this chapter. In practice, this meant that there were regular meetings to present all preliminary results with all the partners. These meetings were held approximately every month. A shared repository space was created to maintain all project documents that were produced as well as all research results as they were obtained. All partners had access to this repository.

Researchers from two of the universities were observers while those from the other universities were insiders (Bartunek & Louis, 1996). The project leader was an insider and the "owner" of a significant amount of the resources to be shared. Data collection was primarily qualitative. One-on-one interviews were conducted with each partner, and group interviews were conducted with the teams tasked with running the university and industrial labs where the research equipment was being used. Interviews were conducted over the course of the full project in an iterative fashion. Each was a semi-structured interview with a clear focus on a single parameter of online collaboration such as trust or leadership model.

The first step of the research was to identify a collaboration model to use (as is or revised as needed). The Shah and Crescendo models were used to evaluate the current and desired phase of collaboration between the project partners. Next key enablers and obstacles to collaboration, especially online collaboration, between participants who had not worked together before were identified, from the academic and practitioner literature. These factors were then identified and their role validated (or not, as the case may be) within the initial interactions among project partners that occurred during the first year of the project (project kickoff and initiation phases). This period of time was selected as this is when the crucial process of learning to work together occurs, including getting to know (and trust) the other participants, define roles and responsibilities, agree on common goals and set the tone for project leadership.

Issues to Deal With

Project participants identified a number of different objectives they had for the project. The industrial partners wanted primarily to solve problems, reduce costs, diminish turnaround time for projects, keep or increase profit margins and standardize processes. The academic partners wanted their students to work on real-life problems and increase their chances of successfully obtaining research grants sustainable funding to continue the project. This difference in objectives only became clear as the project advanced. At first, there appeared to be consensus on project goals and objectives as everyone was using about the same type of language to describe their motivation in participating. However, differences rose to the forefront at the very first project meeting. As participants discussed what they would like to see as project outcomes, it soon became obvious that there were multiple competing project visions. While project participants were surprised, there is ample research evidence that academic-practitioner collaboration is fraught with tension and conflicting objectives (Bartunek & Rynes, 2014).

Roles

Parkinson et al.(2006) note that conflict is most often a result of three common issues: the roles and expectations were not clear, there was an imbalance of power and progress was not made. They recommend writing down clearly what the role of each participant and each organization is with respect to roles, financial rules, and a conflict of interest policy. If there is an imbalance of power in terms of decision-making, then a neutral third party should convene and facilitate meetings in order to ensure that everyone has a voice and that their contributions are taken into account. Finally, disappointing results are difficult to accept. It is important for the leader to refocus the group by reminding them of their goals (Parkinson et al., 2006). A number of conflicts arose during the project's initial phases. At first perplexing, it soon became clear that a large part of these conflicts were due to the fact that roles and responsibilities were definitely not clearly identified. Each member was proceeding according to a set of rules and norms determined by their particular organizational culture which more often than not pitted them directly against competing goals of members from a different organizational culture. As roles were not explicitly identified, everyone proceeded to behave according to their own assumptions – the most important assumption being that "everyone else is just like me." As there was no direct effort made to discuss these differences, participants simply continued work practices they used in their own organizations. The lack of change pointed to the fact that effective collaboration was not taking place. This was best exemplified by the project leader's response to a question as to where the physical resources would reside: "in my lab of course" as he did not want to be inconvenienced.

Ethics

Kenney (1987) was among the first to address the ethical issues that surround these differing cultures and decision-making styles of industrial and academic sectors. The over-arching priority of a profit-driven business is to make a profit. If those accountable do not succeed, they can be fired. The academic sector has a much more complex set of ethical and moral standards that they must meet. These include making the results of research available to all and for researchers to remain unbiased as much as possible. Recent events, such as research being funded by concerned parties, such as a drug being tested and the research being funded by the pharmaceutical company that manufactures that drug, have brought to light the ethical dilemmas that may arise. Academic organizations hold a set of values that emphasize openness, transparency, neutrality or objectivity and healthy skepticism as brought to bear in the form of an organized and systemic inquiry to better understand phenomena. Whereas academic researchers take elaborate steps to ensure they are not in conflict of interest, this is almost expected of industrial employees: they are supposed to reflect the interests of their employer. A specific example is that of the ethics certificate. Academic partners were obligated to apply for and obtain a certificate of ethics from their Ethics Review Board, a lengthy process that can take up to 6 to 8 weeks. This certificate is mandatory for all research involving human research subjects and no data collection can occur before the certificate is issued. Industrial partners found this frustrating and just another delay as they did not have to operate under the same constraints.

Pace and Results

Interviews with participants showed that individuals in competitive profit-driven industries are often used to executive decision making processed – very formalized, top-down decision-making whereas those in the non-profit sectors such as academia are more used to collegial consensus-building approaches to decision making. In the private sector, success is almost always determined by financial return on a financial investment and time periods are typically fairly short (a quarter, a year at most). In the non-profit and academic sectors, benefits include more long-term intangible assets such as the creation of social and/or innovation capital. The measures are much less precise and benefits may take many years to become visible. Academic partners, for example, have their performance assessed partially by their contributions to a body of knowledge, usually in the form of a peer-reviewed publication, which may take over a year to be published. Private sector partners are much more concerned with creating and maintaining a competitive advantage which involved protecting new knowledge and innovations, typically through patents. As a result, academic partners felt the project was proceeding at an appropriate pace whereas industrial partners were much more impatient and wanted to see at least intermediate results sooner.

Lack of Collaboration

These differences were partially responsible for the lack of trust between members. This key finding was very obvious the first-time members of a different organization asked to have access to the resources compiled by another. Ironically, there was shared space that was created for all project partners to make use of. None did. Instead each set up their own repository and there was very strong reluctance to allow others access. While there were a few face-to-face meetings during the initial phases of the project, most interactions took place online. A few members had worked together before, but the majority had not. The result was that there was insufficient time to develop credibility and a positive track record in sharing resources. Instead, the teams from different organizations proceeded to work on tasks on their own. This led to duplication of effort as the roles, and tasks had not been explicitly delineated. Virtual interactions made it more difficult to get to know the others and begin to understand their different ways of working. For example, instead of integrating work done by the different partners, at the initial project meetings, the presentations were clearly "branded" by each organization. There were multiple slide decks, each with their own organizational template and affiliation clearly indicated on the title slide. Each group presented their own work. There was a great deal of overlap and even some direct contradictions in the content presented. It was very obvious that there had been no communication between the teams before the meeting. As a result, the level of collaboration did not evolve beyond teams working on related tasks in a parallel rather than interdependent or more cohesive fashion. An indicator of evolution towards a more mature or advanced level of collaboration would have been, at a minimum, sharing of the respective slide decks before the meeting and, ideally, working on the presentation together to produce one slide deck with the project name on the title slide. The title slide of the final project presentation, in contrast to earlier presentations, contained the project name, the names of all participants as well as the logos of all participating organizations. There was only one slide deck and different people took turns presenting different sections, showing that there was some movement towards the next level of collaboration – cooperation.

Collaboration that brings together very different participants is expected to cause a great deal of cognitive dissonance, to borrow a term from psychology and marketing. Festinger (1957; 1962) defined cognitive dissonance as the psychological stress experienced by people when they are faced with two or more contradictory ideas, something that clashes with previously held beliefs or values. Often this occurs when someone is presented with a differing point of view. When this happens, people experience discomfort as they have been pushed out of their comfort zone in order to confront and somehow integrate these new ideas.

Bonding

Parkinson et al. (2006) also suggest holding bonding activities to foster more personal connections. This was done through a series of lunchtime meetings but could have been done more frequently throughout the duration of the project. Bonding was not very effective as evidenced by the fact that the collaboration was not sustained and participants did not pursue inter-agency interactions to any significant extent once the project was completed. In the project, a very effective method turned out to be to organize field trips to visit each of the partner organizations. In parallel, the location of meetings was rotated so that they were ultimately held in each of the partner organizations. Another useful tool was to create a promotional pamphlet that summarized the objectives, the participants and documented critical information about the collaboration. This is in line with suggestions from Parkinson et al. (2006) to have the public relations group or communications department of the business partners produce "fact sheets for press releases or strategies for announcing a joint project most effectively" *(p. 58).* The pamphlet served as a banner under which everyone could identify (and be proud of). It served to increase the cohesiveness of the group and the sense of belonging of each participant as they used it as a form of "calling card" or project business card when meeting with others and talking about the project.

Leadership Style

The final key finding had to do with leadership style. As discussed earlier, the early stages of the project were marked with conflicting visions, unclear roles, and direct conflicts. In addition, the project was not progressing at a good rate. In terms of disappointing progress, one of the participants made a very inspirational speech about the bigger picture, what everyone was contributing to and how not only the group but the community and even society would benefit in the end. This was very effective and served as a type of "reset" for the group. Participants were able to rise above their specific issues and conflicts and were motivated to try harder to succeed. Unfortunately, this was not the case with the official project leader who was more apt to express his own personal disappointment that some of his specific objectives were not met. The project leader exercised a much more traditional authority-based type of leadership which was not very compatible with the establishment of trust. Simply put, the project leader did not in fact have any formal authority over the other participants. It would have been preferable to have a much more participatory approach to decision-making and the project leader could have benefited from co-managing with the more inspirational participant. In terms of influence, the inspirational participant, who also did not have any formal authority, was much more effective at facilitating trust among project participants which in turn would have led to a consensus on the shared vision for the project.

Some of these results are not unexpected as they have already been well documented in the literature. What was unexpected was that a project with the explicit goal of collaboration could not overcome the key obstacles to collaboration. Collaboration in operation turns out to be much more complex than an agreement in principle. In particular, much as "good fences make good neighbors," good governance makes for successful collaboration. Collaboration does not just happen – it needs guidance and structure to succeed. Nor can collaboration occur by decree – no one individual or organization has the authority or influence to make it happen. Project members engaged in informal sharing which often relies on personal relationships rather than explicit rules and roles. The project interactions indicated that participants were at level 2 in both the Shah and Crescendo models: contribution. At this level, participants have informal relationships with one another, they have agreed to work on a common project and they help each other achieve individual goals. In order for them to be in a better position to achieve some of their goals such as sustainability of the project, increased efficiency, greater innovation for example, they would need to move up at least two more levels in the collaboration models. Last but not least, the major recommendation from all the research studies cited here was to establish very early on the level of collaboration that was the goal. While this was attempted, consensus was not arrived at with respect to the level of desired collaboration required to meet the project goals. The level of commitment was also not uniform amongst the partners – some had more constraints and were not able to devote the necessary time. In addition, there was significant turnover in all the industrial partners as well as in two of the four university partners. As the people were not the same throughout the project, it was very difficult to establish trust, re-explain roles and responsibilities (due to a lack of documentation) and achieve consensus on the degree of desired collaboration.

SOLUTIONS AND RECOMMENDATIONS

The Shah and Crescendo models were used to situate the current and desired level of collaboration for the project. Currently, the participants appear to be at level 2 (Contribution). However, they would like to be at level 5 on both models (Collaboration). Together with the enablers and obstacles that have been identified, these models suggest the succession of stages they should follow to move from a group of independent or barely interactive entities towards a true and effective collaborative partnership. This means that sharing must be reciprocal, that they are working in an interdependent rather than parallel way, that they share a common vision, that their respective roles are clearly identified and explicitly documented and that leadership is more inspirational than authoritative in style.

At level 2, contribution, individual and not collective goals are pursued when accessing and using the shared pool of resources. This model is the traditional model of a lending library, where patron members can check out a book from a shared collection of books. More recently, this model has been replicated with community car or bike sharing applications and tool lending libraries. In the case of the project, partners contribute their research equipment and human resources (expertise), place them in a collective pool whereby each can be reserved for use through a single online interface. The major advantages are: increased capacity (more resources), increased budget (less cost in acquiring collective resources), and shared expertise (collective specialized expertise in the different organizations). In order to reach the next level, coordination (level 3) there would have to be clearly identified collective goals and benefits.

Another requirement to reach the next level would be to foster the emergence of a true community of users. This would require interaction, communication and knowledge sharing not only between the partners but between all users of the collective resources. A true culture of sharing can develop when users begin to help one another out, share what they have learned (e.g. FAQs) and slowly let go of the often dominating need for ownership.

In summary, collaboration levels are meant to be evolutionary in nature so it is expected that if the group is currently at level 2, they can start operating and with time and more collective projects, they will eventually reach level 3. Some additional requirements to attain this level would be:

1. Consensus on shared vision by all participants;
2. Reciprocity and trust (each participant must see a benefit);
3. Participatory and inspirational leadership;
4. Shared knowledge portal that visibly identifies authors and contributors; and,
5. Clear and explicit definition of roles and responsibilities.

In addition, roles and expectations of each participant need to be clearly identified. A safe environment where participants trust each other must be established through explicit attribution and reciprocity throughout all online interactions. The leadership style needs to be closer to the university model than the industrial model in order for online collaboration to be more effective. This means that the leader needs to establish credibility but not authority. The leadership has to be open and transparent. The goal should be to inspire, catalyze and promote conversations as well as to provide a good conversational place so that all voices can be heard and project participants feel they can have a clear impact on the project outcomes.

FUTURE RESEARCH DIRECTIONS

It would be informative to carry out a social network analysis (SNA) to better understand the interactions (and lack thereof) between collaboration partners. This would provide a good means of tracking the development of trust in a dynamic fashion. In particular, it would be a good idea to establish a baseline for their current level of collaboration (2) and then repeat the SNA to see if there is more connecting, more reciprocity and so on which would provide one indicator of evolution towards the next levels of collaboration. While it is not possible to conduct this analysis on the initial project as the partners have disbanded, it would be useful to assess the SNA baseline in future collaborative partnerships and repeat the SNA to assess whether or not the collaboration level was improving and a greater cohesion was being formed. The same partners are continuing their interaction but this is limited to seeking new sources of funding in order to continue the research project.

Another interesting avenue to pursue would be to develop a simulation model that could test the sharing of resources between project partners. A similar study was done by Pasin et al. (2002) for resource sharing among health-care organizations. They designed a simulation for the pooling of healthcare resources between different organizations to test out different sharing scenarios to ensure fairness. For example, assigning priorities to certain users for example and then verify whether any of the participants were penalized. The simulation tool helped overcome initial resistance by showing that all participants would benefit from pooling their resources and that the more they used the common resources, the less their costs and the greater their benefits.

Finally, the development of a knowledge sharing portal would be the next step in our research study. A similar approach was described by McAdam et al. (2008) who also looked at university-industry collaboration. They found that there were positive and negative aspects to sharing knowledge and information. McAdam et al. (2008) found that a multi-level framework of sharing regulations tied to knowledge type to be shared and degree of collaboration required led to effective collaboration. They were better able to resolve conflicts of interest between academic and industrial partners (e.g. scientific research vs. product development). They also developed a knowledge portal to provide a more systematic way of integrating knowledge that was collaboratively created. This is a much more structured shared space where knowledge contributors are clearly identified and content is well classified to facilitate its reuse by others. The clear attribution of authors and contributors of resources is expected to also help foster more trust and therefore more sharing among participants.

The collaboration model proposed in this paper can be applied by researchers and practitioners to improve the efficiency and effectiveness of information and knowledge sharing between disparate and geographically dispersed organizations. The model can also help diagnose the current and desired level of collaboration and provide concrete next steps as well as milestones to assess as progress is made. The degree to which a vision is shared by all participants, the degree to which they trust one another and the appropriateness of the leadership style can also be benchmarked against the different levels of collaboration in the model. However, there are also limitations to this approach and to the model itself. There is a solid foundation of research that highlights the role played by organizational culture and individual differences in sharing, whether it be sharing of physical resources or sharing of information online. There is a need to assess or diagnose organizational culture to come up with a profile. In parallel, a profile is needed for each participant as they will differ in how they prefer to share, both as the giver and the receiver, in addition to other parameters such as language, seniority, motivation, skill level, time constraints to name a few (Durugbo et al., 2011).

CONCLUSION

The research project consisted of an extensive review of the literature, combined with an in-depth analysis of similar sharing taking place in organizations around the world to propose an adapted model of sharing. The Shah and Crescendo models were chosen as they shared many features and were applicable to industry-university collaborations that occurred primarily online. These models should not be seen as a finality, but rather as a diagnostic tool that can be used to track the evolution of the sharing that can occur amongst industrial and academic partners in the aerospace industry. All partners were interviewed in order to gain a better understanding of what sharing level exists today and what level can be targeted in the next phase of the project. These interviews showed that the current status is at level 2 – contribution. Level 3, coordination, would be the next step which would entail more collective cohesion, interdependency and more explicit definition of sharing rules and roles. This would require that participants spend more time "bonding" with one another, getting to know one another and their respective institutions. There needs to be a minimum amount of time in order to establish a good credibility and track record of reciprocity. The ability to articulate flexible, adaptable and shared goals, vision and leadership with continuous communication throughout this community of users would then be required in order to attain the next level, 4, cooperation. Clear guidelines on how to resolve any disputes or conflicts that may arise will be needed. In order to attain the level of collaboration (level 5), a new entity would need to be created that will manage the collective pool of resources, ideally with dedicated resources (staff, budget) in order to be sustainable.

REFERENCES

Abrams, L., Cross, R., Lesser, E., & Levin, D. (2003). Nurturing interpersonal trust in knowledge-sharing networks. *Academy of Management Review*, *17*(4), 64–78.

Allan, J., & Lawless, N. (2005). Learning through online collaboration by SME staff: A scoping investigation into likely team-role stressors. *Education + Training*, *47*(8/9), 653–664. doi:10.1108/00400910510633189

Altrichter, H., Kemmis, S., McTaggart, R., & Zuber-Skerritt, O. (2002). The concept of action research. *The Learning Organization*, *9*(3), 125–131. doi:10.1108/09696470210428840

Bailey, D., & Koney, K. (2000). *Strategic alliances among health and human services organizations: From affiliations to consolidations*. Thousand Oaks, CA: Sage. doi:10.4135/9781483328546

Bardach, E. (1998). *Getting agencies to work together: The practice and theory of managerial craftsmanship*. Brookings Institution Press.

Bartunek, J., & Louis, M. (1996). *Insider/outsider team research* (Vol. 40). Sage Publications.

Bartunek, J., & Rynes, S. (2014). Academics and practitioners are alike and unlike. The paradoxes of academic-practitioner relationships. *Journal of Management*, *40*(5), 1181–1201. doi:10.1177/0149206314529160

Biocca, F., Harms, C., & Burgoon, J. K. (2003). Toward a more robust theory and measure of social presence: Review and suggested criteria. *Presence (Cambridge, Mass.)*, *12*(5), 456–480. doi:10.1162/105474603322761270

Borgatti, S., & Cross, R. (2003). A social network view of organizational learning: Relational and structural dimensions of "know wh0.". *Management Science*, *49*(4), 432–445. doi:10.1287/mnsc.49.4.432.14428

Bradbury Huang, H. (2010). What is good action research? Why the resurgent interest? *Action Research*, *8*(1), 93–109. doi:10.1177/1476750310362435

Cranshaw, J., & Kittur, A. (2011, May). The polymath project: lessons from a successful online collaboration in mathematics. In *Proceedings of the SIGCHI Conference on Human Factors in Computing Systems* (pp. 1865-1874). ACM. doi:10.1145/1978942.1979213

CRESCENDO Consortium. (2009). *Enterprise collaborative state-of-art review report*. CRESCENDO FP7– 234344. Retrieved from http://cordis.europa.eu/docs/results/234344/final1-crescendo-d017-final report-20130430-r1.pdf

Daft, R., & Engel, R. (1984). Information richness: A new approach to managerial behaviour and organizational design. *Research in Organizational Behavior*, *6*, 191–233.

Dalkir, K. (2011). *Knowledge management in theory and practice* (2nd ed.). Boston, MA: MIT Press.

Dirks, K., & Ferrin, D. (2001). The role of trust in organizational settings. *Organization Science*, *12*(4), 450–467. doi:10.1287/orsc.12.4.450.10640

Durugbo, C., Hutabarat, W., Tiwari, A., & Alcock, J. R. (2011). Modelling collaboration using complex networks. *Information Sciences*, *181*(15), 3143–3161. doi:10.1016/j.ins.2011.03.020

Evans, M. (2012). *Knowledge Sharing: An empirical study of the role of trust and other social-cognitive factors in an organizational setting* (Doctoral thesis). Faculty of Information Studies, University of Toronto, Toronto, Canada.

Evans, M. M., & Wensley, A. K. P. (2009). Predicting the influence of network structure on trust in knowledge communities: Addressing the interconnectedness of four network principles and trust. *Electronic Journal of Knowledge Management*, *7*(1), 41–54.

Evans, M. M., Wensley, A. K. P., & Frissen, I. (2015). The mediating effects of trustworthiness on social-cognitive factors and knowledge sharing in a large professional service firm. *Electronic Journal of Knowledge Management*, *13*(3), 240–254.

Festinger, L. (1957). *A Theory of Cognitive Dissonance*. Palo Alto, CA: Stanford University Press.

Festinger, L. (1962). Cognitive dissonance. *Scientific American*, *207*(4), 93–107. doi:10.1038/scientificamerican1062-93 PMID:13892642

Fichter, D. (2005). The many forms of e-collaboration: Blogs, wikis, portals, groupware, discussion boards, and instant messaging. *Online (Bergheim)*, *29*(4), 48–50.

Frey, B. B., Lohmeier, J. H., Lee, S. W., & Tollefson, N. (2006). Measuring collaboration among grant partners. *The American Journal of Evaluation*, *27*(3), 383–392. doi:10.1177/1098214006290356

Gajda, R. (2004). Utilizing collaboration theory to evaluate strategic alliances. *The American Journal of Evaluation*, *25*(1), 65–77. doi:10.1177/109821400402500105

Gambetta, D. (1988). *Trust: Making and breaking cooperative relations*. Oxford, UK: Basil Blackwell.

Gartner Group (2002). *Gartner survey on managing information*. Note number COM-15-0871. Author.

Gartner Group. (2007). *Enterprise 2.0: Collaboration and Knowledge Management Renaissance*. Report number G00203367. Author.

Greenwood, D. J., & Levin, M. (2006). *Introduction to action research: Social research for social change*. SAGE publications.

Hardin, G. (1968). The tragedy of the commons. *Science*, *162*(3859), 1243–1248. doi:10.1126/science.162.3859.1243 PMID:5699198

Herbsleb, J. D., & Mockus, A. (2003). An empirical study of speed and communication in globally distributed software development. *IEEE Transactions on Software Engineering*, *29*(6), 481–494. doi:10.1109/TSE.2003.1205177

Hogue, T. (1993). *Community-based collaboration: Community wellness multiplied*. Bend, OR: Chandler Center for Community Leadership. Oregon Center for Community Leadership, Oregon State University.

Kahai, S. S., & Cooper, R. B. (2003). Exploring the core concepts of media richness theory: The impact of cue multiplicity and feedback immediacy on decision quality. *Journal of Management Information Systems*, *20*(1), 263–299. doi:10.1080/07421222.2003.11045754

Kemmis, S., McTaggart, R., & Nixon, R. (2013). *The action research planner: Doing critical participatory action research*. Springer Science & Business Media.

Kenney, M. (1987). The Ethical Dilemmas of University – Industry Collaborations. *Journal of Business Ethics*, *6*(2), 127–135. doi:10.1007/BF00382026

Kramer, R., & Tyler, T. (Eds.). (1996). *Trust in organizations: Frontiers of theory and research*. Thousand Oaks, CA: Sage.

Larsson, A. (2005). *Engineering know-how: Why social connectedness matters to global design teams* (Doctoral thesis). Department of Applied Physics and Mechanical Engineering, Lulea University of Technology, Lulea, Sweden.

Lave, J., & Wenger, E. (1991). *Situated learning: Legitimate peripheral participation*. Cambridge, UK: Cambridge University Press. doi:10.1017/CBO9780511815355

Lawrence, T., Hardy, C., & Phillips, N. (2002). Institutional effects of interorganisational collaboration: The emergence of proto-institutions. *Academy of Management Journal*, *45*(1), 281–290. doi:10.2307/3069297

Levesque, V. R., Calhoun, A. J., Bell, K. P., & Johnson, T. R. (2017). Turning contention into collaboration: Engaging power, trust, and learning in collaborative networks. *Society & Natural Resources*, *30*(2), 245–260. doi:10.1080/08941920.2016.1180726

Lock, J. V., & Redmond, P. (2006). International online collaboration: Modeling online learning and teaching. *Journal of Online Learning and Teaching / MERLOT*, *2*(4), 233–247.

Mattessich, P. W., Murry-Close, M., & Monsey, B. R. (2001). *Collaboration: What makes it work* (2nd ed.). St. Paul, MN: Wilder Publishing Center.

Mayer, R., Davis, J., & Schoorman, S. (1995). An integration model of organizational trust. *Academy of Management Review*, *20*, 709–734.

McAdam, R., O'Hare, T., & Moffett, S. (2008). Collaborative knowledge sharing in composite new product development: An aerospace study. *Technovation*, *28*(5), 245–256. doi:10.1016/j.technovation.2007.07.003

McLaughlin, K. (2004). Towards a "modernized" voluntary and community sector? *Public Management Review*, *6*(4), 555–562. doi:10.1080/1471903042000303337

McNiff, J. (2013). *Action research: Principles and practice*. Routledge.

Nahapiet, J., & Ghoshal, S. (1998). Social capital, intellectual capital, and the organizational advantage. *Academy of Management Review*, *23*(2), 242–266.

Parkinson, C. (2006). Building successful collaborations: A guide to collaboration among non-profit agencies and between nonprofit agencies and businesses. Cambridge, Canada: Cambridge & North Dumfries Community Foundation.

Pasin, F., Jobin, M. H., & Cordeau, J. F. (2002). An application of simulation to analyse resource sharing among healthcare organisations. *International Journal of Operations & Production Management*, *22*(4), 381–393. doi:10.1108/01443570210420395

Payne, M. (2000). *Teamwork in multiprofessional care*. Basingstoke, UK: Macmillan.

Peterson, N. (1991). Interagency collaboration under part H: The key to comprehensive, multidisciplinary, coordinated infant/toddler intervention services. *Journal of Early Intervention*, *15*(1), 89–105. doi:10.1177/105381519101500111

Rao, R., Dalkir, K., & Wiseman, E. (2004). Public access to information and the creation of an 'information commons.' In *ASIST 2004: Proceedings of the 67th ASIST annual meeting*. Chicago, IL: Information Today, Inc.

Schrage, M. (1990). *Shared minds*. New York, NY: Random House.

Shah, C. (2009). Toward collaborative information seeking (CIS). *Proceedings of JCDL 2008 Workshop on Collaborative Exploratory Search*.

Shah, C. (2012). *Collaborative information seeking: The art and science of making the whole greater than the sum of all* (Vol. 32). Springer Information Retrieval Series. doi:10.1007/978-3-642-28813-5

Short, J. A., Williams, E., & Christie, B. (1976). *The social psychology of telecommunications*. London, UK: Wiley.

Sun, P. C., & Cheng, H. K. (2007). The design of instructional multimedia in e-Learning: A Media Richness Theory-based approach. *Computers & Education*, *49*(3), 662–676. doi:10.1016/j.compedu.2005.11.016

Tseng, H., Wang, C. H., Ku, H. Y., & Sun, L. (2009). Key factors in online collaboration and their relationship to teamwork satisfaction. *Quarterly Review of Distance Education*, *10*(2), 195.

Wenger, E. (1998). *Communities of practice: Learning, meaning and identity*. Cambridge, UK: Cambridge University Press. doi:10.1017/CBO9780511803932

KEY TERMS AND DEFINITIONS

Action Research: A type of research methodology in which the human subjects who form part of the study are not treated as "subject" but rather as active participants who play a significant role throughout the research. They contribute to the design and implementation of the research study and in turn they receive results as they are obtained in order to make use of these to improve their organization in general or specific work practices in particular.

Cognitive Dissonance: This is a type of mental discomfort that occurs when previously well understood and comfortably assimilated knowledge is directly contradicted by newly acquired knowledge. Human beings strive to make sense of their world at all times, so when this occurs, there is some discomfort, but ultimately, the new knowledge is assimilated and existing mental conceptual frameworks are revised or updated in order to take this new knowledge into account.

Collaboration: A form of relationship between at least two people where everyone understands, shares, and is working toward a common goal. Synergy is present as the group is often more effective at organizational tasks such as problem solving then would be possible by just adding up the contributions of each participant. There is a great deal of interdependency and participants should feel they are in a mutually respectful, trusting, and reciprocal relationship with everyone else.

Leadership Style: The way in which someone who holds a position of authority manages those who report to him or her. This includes the ways in which they make decisions (e.g., authoritative vs. participatory), how they convey their plans and the roles everyone will play in them to their staff, how they motivate people (e.g., intrinsic vs. extrinsic rewards), and other factors such as coaching and/or mentoring.

Organizational Culture: The system of shared assumptions, values, and beliefs that lie behind why things are done the way are done in a particular organization. Culture is a complex interaction of system rewards and censures which when well understood can explain why people behave the way they do. Organizational culture will influence how people interact, communicate, and work in their organization (e.g., preference for face-to-face vs. texting, dress code, and work/life balance).

Power: The ability to directly influence something or someone; to control something; to make something happen or to prevent something from happening. An individual has the "power of life and death" when they are a court judge about to pronounce a sentence. A manager has the power to assign a given staff member to a specific project. Power may have diverse sources other than formal authority.

Trust: Trusting someone means that you are confident you can rely upon them if you need them to do something for you. A trusted person is someone who has a positive track record with you (e.g., they always reciprocated in the past), someone who is credible, who is honest, who has a positive reputation. Trust may develop through direct interactions or through a third party (e.g., someone vouching for someone else – as long as the voucher is someone you trust).

Vision: The ability to form a picture in your mind of what you hope your organization (project or team) can achieve in the foreseeable future. Being able to clearly describe what the desired state should be within a certain time horizon. An explicitly articulated intention of what you expect to happen in the future.

Chapter 8
Power in Online Leadership

Morten Fogsgaard
University of Aalborg, Denmark

Claus Elmholdt
University of Aalborg, Denmark

Rikke Lindekilde
Virtuel-Ledelse.com, Denmark

ABSTRACT

In this chapter, the authors explore how power processes can promote or hinder effectiveness in online leadership. The chapter offers a theoretical and empirical analysis of the power processes in online leadership, using the triangle of structural, personal, and discursive power as an analytical lens. The theoretical outset for understanding online leadership is the literature on boundaries and boundary-spanning leadership. The authors define online leadership as social influence processes mediated by technology to produce individual and collective efforts to accomplish shared objectives. The chapter is empirically based on material from one of the authors' PhD work on online leadership in a global Danish manufacturing company. The chapter illustrates how power can be used both constructively and destructively in online leadership.

INTRODUCTION

Alignment of expectations across locations is crucial. I think this part is generally underestimated. For instance, when managing introverted engineers at a distance, you need to know their individual expectations towards work and to you as a manager. Then you have to communicate your managerial expectations and align these in order to shape work and prevent deviance, no tool or groupware can help you on this.

These words were spoken by the manager of a global virtual team approximately 12 months after the team was established. The manager expressed his realization that, when managing online knowledge workers, his previous methods of supervising were not sufficient to achieve success and show that, given the constraints of technological interaction, online leaders and employees must develop team agreements for

DOI: 10.4018/978-1-5225-4094-6.ch008

how to interact and conduct their distributed collaboration. These agreements apply to various essential aspects of collaboration, such as the distribution of power in decision-making processes, when to send a query, how to communicate sensitive and personal feedback and even how to develop and maintain trustful relationships across distances. Whereas, in collocated collaboration, a number of social forces operate to ensure that both social information and activity information are exchanged, these forces are weak or absent in virtual teams. Therefore, leading a geographically distributed online workforce, such as those found in virtual teams, creates new demands for both team leadership and organization (Lipnack & Stamps, 1997). The team leader no longer has close interactions with individual team members, and it is no longer possible to oversee employees' everyday work or to have full knowledge of a given situation (Gibson & Cohen, 2003). The authors suggest that one approach for a geographically dispersed team to overcome these issues is to rely on a more empowering approach to leadership: an approach in which both the team members and the team take part together in the leadership activities (Lipnack & Stamps, 1997).

In this chapter, the authors explore the research question of how power processes can promote or hinder the accomplishment of effectiveness in online leadership of virtual teams. They define online leadership as the social influence processes in virtual teams that produce individual and collective efforts to accomplish shared objectives. Research on effectiveness in online leadership has mostly addressed issues of technology, communication and leadership styles, largely overlooking the question of power. A recent search (31 May 2017) on the words "online leadership" and "power" using the search engine "Primo" revealed zero hits. Managers need to understand the role of power in organizations better in general, and in particular they need to understand the role of power in virtual teams, which is a different way of organizing and managing work activities. This chapter seeks to contribute to the sparse research on the topic of power in online leadership. A theoretical and empirical analysis of power processes in online leadership is conducted using the triangle of structural, personal and discursive power (Fogsgaard & Elmholdt, 2016) as an analytical lens. The authors' theoretical outset for understanding online leadership is the literature on boundaries (Akkerman & Bakker, 2011) and boundary-spanning leadership (Ernst & Chabot-Mason, 2011). The chapter is empirically based on material from one of the authors' PhD work on online leadership in a global Danish manufacturing company. The chapter illustrates how power can be used both constructively and destructively in online leadership in a way similar to that used in face-to-face leadership. We will explore and discuss the potentials, challenges and pitfalls of power in relation to online leadership in virtual teams.

The aim is to create the basis for a constructive practical development of online leadership in virtual teams. The authors illustrate how increased awareness, analysis and discussion of power issues in online leadership can be constructive. It is thus a central point that power is not necessarily something that breaks down and represses, but that an explicit focus on power can be productive for leadership in virtual teams.

THE CHALLENGE OF BOUNDARIES IN ONLINE LEADERSHIP

Over the last 20 years, there has been an increase in people working physically distributed in virtual teams in knowledge-intensive industries. Teams working with, for example, research and development (R&D), product development or innovation are often distributed across the world to meet the demands for knowledge and expertise. They work in virtual teams in which team members use technology to interact with one another across geographic, organizational and other boundaries to share knowledge and work towards common goals (Martins et al., 2004).

The new layouts of boundaries in online leadership of virtual teams is interesting to study for at least two reasons. (1) Most (if not all) organizations today are virtual to some degree, as virtuality cannot be limited to a specific state or certain parameters. Information technology (IT) does not dictate virtuality; rather, it should be considered in a way that makes virtuality multidimensional based on several parameters, such as the degree of temporal, cultural, spatial and organizational dispersion (Panteli & Chiasson, 2008). (2) Although arguments have been made for the great benefits of cross-boundary and virtual teams, research has found that between 60 and 80% of virtual teams fail to meet their strategic or operational goals (Govindarajan & Gupta, 2001). Cross-boundary teaming is an increasing popular strategy for innovation (Edmondson & Harvey, 2016). Yet organizational case studies reveal that teaming across knowledge boundaries, which can be thickened by differences in language, culture and spatial disparity, can be difficult in practice, and innovation is not always realized (Carlile, 2002, 2004). Research shows that the reason behind this high failure rate lies in leaders' inability to manage the distributed workforce (Kanawattanachai & Yoo, 2002). Binsbergen (as cited in Pantelli & Chiasson, 2008) states that: "globalization is not about the absence or dissolution of boundaries, but about... the opening up of new spaces and new times within new boundaries that were hitherto inconceivable" (p. 875). All of these factors change the layout of the organizational boundaries to bridge and work across and force online leaders to master the skills of boundary-spanning leadership, defined as "the ability to create direction, alignment and commitment across group boundaries in service of a higher vision and goal" (Ernst & Chrobot-Mason, 2011, p. 2).

Paradoxically, while technological innovations and globalization have made doing business across spatial and temporal borders easier, social psychological and relational barriers remain obstacles to productive human interactions across professions, departments and organizations (Edmondson, 1999). As stated by Ernst and Chrobot-Mason (2011, p. 1)" while physical barriers have been diminished, relational barriers – obstacles to productive human interaction – not only remains largely unchanged but in some cases, has deepened". The term 'boundary' indicates both a difference and an encounter, in which two (or more) "sites", "things", "arenas" or "ideas" are collated and perceived as both relevant to each other *and* different (Akkerman & Bakker, 2011). In line with practice theory (Corradi, Gherardi & Verzelloni, 2010), the authors view boundaries not as stable entities—things that are set and given—but as things that are actively negotiated and renegotiated in organizational practice. Hernes (2004) suggests a useful framework for studying boundaries differentiating three categories: "physical"," social" and "mental". Physical boundaries refer to material boundaries, such as spatial and temporal dispersion. Social boundaries are associated with the social bonds among people and the creation of a group identity in relation to other groups. Finally, mental boundaries relate to such mechanisms as core ideas, understandings and beliefs that tend to guide actions within a group or organization. Case studies in tradition of practice-based theory reveal that boundaries are not stable entities but ambiguous, volatile and embedded with power processes in social practice (Corradi, Gherardi & Verzelloni, 2010). In an early study on boundaries discussing the setting up of Berkeley's Museum of Vertebrate Zoology, Star and Griesemer (1989) developed the concept of boundary objects (in this case, animals and classification records), and proposed that they play a crucial role in enabling cooperation in temporally, spatially and culturally dispersed groups.

Boundary objects are objects, which are both plastic enough to adapt to local needs and the constraints of the several parties employing them, yet robust enough to maintain a common identity across sites... These objects may be abstract or concrete (Star & Griesemer, 1989, p. 393).

The research on boundary objects has mainly focused on the positive and performative aspects, which were highlighted from the beginning in the description of boundary objects as "anchors and bridges" (Ibid., 414). More recently research has acknowledged that boundary objects may also generate conflicts and tension in dispersed groups, "acting as barricades and mazes" (Oswick & Robertson, 2009, 179). In a case study conducted by the authors the virtual work room used as collaboration interface by the dispersed team of engineers may be defined as a boundary object. Yet, as the authors will show later in this chapter, the boundary object of a virtual work room, does not by itself guarantee positive and prosperous collaboration in the team. They argue that a power conscious leadership facilitating the negation of expectations and the formulation of a shared work agreement is needed in order to create common ground for collaboration in the virtual team. The point that boundary objects does not guarantee smooth collaboration across boundaries is supported by the case studies on cross-boundary work in innovation teams by Carlile (2002, 2004), who described three major barriers to knowledge sharing in dispersed groups. The syntactic (differences in language), the semantic (differences in meaning) and the pragmatic (differences in practice). Carlile described the pragmatic barriers as the most difficult to overcome as they necessitate the development of new practices across dispersed communities of practice. This is in line with the case study in this chapter, discussing the managerial role in building new practices in virtual teams.

ORGANIZATIONS AND POWER

Power is about exercising direct and indirect pressures on individuals or groups that give way (Yukl, 2013). Though many agree with this classic definition, it can also be challenged. Is power nothing more than enforcing one's authority over others, giving orders and directly compelling someone to do something? Is it not also powerful to be able to motivate and direct organization members' behaviors toward specific goals? Is it not also powerful to create a loving" container" that might eliminate the importance of the asymmetric relationship between leader and employee, so that the usual insecurity, incongruence and distrust disappear and are replaced by feelings of equality, openness and cohesion?

In traditional hierarchical organizations, the distribution of power and its basic conditions and consequences are relatively clear. This differs somewhat in more dialogically oriented and "flat" organizations. There has occurred an atomization of power, such that power no longer has a unique center or a clear structure. This tendency is particular evident in the online leadership of spatially, temporally and culturally dispersed virtual teams. Together, the physical, social and mental boundaries of online leadership blur the concept of direct positional managerial power (Ernst & Chrobot-Mason, 2011).

Organizations emerge when people coordinate and cooperate to achieve certain goals. The organizational theorist Mary Jo Hatch (2011, p.1) describes the concept as follows: "Organizations emerge when people work together to achieve a particular desired state or goal. " Organizations often consist of many individuals, each with unique backgrounds and interests. Encouraging all of these individuals to work together to achieve a common organizational goal is, essentially, a form of power (Clegg et al., 2006). The authors consider power to be an analytical concept that is neither negative nor positive.

Rather, it is the way in which power is used that can be negative or positive. As such, an integrative view of power and influence in organizations that incorporates their various negative and positive forms and consequences is conceptualized. Central to this approach is Foucault's (1982) understanding of power as a relational and discursive phenomenon that creates subjects. Power and influence are understood, therefore, not as separate processes, but as two sides of the same coin: power is a tool of influence, and influence is a source of power.

THE TRIANGLE OF POWER

The authors suggest an integrative conception of power using the *triangle of power* as an analytical tool grounded in three different power perspectives (Fogsgaard & Elmholdt, 2014). The triangle of power is an attempt to develop an integrative and analytical framework of power in online leadership, referring to a wide spectrum of power and influence dimensions. The triangle of power is elaborated in relation to online leadership throughout the rest of this chapter. The strength of not limiting power to a narrowly defined concept is that this approach secures sufficient openness to what truly goes on in the field of power and leadership in virtual teams. A multiple power concept can help sensitize one to the diversity of the field and, thus, sharpen one's awareness of opportunities for change in power relations. The triangle of power is a model grounded in three different power perspectives that mutually contribute to an understanding of the dynamics of power within online leadership. We argue that an analysis of power in online leadership should include all three perspectives and that successful leadership is about precisely balancing and using all of these power dimensions. The three perspectives of the power triangle are internally related, as shown in the following figure:

- Structural power focuses on an organization's positional and formal power bases.
- Personal power focuses on individual properties and resources.
- Discursive power focuses on the organizational play of language and discursive power struggles in informal processes.

Structural Power

The structural perspective on power is based on a zero-sum assumption. This means that more power to A means less power to B. This forms the basis for a conflict of interest between A and B regarding who should have the power and whose interests are to be protected. In keeping with Dahl (1957), we can describe power as a relation of influence between two individuals, in which A gets B to do something in accordance with A's wishes that B would otherwise not have done. This is a "power-over" perspective, in which power is viewed as a resource for the few: a repressive phenomenon that dominates, oppresses and robs one of freedom. It is primarily a case of a mono-centered understanding of power: a top-down perspective on the distribution of power. In this perspective, the organization contains various positions, each with a certain amount of respective associated power. Power is linked to specific actions and positions, such as the board, the executive, the middle manager, etc., and these each possess different degrees of power, depending on where in the hierarchy they find themselves. It is critical to determine the positions of and power distances among the various actors in the organization, since these will have

Figure 1. Perspectives of the triangle of power

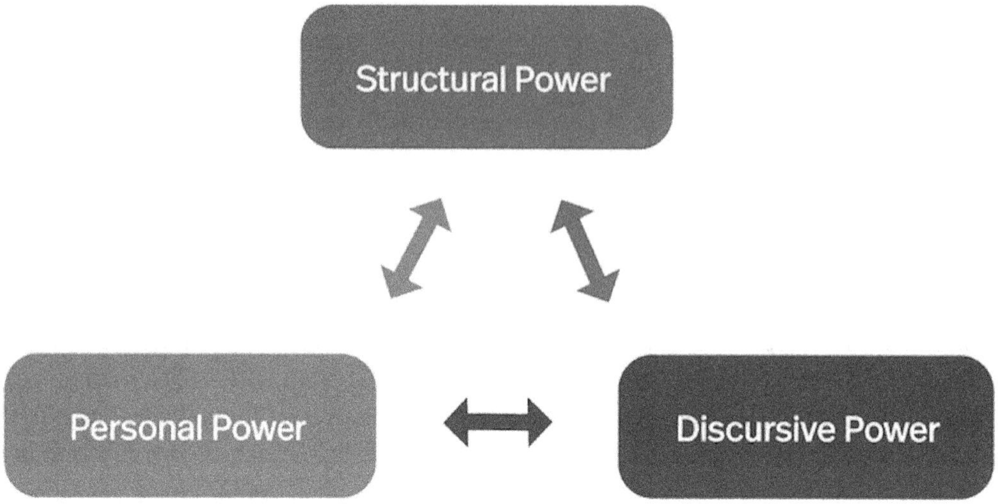

a major influence on their behavior. The structural composition of the organization (e.g. the management hierarchy, the composition of teams, etc.) all contain mechanisms that allow for certain behavioral patterns. Power must, therefore, be understood as something possessed in relation to certain organizational roles, such that a person's source of power is implied by his or her position within the organization.

Personal Power

In an analysis of power, one should consider that individuals participate with a certain degree of free will and a variety of choices. The consequence is that, while organizational structures influence the individual, the individual also influences the organizational structures (Buch-Hansen & Nielsen, 2005). In their own ways, all persons within an organization possess personal power bases related to, for example, specific skills, previous experience, personal dispositions, personal background, etc. (Yukl, 2013). As early as 1959, French and Raven presented a theory of power bases in which they defined power bases as power resources, understood as individual or group capacities within the organization. French and Raven (1959) emphasize that these power resources function by others' acknowledgement of them. They list five types of power:

1. **Reward Power:** The ability to give and withhold rewards. In the organizational context, reward power can manifest in the form of salary increases, employee perks, fringe benefits, promotion or offers to work on more attractive and exciting tasks.
2. **Coercive Power:** The ability to punish or sanction. Within organizations, this could be the ability to demote an employee, issue warnings, deprive an employee of bonuses or even fire the employee.
3. **Legitimacy Power:** The power attached to a specific position in a certain social context. Employees internalize their specific positions and associated powers via socialization. Legitimacy power is exercised from a position in a social structure where it is customary or agreed that the holder of the position has the right to exercise influence over people in other positions. In an organizational context, the leader is an example of an individual with legitimacy power.

4. **Reference Power:** The power that enables individuals or groups to act favorably on their surroundings due to inherent characteristics. Identification with an individual or group is the key element of this power. People with a lot of charisma are a clear example of reference power.
5. **Expert Power:** The power that a person has due to others' recognition of their extensive knowledge or expertise in an area.

The first three forms may be described as "hard power", tied to a power over others, where as the latter two may be described as "soft power" – the power to make something happen through others. These power resources should be seen in relation to both individuals (e.g. an experience or acquired knowledge) and structures (i.e. institutional features; e.g. certain professions and expertise are valued higher than others). Person-based power derives from recognition from others. It is not only the role, but also the person in the role around whom others rally. A leader must make him/herself "deserving enough" to exert influence, and both conscious and unconscious individual and collective assumptions about good leadership play a role in what kind of managerial behavior obtains authority in a local setting. In this sense, online teams can be seen as local settings with specific individual and cultural assumptions about good leadership that operate in an evolving culture. A leader's personal trajectory of making him/herself deserving enough to exert influence evolves though his/her interplay with the characteristics of the online context. Several management and leader scholars have emphasized that the personal power dimension is crucial for creating adherence to an organization's mission and objectives (Yukl, 2013). At the same time, it is important to emphasize that, regardless of the increasing importance of personal power, leader can never escape their structural positional power. Similarly, while a moderate amount of positional power increases opportunities for personal power, positional power never guarantees personal power (Yukl, 2013). It is, therefore, important to analyze the power strategies used by various leaders, employees and groups and to view them in relation to the structural conditions and dominating beliefs of good leadership in the local culture.

Discursive Power

Discourse is the third leg in the triangle of power. The starting point for the understanding of discursive power is a reading of Michel Foucault's concept of power. Foucault opposes the traditional Hobbesian understanding of power as something one can own, distribute or conquer (Foucault, 1982). He emphasizes that power produces and creates, stating that "Power ... is the name given to a complex strategic situation in a given society" (Foucault, 1980 p. 99). Power is, by nature, processual and operational. With his definition, Foucault seeks to direct our gaze towards how power is exercised, the mechanisms power uses, and its intended and unintended consequences.

Power and knowledge come together in discourses to be understood as socially embedded structuring principles of what can be said about a given subject at a given moment in history. As a result, power often develops between people and operates unnoticed, as discourses and as the structuring principle of modes of conduct. Foucault describes power as something that comes from everywhere, meaning that it operates at the macro-, meso- and micro- levels (Heede, 2002). Through discourse, language plays a central role the grasping of power (Jensen, 2005). Foucault's concept of power, thus, involves a ubiquitous and discursive focus on relations and language. He argues that it is never possible to reach the truth because one can never speak from a position outside the reigning discourse, while the truth is embedded in and produced by power systems. Since one cannot ever reach the truth, one should give up asking whether something is true or false and instead focus on how truth effects are created. What specific power technologies and techniques are we talking about? This will be addressed below.

In this chapter, "power technologies" are "the technologies that seek to determine the behavior of individuals and submit them to goals and forms of discipline and control" (Mik-Meyer & Villadsen, 2007, p. 28). The prominent example of modern power technologies derives from Foucault's analysis of the Panopticon prison designed by the English philosopher Jeremy Bentham (1748-1832). The prison is built so that the prisoners, due to the backlight and the building construction, cannot see those who are monitoring them (Raffnsøe, Gudmand & Thaning, 2008). The Panopticon has become a metaphor for power technologies, which, by positioning the individual in a conscious and permanent state of visibility, ensure that power and social order are maintained (Foucault, 2002). This leads Foucault to the theory of modern management discipline. Foucault's work explains the mechanism influencing behavior via surveillance at a distance: a mechanism that is also at work in modern management tools, such as digitalized performance measurement systems that log employees' billable time. Such power technologies rely on an awareness of constant and distanced surveillance, which makes the employees monitor and discipline themselves in relation to set targets (e.g. a billing rate of 80%). Because of the ubiquitous surveillance built into these kinds of management tools, control may become automated: In other words, the individual may internalize the external power technology and take on the responsibility of monitoring and disciplining himself. Once this happens, the leader's gaze works even at a distance, producing certain subjects with certain behavioral strategies. Management's power to define standards (e.g. a billing rate of 80%) and apply distanced surveillance tools in online leadership creates a situation of bounded autonomy, in which the online employee is given a significant degree of self-leadership, but is also expected to do what is necessary to reach the target goal. Unintended consequences might be darker self-disciplining strategies, such as not time-logging extra hours to keep the target 80% billing rate or to appear more successful in relation to peers. Such darker power dynamics in modern performance management tools have been related to the alarmingly increased frequencies of stress in the workforce, especially in knowledge-intensive work, where the use of online leadership is widespread (Andersen & Prætorius, 2016). This example shows how the distanced power technology of performance measurement through self-governed time-logging creates subjects who monitor and control themselves by adapting, in one way or another, to the prevailing discourse (e.g. an 80% billing rate).

Today, says Foucault (1982), we tolerate power only if its direct expression is masked. The methods of power have moved from explicit laws and threats of corporal punishment to the use of what he calls "normalization" and "technologies of the self" (Foucault 1980) Normalization refers to how different strategies for self-discipline lead individuals to act and experience themselves and each other in ways that are consistent with the prevailing discourse. Technologies of the self are specialized techniques that individuals use on their own bodies, mind and conduct to regulate themselves in accordance with specific interpretations of the world, as in the above example of employees' self-discipline strategies developed in relation to performance measurement using a self-governed time-logging tool. There is no doubt that Foucault's main contribution to the understanding of power in organizations is his demonstration of how power shifts from being exercised by an external authority to being supplemented by inner control.

However, another important aspect of Foucault's notion of power is that it allows us to probe more deeply into the linguistic mechanisms that help to constitute the historical and, therefore, arbitrary limits for a wide range of organizational acts and forms of consciousness. The most important point here is Foucault's insistence on power relations containing not only repressive, but also productive aspects. The discursive understanding of power opens a path toward applying power constructively in online leadership. From a discursive perspective, management is a social construct, like all other phenomena whose contents are determined locally in social interactions.

Thus, the online leader's role is the result of subject positions that the leader is invited to take and that the leader invites others to take in discursive interaction with both subordinate employees and superiors. The online leader's role is, thus, both self-created and contingent. As such, the online leader's task is to establish a sense of meaning and direction within the virtual team by co-operating with employees to align common discourses and narratives (Holmgren, 2006). These discourses and narratives unite employee experiences into a coherent and meaningful understanding of themselves and thc world. These discourses, therefore, are critical in giving meaning to organizational life and constitute the basis upon which the employees carry out their self-managing roles.

WORKING ONLINE: WHAT ARE THE POWER CHALLENGES OF LEADERSHIP?

Virtual teams represent a widespread way of working in today's knowledge-intensive companies (Leonardi & Bailey, 2008). They allow knowledge workers from different locations to operate on the same team, thus potentially reducing costs, travel, time and stress (Hinds et al., 2011). The term "virtual teams" is often used synonymously with the term "distributed teams", which refers to a way of working together while in different geographic locations (Lipnack & Stamps, 1997). Research on virtual teams has increased significantly since the 1990s (Watson-Manheim et al., 2012). The field research has also shifted from a dichotomous definition that compares face-to-face teams with virtual teams to a more complex hybrid definition, in which such factors as geographic dispersion, technology, time zones, language difficulties and national and organizational cultures are all identified as potential constraints to global virtual collaboration (Gibson & Cohen; 2003; Gilson et al., 2015; Hinds & Kiesler, 2002; Klitmøller & Lauring, 2013; Maznevski & Chudoba, 2000).

Both existing research and practical experience show that managing and rendering effective results often proves difficult due to the problem of finding "common ground" in distributed collaboration (Cramton, 2002; Olson & Olesen, 2000). Common ground here refers to the knowledge that people share and know that they share with others (Clark & Brennan, 1991). Because remote leader s and online workers do not "stand on common ground", they tend to lack mutual knowledge of important aspects of one another's situations; this makes it difficult for leader s to supervise effectively. More specifically, online leaders tend to run into problems when they assume that distributed employees have salient information and that they share this information with their distributed collaborators or when they attempt to interpret the meaning of their distributed employees' (lack of) communication. For these reasons, it is difficult for online leaders to supervise and influence the workflows of online employees. In other words, online knowledge workers tend to work on what they choose to or think they should focus on, without direction from a physical task environment or the structural power of a physically present direct manager. Management's structural power becomes less powerful in the absence of physically present personal power (Yukl, 2013). The positional power-over is essentially still at play in the leader-employee relationship; however, the physical distance might blur this fact in the employees' perception, making directive management techniques even less effective. Moreover, from a functionalistic perspective, directive management techniques that draw heavily on the direct positional power of a physically present manager are not easily applicable in a physically distanced working environment. Most importantly, online employees expect a huge degree of autonomy and self-leadership. As such, managerial attempts to supervise and overview the process of employees' work (e.g. work hours/place/focus) may be perceived as illegitimate and abusive uses of power that may threaten their legitimate autonomy and self-leadership. As a result, such managerial efforts are counterproductive and may be perceived as signals of managerial mistrust. For these reason, online leaders must find alternative ways of influencing employees in the absence of direct physical control and information channels.

METHODS AND DATA

The analysis in this chapter is based on empirical material from one of the authors' PhD work on online leadership in a global Danish manufacturing company. By applying a power analysis to cases from the empirical material, the authors seek to show that an examination of power dynamics in online leadership has the potential to help us understand new aspects of the challenges of online leadership and ways to manage these challenges successfully. Given the nature and conditions of the online leadership of knowledge workers, the authors argue that such leadership should be considered a two-way exchange.

The purpose of the PhD project on which this chapter builds was to both improve the practice of online leadership and develop new theoretical understandings of how to create enabling conditions for managing distributed work. Existing research talks about organizational change and impact as what distinguishes action research from other research areas (Baskerville & Myers, 2004). Thus, opposed to other research methods, where the researcher seeks to study organizations but not to change them, the action researcher is concerned to create organizational change while studying the process (Baskerville & Myers, 2004). The dual goal of both providing scientific knowledge and supporting organizational work made action research an applicable choice as the main research approach. The authors have used the definition of action research proposed by Hult and Lennung (1980):

Action research simultaneously assists in practical problem-solving and expands scientific knowledge, as well as enhances the competencies of the respective actors (p. 247)

Attention to the two goals and their interaction helps researchers produce knowledge by applying a theoretical framework to respond to the practical problems at hand and by using practical insights from real-world settings and interventions taken to discover new theoretical knowledge to inform future research (Chiasson et al., 2009).

The action research project was conducted from January 2012 to October 2013 in a large multinational Danish engineering manufacturer (DEM) with 17,000 employees. The DEM is headquartered in Denmark and has subsidiaries in more than 55 countries. In recent decades, the DEM has undergone continuous globalization and is aiming to increasingly empower and involve its subsidiaries in the development of new products in the R&D department. The distributed organization of the R&D department demands a collaborative working culture to ensure optimal utilization of the global employees. In this research project, the DEM participated in the action research with teams and leaders from the R&D department. During and after the data collection, the research findings were reflected on with the participating DEM leaders through seminars and online sessions.

One of the primary strengths of action research is how it builds bridges between research and practice and facilitates the engaged involvement of both worlds through cyclic and iterative research processes (Baskerville & Wood-Harper, 1998). In action research literature, there is a broad variety of approaches, but nearly all follow the same general process: first, a diagnostic of the practical circumstances; then, the execution of interventions based upon theoretical considerations; next, the collection of empirical data about the intervention; and, finally, a reflection upon the results (e.g. Baskerville & Wood-Harper, 1998; Iversen et al., 2004). In this study, the authors have been inspired by the collaborative action research (CPR) approach of Mathiassen et al. (2002). The CPR approach is characterized by the goal of understanding and improving professional practices through a close collaboration between researchers and practitioners. The CPR research process is guided by a pluralist methodology (Mingers, 2001) that uses action research as the dominant approach and other conventional methods (e.g. case studies) as supplementary approaches. Action research requires the use of participative methods for data collection, diagnosis and intervention. In the next section, we'll present the methods for data collection and analysis, followed by a case example from the research project.

Action research stresses the need to understand an organization before intervening to improve the current situation. Thus, in order to build a solid action research base within the research setting, the intention was to develop an initial understanding of both the DEM itself and its R&D department. The data- collection-approach involved systematic and comprehensive data collection that built on multiple sources of data as the foundation for the action research interventions. The multiple data sources supported triangulation, systematic gathering and reliable data, which, in turn, helped to ensure the validity of the study. The data collection during the problem diagnosis phase involved semi-structured interviews, observations and documentary sources.

Semi-Structured Interviews

During the 21 months of study, 50 semi-structured interviews were conducted. Semi-structured interviews were chosen as a method for data collection because it brought a rich telling of the participants' experiences with online leadership. The semi-structured interviews were primarily conducted during the first half of 2012; however, 10 were conducted as repeat interviews during the summer of 2013. Data were collected from 50 interviews (29 with managers and 21 with employees) from both the Danish headquarters and the other three R&D locations worldwide (the US, China and Hungary). The semi-structured interviews were facilitated as a "guided" dialogue between the researcher and the respondent in order to both address the themes outlined in the research questions and allow the respondents to break away from the questions to raise new insights and connections. In addition to the formalized interviews, also informal unstructured interviews were conducted during lunch-breaks, hallways, offices, cars, flights, etc. The significance of these interviews was high, as they offered insight into the informal part of everyday life in the R&D department. Thus, in the more relaxed and informal settings, the interviewees disclosed more frustrations related to the challenges of the distributed online set-up. The interviews were either recorded or, in the informal settings, recalled through extensive same-day note-taking (cf. Eisenhardt, 1989). Both the informal interviews and the formal ones were conducted in either Danish or English. The quotes reported in the results have all been translated into English, omitting all information that would allow the identification of the informants.

Participant Observation

To gain an inside perspective into the daily life of the R&D department, one of the authors also occupied a workspace at the Danish headquarters for the duration of the 21 months of study. The purpose of this participant observation was two-fold: First it allowed the researcher to immerse herself into the organization under study and to obtain an insider's perspective on organizational life. Second, it allowed the researcher to change her prior perceptions (Spradley, 1980). In general, participant observation seeks to continuously challenge researchers' subjective pre-perceptions with the ultimate aim of helping them obtain more advanced understandings of the organization under study (Bourdieu, 2003).

The managers involved in the data collection operate in a working arena that includes virtually the whole world. The consequence is that they travel often and far and that they have numerous virtual meetings. The growing complexity of professional work demands study methods that can cope with such complexity (Czarniawska, 2007). Hence, the data in this study have been collected through activities involving either physically collocated observation or the use of communication technology. All observations on activities related to research aspects were written down in full field notes either during or after the workday (Czarniawska, 2007).

Interventions

Three different workshops were developed and conducted as specific action research interventions. All interventions were designed based on the empirical findings and with the aim of supporting the managers of the R&D department in improving their online leadership. Each workshop was planned as an intervention activity, and data were collected about each intervention. The three workshops represent the action research cycles comprising the action research phases proposed by Checkland (1991): diagnosing, action planning, action taking, evaluating and specifying learning. Building on the previous data collection and problem definition, the design was aligned with the interventions specifically for the target audience, taking into consideration the organizational culture of the particular collaborators in the R&D department. The interventions were explicitly built on and combined existing research in order to both inform future research and break with and challenge existing work practices in online leadership. More specifically, the interventions were guided by theory on transformations of perceptions and work practices in global work (Bjørn et al., 2006; Tyre & Orlikowsky, 1994). Building on this literature and the researchers' problem diagnosis, the workshops covered the following content: opening with plenary presentations, alignment of global work practices, casework and contracts of collaboration. By the end of the workshops, considerable commitment was established among the managers regarding improving collaboration in global teams.

Data Analysis

The case example included in this chapter illustrates how the problem diagnosis obtained through the semi-structured interviews, the participatory observations and the document analysis were used as a basis for developing workshops to improve the managers' online leadership. As a basis for this chapter, the data analysis followed an iterative process deductively guided by theories on globally distributed collaboration, online leadership, power and leadership transformation. The approach recommended by Miles and Huberman (1994) served as the inspiration and benchmark for coding and analyzing the data.

While the data analysis covered the entire period of interaction with the DEM, this chapter focuses particularly on the specific activities of evaluating the results of the workshop activities. Half a year after the last workshop was held, we investigated these activities in more detail and re-examined the empirical data from the workshops in order to understand whether, why and how the managers experienced transformations in their online leadership practice. The case presented below illustrates how, through repeat interviews, the researchers returned to the involved managers six months after the interventions and evaluated significant learnings related to the organization's capability to manage globally distributed teams. The case further explores how the researchers identified the long-term impact of the action research interventions, including, namely, the power of aligning expectations in online leadership, which was identified in the future ripples of the past interventions.

CASE: THE POWER OF ALIGNING EXPECTATIONS IN ONLINE LEADERSHIP

The following empirical case illustrates how divergent expectations of online leadership must be aligned to create a power base for exerting influence and accomplishing shared objectives. The case illustrates how this can be accomplished in practice. It further shows the importance of paying specific attention to the discursive aspects of power in online leadership.

With continuous annual growth in recent decades, the studied DEM has been expanding its global market share on engineering products. Due to organizational changes in 2013, the need for online leadership has become vital, especially in the increasingly global R&D department. The R&D managers are concerned about how to stay in touch with and supervise their physically and culturally distributed team members. The managers define the challenge as a question of "how to lead the virtual team—align team tasks, expectations and work practices across locations—making the daily team collaboration succeed and accomplish shared goals". When asked, 80% of the managers in the R&D department stated that the difficulties lie not only in the dispersion of the geographical locations, but, more importantly, in the fact that individuals approach the teamwork significantly differently, with diverse views and experiences of responsibilities, roles, work processes, local customs, etc. This diversity is experienced as a significant challenge for the online leaders, since their global employees have several different expectations of work, teamwork and the question of "what is good online leadership?"

Consequently, it was decided that organizational workshops should be created to address the development of online leadership competences among the R&D managers. These workshops focus specifically on assisting the managers in creating agreements to support the alignment of expectations and coordination. Thus, the workshops are directed at supporting global managers in facilitating "contracts of collaboration" with their distributed team members. The aim in this chapter is to discuss one important insight gleaned from the intervention that was pertinent for the managers' success in mitigating the challenges of creating enabling conditions for online leadership, namely, the "contract of collaboration" should address work coordination (e.g. scheduling, organizing, and structuring work), and also clarify responsibilities and power distributions. One Danish manager said:

I had a constructive dialogue in the team about the "contract of collaboration", and the feedback confirmed the need for this type of document. It is important for me to know that everyone in the team is aligned on specific topics. Besides the daily use, it is also very useful in relation to the introduction of new team members. [Most recently], I have used it when introducing new team members from China. (manager, August 2013)

During the intervention, it became clear to the managers that, by negotiating and determining obligations and roles within the team, they created a platform of shared expectations of online leadership. The dialogues also provided the managers with additional knowledge about remote employees, which supported the development of shared meaning. A manifest outcome of the workshop intervention was a common" agreement template" and process guidelines by which to negotiate this contract in global teams. Another global manager said:

In creating contracts of collaboration and, thereby, getting inputs from my employees, I really got an insight into what thoughts everybody has, not just in relation to collaboration within our team, but a lot about the collaboration in the global setting. (manager, June 2013)

As explained by this manager, the coordination agreements were to be implemented through the mutual collaboration of all team members, including explicit alignment regarding central elements, such as remote meetings, emails, coordination, communication tools, frequency of communication, content of communication, etc. Managers participating in the program reported that the clear and explicated form of collaborative structure was well received and helped to make implicit expectations more explicit, thereby giving the managers a greater sense of power and control. Using the coordination agreements, team members could use information about one another's roles, tasks, and priorities to guide their concrete activities.

Summing up, it was clear that the alignment of expectations introduced by the action research interventions directed at improving conditions for online work had been embraced by the organizational members and helped them to further develop their organization as ripples in the water. Several managers had posted on the online portal how they either implemented contracts of collaboration together with new ways of facilitating these with the purpose of engaging their global team members in productive collaborative work practices. Thus, linking back to the action research interventions, the researchers detected a growing capability among the organizational members to ensure aligned expectations through team agreements. While the action research project was helping to ensure the managers capability to create successful conditions for online work, a manager told how an employee in his team expected the manager to be physically present in order to be a good and successful manager. As the manager was leading an online team distributed across 5 locations the manager would never succeed to fulfil this specific expectation from his employee. Thus, the expectations employees bring to the work setting influence the perception of fulfilment and/or breaches.

DISCUSSION AND IMPLICATIONS FOR PRACTICE

Reporting from the action research study, the researchers wanted to examine the ways in which leaders are able to lead distributed work processes and online team members successfully. In doing so, the study closely examined how power processes can promote or hinder the accomplishment of effectiveness in online leadership. Existing research agrees that managing online workers is difficult, thus on-line leaders must consider new understandings as well as new power strategies supporting online work. In this section, the authors discuss how the use of *empowerment* and *social power* provides the possibility to detect successful online leadership that manage to establish a shared practice around the boundary object of a virtual work space.

The Empowerment of Online Workers

Dacher Keltner (2016) argues in favor of a new understanding of power, in which the leader acts as a collaborator who wins over others by making everyone stronger. This concept was exemplified by the manager in the case who implemented coordination agreements through the mutual collaboration of all team members. Such an approach allows power to become a means for strengthening the common good, rather than a goal *per se*. If one analyzes the way in which the successful online leader operates in the case, one finds that, to him, power is not a zero-sum game, but, rather, a process of mutual reinforcement. In other words, power is more about empowerment than the raw execution of power. Thus, the empowerment of the distributed team members and the establishment of mutual norms and agreements has, in the above case, been shown to be important for the successful leadership of online teams. The reason for this is that both personal and managerial power forces tend to be weaker in distributed teams.

As opposed to the image of the physically present charismatic leader, the above case has shown how the empowerment of team members is important for the successful leadership of online teams. Empowerment is here understood as a qualification of others and a delegation of one's own power (Fogsgaard & Elmholdt, 2014). This approach sees power from a "power-to" perspective, in which what the power does or how it is exercised is most meaningful. This approach also considers the effects that power has on culture, individual consciousness and the organization (Clegg et al., 2006). In empowered online leadership, the employees' ability to master is enhanced, and the leader facilitates employees' development, delegates responsibilities and provides opportunities for ownership. Empowered leadership alters the dynamic between the leader and the employee and blurs the asymmetry obvious in the "power-over" perspective. It is important to note that this does not mean that the power disappears; rather, the power achieves a new purpose and takes different forms. Therefore, the "power-to" perspective often refers to a soft and hidden power that expands a leader's actual opportunities of influence. Based on this idea of empowerment, Keltner (2016) formulates the following "power principle": "We keep and grow power by giving it away. [...] enduring power is found in giving it away. Your power expands as you empower others."

Social Power Among Online Workers

As illustrated in the empirical case, the process of empowerment builds on aligning expectations and establishing common ground in the team. However, it further involves a significant social power pressure among team members, which together reinforce employees' commitment to the team agreements. In the empirical case, the managers did not persuade team members to accept specific collaborative rules or norms as part of their agreements; however, they did make sure to communicate that the team members should align expectations and that this should result in a specific agreement among all team members.

This case shows that the traditional "power-over" model has not been completely sidelined, however the "power-over" here has a more socially grounded form. The case further shows that in successful online leadership it is no use to follow only one form of power. Rather, the leader must integrate hard structural based power (goalsetting, evaluation, deadlines etc.) with soft, more personal and social power and reflect upon how it is possible as a leader to show direction, while remaining in the background in the team.

Agreements in collocated teams can often be formed in a natural and implicit way (Mark, 2002), since team members and managers more or less consciously regulate and model one another's behaviors. However, in distributed teams, in which perspective, experiences and cultural backgrounds can be quite diverse, articulation is needed to achieve a common understanding. Agreements serve to establish common ground, to strengthen social power pressure. Leaders must here consider how to be both invisible and visible at the same time. Therefore, the leader, must reflect upon his/her own leadership behavior, upon the power that is bound to his/her actions and upon how this power and these actions affect employees. It is essential that leaders understand the complexity and contradictions of the phenomenon. They must relate to the tension they experience so that they can exist and act in a manner appropriate for themselves, for their employees and for the organization. Only then, an online leader is able to reflect upon when what types of practice are most appropriate (e.g. When do I set direction, and when do I delegate?). Different leadership roles have their own function and legitimacy, and these roles are rarely clear in practice. Thus, an online leader must reflect on what he/she does, why he/she does it and what consequences these actions have. To be capable of navigating as an online leader, he/she must balance within this tension field and continually integrate hard and soft power.

CONCLUSION

In this chapter, the authors have focused on how online leadership is linked to power. They have argued that online leadership involves both hard power, which determines, sets direction and decides, and soft power, which focuses on empowerment, involvement and recognition. Thus, the authors have shown that online leadership is a practice that covers many conflicting managerial demands and expectations. To balance these many demands and help leaders focus on their overall leadership task, the power triangle has been introduced as an analytical tool and guideline.

The authors wish to emphasize that online leadership is always a power-based relationship. Based on the formal structural understanding of power, they have argued that leadership is linked to various positions of vertical organizational power. However, it is not sufficient only to focus on the structural and formal aspects of the organization. Personal characteristics and relational interaction also play a key role, both positively (i.e. by strengthening the leader's position) and negatively (i.e. by causing a loss of legitimacy). The authors thus suggest that an essential approach to understanding leaders' actions and organizational legitimacy is to examine leaders' personal power bases and how they are used. The goal of a power-oriented approach to online leadership could, thus, be to strengthen the leader's ability to act through enhanced insight into the dynamics of power in the field of tension, constituted by hard and soft types of power. Foucault (1982) contributed by focusing on the linguistic and less visible forms of power, through which the leader becomes part of the discursive practices of the organization and different perceptions of the organization thrive side-by-side, without the involved persons being aware that there are versions of reality other than their own. This makes it interesting to study how some forms of leadership can have a greater impact than others.

The authors argue that the online leader plays an important role in clarifying and reshaping the concept of leadership to achieve success in online leadership. The leader must negotiate and determine team obligations and roles to create a platform of shared expectations. Thus, an important task for the online leader is to communicate with employees about the psychological contract of a distant employee–leader relationship. The transaction between the manager and the employee (i.e. What are your responsibilities at a distance, and what are mine?) is particularly important. Power in online leadership is a sparsely research topic, and there is a need for more research addressing the question of power bases and power processes for exerting influence and obtaining shared objectives in online leadership.

REFERENCES

Akkerman, S. F., & Bakker, A. (2011). Boundary crossing and boundary objects. *Review of Educational Research, 81*(2), 132–169. doi:10.3102/0034654311404435

Andersen, M. F., & Prætorius, N. (2016). Går skønne kræfter til spilde? In *Tæller vi det der tæller? Målstyring og standardisering i arbejdslivet.* Aarhus: Klim.

Baskerville, R., & Myers, M. D. (2004). Special issue on action research in information systems: Making IS research relevant to practice: Foreword. *Management Information Systems Quarterly, 28*(3), 329–335. doi:10.2307/25148642

Baskerville, R., & Wood-Harper, A. T. (1998). Diversity in information systems action research methods. *European Journal of Information Systems, 7*(2), 90–107. doi:10.1057/palgrave.ejis.3000298

Bjorn, P., Scupola, A., & Fitzgerald, B. (2006). Expanding technological frames towards mediated collaboration. *Scandinavian Journal of Information Systems, 18*(2), 1.

Bourdieu, P. (2003). Participant objectivation. *Journal of the Royal Anthropological Institute, 9*(2), 281–294. doi:10.1111/1467-9655.00150

Buch-Hansen, H., & Nielsen, P. (2005). *Kritisk realisme.* Frederiksberg: Samfundslitteratur/Roskilde Universitetsforlag.

Carlile, P. R. (2002). A pragmatic view of knowledge and boundaries: Boundary objects in new product development. *Organization Science, 13*(4), 442–455. doi:10.1287/orsc.13.4.442.2953

Carlile, P. R. (2004). Transferring, translating and transforming knowledge: An integrative framework for managing knowledge across boundaries. *Organization Science, 15*(5), 555–568. doi:10.1287/orsc.1040.0094

Checkland, P. (1991). From framework through experience to learning: The essential nature of action research. *Information Systems Research: Contemporary Approaches and Emergent Traditions,* 397–403.

Chiasson, M., Germonprez, M., & Mathiassen, L. (2009). Pluralist action research: A review of the information systems literature. *Information Systems Journal, 19*(1), 31–54. doi:10.1111/j.1365-2575.2008.00297.x

Clark, H. H., & Brennan, S. E. (1991). Grounding in communication. *Perspectives on Socially Shared Cognition*, *13*, 127–149. doi:10.1037/10096-006

Clegg, S., Courpasson, D., & Phillips, N. (2006). *Power and organizations*. Thousand Oaks, CA: Sage Publications.

Corradi, G., Gherardi, S., & Verzelloni, L. (2010). Through the practice lens: Where is the bandwagon of practice-based studies heading? *Management Learning*, *41*(3), 265–283. doi:10.1177/1350507609356938

Cramton, C. D. (2002). Finding common ground in dispersed collaboration. *Organizational Dynamics*, *30*(4), 356–367. doi:10.1016/S0090-2616(02)00063-3

Czarniawska, B. (2007). *Shadowing: And other techniques for doing fieldwork in modern societies*. Copenhagen: Copenhagen Business School Press DK.

Dahl, R. A. (1957). The concept of power. In Behavioural science (vol. 2). Academic Press. doi:10.1002/bs.3830020303

Edmondson, A., & Harvey, J. F. (2017). Cross-boundary teaming for innovation: Integrating research on teams and knowledge in organizations. *Human Resource Management Review*. doi:10.1016/j.hrmr.2017.03.002

Eisenhardt, K. M. (1989). Building theories from case study research. *Academy of Management Review*, *14*(4), 532–550.

Ernst, C., & Chrobot-Mason, D. (2011). *Boundary Spanning Leadership*. New York: McGraw-Hill.

Fogsgaard, M. K., & Elmholdt, C. W. (2016). *Power and leadership. In Global Encyclopedia of Public Administration, Public Policy, and Governance* (pp. 1–8). Switzerland: Springer Publishing Company. DOI.

Foucault, M. (1980). Two lectures. In M. Foucault (Ed.), *Power/knowledge*. Harlow: Pearson Education. (Original work published 1977)

Foucault, M. (1982). The subject and power. In H. L. Dreyfus & P. Rabinow (Eds.), *Michel Foucault—Beyond structuralism and hermeneutics*. Chicago: The University of Chicago Press. doi:10.1086/448181

Foucault, M. (2002). *Overvågning og straf – fængslets fødsel*. Frederiksberg: Det lille Forlag.

French, J. R. P., Jr., & Raven, B. (1959). The bases of social power. In D. Cartwright (Ed.), Studies in social power. Ann Arbor, MI: University of Michigan.

Gibson, C. B., & Cohen, S. G. (Eds.). (2003). *Virtual teams that work: Creating conditions for virtual team effectiveness*. San Francisco, CA: John Wiley & Sons.

Gilson, L. L., Maynard, M. T., Young, N. C. J., Vartiainen, M., & Hakonen, M. (2015). Virtual teams research 10 years, 10 themes, and 10 opportunities. *Journal of Management*, *41*(5), 1313–1337. doi:10.1177/0149206314559946

Govindarajan, V., & Gupta, A. (2001). Building an Effective Global Business Team. *MIT Sloan Management Review*, *42*(4), 63–71.

Hatch, M. J. (2011). *Organization theory – modern symbolic and postmodern perspectives*. Oxford, UK: Oxford University Press.

Heede, D. (2002). *Det tomme menneske – Introduktion til Michel Foucault* (2nd ed.). København: Museum Tusculanums Forlag.

Hernes, T. (2004). Studying composite boundaries: A framework of analysis. *Human Relations*, *57*(1), 9–29. doi:10.1177/0018726704042712

Hinds, P., Liu, L., & Lyon, J. (2011). Putting the global in global work: An intercultural lens on the practice of cross-national collaboration. *The Academy of Management Annals*, *5*(1), 135–188. doi:10.1080/19416520.2011.586108

Hinds, P. J., & Kiesler, S. (2002). *Distributed work*. Cambridge, MA: MIT Press.

Holmgren, A. (2006). *Ledelse med et narrativt og poststrukturalistisk perspektiv. I: Helth, P. Lederskabelse – det personligevlederskab*. Roskilde: Samfundslitteratur.

Hult, M., & Lennung, S.-Å. (1980). Towards a definition of action research: A note and bibliography. *Journal of Management Studies*, *17*(2), 241–250. doi:10.1111/j.1467-6486.1980.tb00087.x

Iversen, J. H., Mathiassen, L., & Nielsen, P. A. (2004). Managing risk in software process improvement: An action research approach. *Management Information Systems Quarterly*, *28*(3), 395–433. doi:10.2307/25148645

Jensen, A. F. (2005). Mellem ting – Foucaults filosofi. Det lille Forlag.

Kanawattanachai, P., & Yoo, Y. (2002). Dynamic nature of trust in virtual teams. *The Journal of Strategic Information Systems*, *11*(3), 187–213. doi:10.1016/S0963-8687(02)00019-7

Keltner, D. (2016). *The Power Paradox: How We Gain and Lose Influence*. New York: Penguin Press.

Klitmøller, A., & Lauring, J. (2013). When global virtual teams share knowledge: Media richness, cultural distance and language commonality. *Journal of World Business*, *48*(3), 398–406. doi:10.1016/j.jwb.2012.07.023

Leonardi, P. M., & Bailey, D. E. (2008). Transformational technologies and the creation of new work practices: Making implicit knowledge explicit in task-based offshoring. *Management Information Systems Quarterly*, *32*(2), 411–436. doi:10.2307/25148846

Lipnack, J., & Stamps, J. (1997). *Virtual teams: reaching across space, time and organization with technology*. London: John Wiley & Sons.

Mark, G. (2002). Conventions and commitments in distributed CSCW groups. *Computer Supported Cooperative Work*, *11*(3), 349–387. doi:10.1023/A:1021289427473

Martins, L. L., Gilson, L. L., & Maynard, M. T. (2004). Virtual teams: What do we know and where do we go from here? *Journal of Management*, *30*(6), 805–835. doi:10.1016/j.jm.2004.05.002

Mathiassen, L. (2002). Collaborative practice research. *Scandinavian Journal of Information Systems*, *14*, 57–76.

Maznevski, M. L., & Chudoba, K. M. (2000). Bridging space over time: Global virtual team dynamics and effectiveness. *Organization Science, 11*(5), 473–492. doi:10.1287/orsc.11.5.473.15200

Mik-Meyer, N., & Villadsen, K. (2007). *Magtens former: Sociologiske perspektiver på statens møde med borgeren*. København: Hans Reitzel.

Miles, M. B., & Huberman, A. M. (1994). *Qualitative data analysis: An expanded sourcebook*. London: Sage.

Mingers, J. (2001). Combining IS research methods: Towards a pluralist methodology. *Information Systems Research, 12*(3), 240–259. doi:10.1287/isre.12.3.240.9709

Mulder, M. (1977). *The daily power came*. Leiden, The Netherlands: Maritinus Nijhoff Social Science. doi:10.1007/978-1-4684-6951-6

Olson, G. M., & Olson, J. S. (2000). Distance matters. *Human-Computer Interaction, 15*(2), 139–178. doi:10.1207/S15327051HCI1523_4

Oswick, C., & Robertson, M. (2009). Boundary Objects Reconsidered: From Bridges and Anchors to Barricades and Mazes. *Journal of Change Management, 9*(2), 179–193. doi:10.1080/14697010902879137

Panteli, N., & Chiasson, M. (2008). Rethinking virtuality. In Exploring Virtuality within and beyond Organizations (pp. 1-17). Palgrave Macmillan UK. doi:10.1057/9780230593978_1

Raffnsøe, S., Gudmand-Høyer, M., & Thaning, M. (2008). *Foucault*. Frederiksberg: Forlaget Samfundslitteratur.

Spradley, J. P. (1980). *Participant observation*. New York: Holt Rinehart and Winston.

Star, S. L., & Griesemer, J. R. (1989). Institutional ecology, 'translations' and boundary objects: Amateurs and professionals in Berkeley's Museum of Vertebrate Zoology. *Social Studies of Science, 19*(3), 387–420. doi:10.1177/030631289019003001

Tyre, M., & Orlikowski, W. (1994). Windows of opportunities: Temporal patterns of technological adaptation in organizations. *Organization Science, 5*(1), 98–118. doi:10.1287/orsc.5.1.98

Watson-Manheim, M. B., & Chudoba, K. (2002). Discontinuities and continuities: A new way to understand virtual work. *Information Technology & People, 15*(3), 191–209. doi:10.1108/09593840210444746

Yulk, G. (2013). *Leadership in organizations*. Albany, NY: Prentice Hall.

KEY TERMS AND DEFINITIONS

Action Research: Simultaneously assists in practical problem-solving and expands scientific knowledge, as well as enhances the competencies of the respective actors.

Boundary: Indicates both a difference and an encounter, in which two "sites," "things," "arenas," or "ideas" are collated and perceived as both relevant to each other *and* different.

Discursive Power: Power and knowledge come together in discourses to be understood as socially embedded structuring principles of what can be said about a given subject at a given moment.

Empowerment: An approach to leadership, in which the leader acts as a collaborator who wins over others by making everyone stronger.

Online Leadership: The social influence processes in virtual teams that produce individual and collective efforts to accomplish shared objectives.

Personal Power: All persons within an organization possess more or less power related to, for example, specific skills, previous experience, and personal dispositions.

Power: A tool of influence in virtual leadership that is neither negative nor positive. Rather, it is the way in which power is used that can be negative or positive.

Structural Power: Formal power bases linked to specific actions and positions, such as the board, the executive, the middle manager.

Chapter 9
Improving Online Collaboration in Contemporary IT Development Teams

Jan Pries-Heje
Roskilde University, Denmark

Lene Pries-Heje
The IT University of Copenhagen, Denmark

ABSTRACT

An interview study focusing on online collaboration in geographically distributed IT development teams in Danske Bank revealed seven problem areas. To cope with the problems the authors applied a design science research approach to construct a conceptual framework for improving online collaboration. The conceptual framework combines a six-phase teambuilding model with six elements of social capital. Thus, in each phase of teambuilding, the online collaborators aim at building up all six elements of social capital. The complete six-by-six framework was successfully tried and diffused throughout Danske Bank. This chapter gives an account of the framework content and the results from the evaluation.

INTRODUCTION

Future collaborative projects will be undertaken everywhere; globalisation is here now (cf. Friedman, 2006). This means that work will be assigned to anyone who does it better, cheaper or faster. It also means that a company will have many projects, characterised by rapidly assembled project teams, geographically dispersed but with highly specialised professionals who perform specific tasks. Individual project teams will gather and collaborate (online) for varying periods of time, depending on the type of work, and will be dissolved as soon as the project task is completed. For an individual, based somewhere in the world, work and career will be about participating in a series of individual projects, characterised by progress and achievements.

DOI: 10.4018/978-1-5225-4094-6.ch009

Thus, online collaboration and projects with geographically distributed team members will be common in the future. In our understanding, a distributed online collaboration team is separated by geography, time zone and/or culture; nevertheless, its members have to work together as a team.

In this chapter, the authors examine a concrete process improvement effort in Danske Bank. Project work in Danske Bank is characterised by many teams collaborating online, with members based in both India and Denmark, but these teams need to improve their collaboration. Danske Bank realised this issue some years ago, and the process improvement effort reported here was initiated from that realisation.

The remainder chapter is organized as follows. First, the chapter carefully explains the research method, specifically, design science research. Second, the chapter give a short account of existing knowledge related to online collaboration, distributed teams and management, as well as the problems at hand. Third, the chapter lays out all the details of an interview conducted in Danske Bank. Fourth, the chapter gives the details of the design of a conceptual framework for solving the problem faced in Danske Bank. Fifth, the diffusion and successful adoption of a six-by-six conceptual framework (i.e., the design)is explained. Finally, the chapter ends with a conclusion.

RESEARCH METHOD

In this section, the authors carefully explain the design science research approach taken, the business needs being addressed and the knowledge applicable to the problem at hand.

Benbasat and Zmud (1999) argue that much information system (IS) research today is irrelevant, and they recommend research that is more relevant but without fundamentally challenging the existing academic value system. The authors believe that design science research offers the practical relevance and utility required because it emphasises that a design should address a need or a problem and at the same time, should stand on the shoulders of existing research on the problem area. Probably the first journal paper on design science research about IS is that of Walls et al. (1992), who argue that design is both a product and a process. Thus, a design theory must on one side, handle the design product; on the other side, it should handle the design process. In another influential paper on design science research (March & Smith, (1995), one of the key points is that in design science, a researcher can build and evaluate four elements: constructs, models, methods and instantiations.

Continuing the work of March and Smith (1995), Hevner et al. (2004) present a design science research framework that also enhances the study of Walls et al. (1992). Figure 1 shows an overview of the framework. At the core are such elements as build and justify. The arrows back and forth symbolise the iterative nature of design science research. The left hand side of the figure shows the business needs stemming from people, an organisation or technology. The right hand side of the figure shows the common knowledge base consisting of foundations, methodologies and technology. The bottom two arrows show the main outcomes of design science research – applications in a concrete setting and additions to the general knowledge base.

The Case and How the Research Was Initiated

The research reported in this chapter was initiated when Linda Olsen, the First Vice President for Danske Bank's outsourcing setup called Development Center India (DCI), realised that the firm needed improvement in online collaboration and management.

Figure 1. Components of a design theory according to Hevner et al. (2004)

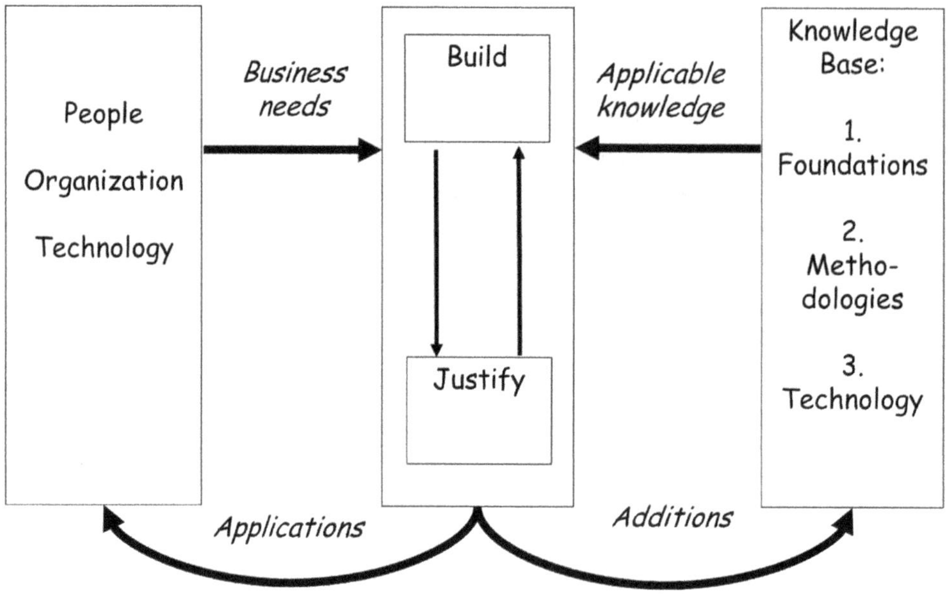

The setup between DCI and Danske Bank in Denmark was at that time built on an outsourcing strategy called staff augmentation. This strategy meant that an increasing number of projects would have people from different sites (e.g., Bangalore in India, and Ejby, Brabrand and Lyngby in Denmark) working together in the same project team. The main advantage of the staff augmentation approach was that it could leverage existing resources in Denmark, as well as utilise outsourced services and contract workers in Bangalore. Specifically, Danske Bank had close to 500 people working in Bangalore as part of project teams using online collaboration of both Danish and Indian team members. Most shared project teams did not create totally new information technology (IT) systems but were responsible for updates and maintenance of existing systems, that is, system management projects.

For a start, the authors interviewed Olsen to obtain a better understanding of the business need for better online collaboration management. She told the researchers that Danske Bank had two types of projects. One type comprised new development projects where something new would be developed, often as an add-on to existing applications or from scratch. The other type involved system management projects where development would consist of smaller changes, additions and defect corrections. The online collaboration projects (across Denmark and India) were mainly of the latter type. Additionally she also explained that in her opinion, enough technology was available to the online collaboration project teams. Danske Bank had built telepresence rooms in all main sites, including Bangalore. They had eMeeting and chat software at all workstations in Denmark and India. They also had several video conferencing facilities in both Bangalore and Danish sites. Thus, Olsen emphasised that the need for better online collaboration management was in her view, a *management* problem.

A series of interviews was planned and conducted to obtain a deeper and more thorough understanding of the management problem and thus to address it properly.

Applicable Knowledge From The Knowledge Base

Design science research requires standing on the shoulders of all the other researchers who have investigated a similar problem, as indicated by the upper right arrow labelled "Applicable knowledge" in Figure 1. Many researchers have examined online collaboration (project) management. According to Lacity et al., "Much depends on experiential learning and sheer hard work" and "[...] outsourcing is not about giving up management but managing in a different way" (2008, p. 32).

What then should this "different way" be? The literature review of the knowledge base reveals an interesting paper (Bhat, Gupta, & Murthy, 2006), building on case studies from an Indian IT services firm where the authors identify five factors essential for success, using a root-cause analysis. The five factors include team collaboration or shared goals, shared culture, shared processes, shared responsibility and trust in one another. These five key factors have been called *social capital*, which forms a major part of the literature describing this theme.

Social capital can be defined as the people you know, how well you know them, and the mutual knowledge and concepts shared between you. Social capital brings together a number of concepts, such as informal organisation, trust, culture, social support, social exchange, social resources, rational contracts, social networks and interfirm networks (Adler & Kwon, 2002). As a construct, social capital can be defined as "the goodwill available to individuals or groups. Its source lies in the structure and content of the actor's social relations. Its effects flow from the information, influence, and solidarity it makes available to the actor" (Adler & Kwon, 2002, p. 23).

The concept of social capital originates from the French philosopher Pierre Bourdieu (Bourdieu, 1983), who distinguishes among three kinds – classical monetary, cultural and social capital. Bourdieu focuses on the benefits gained from having social capital and shows how the ruling classes in society exploit their networks to maintain inequality among social classes. Robert Putnam is the other major figure who has researched on social capital. In his study of American cities, he shows how the relocation to suburbs means that the social networks, where people could avail of help, support, friendship and services, have now disappeared or at least become a lot smaller. He defines social capital as the collective value of all "social networks" and the inclinations created by these networks to do things for one another (Putnam, 1995, 2000).

Putnam (2000) also uses the concepts of bonding and bridging in social networks. *Bonding* is about close ties among people who know one another well, such as family members, close friends, neighbours and members of groups with the same socioeconomic position and demographic characteristics. *Bridging* is about more loose relationships among people across groups or in collaborations. Some scholars mention a third type of relation-oriented social capital, *linking*, which is about creating links outside of another group or collaboration, for example, up to the leadership of an organisation (Gittell & Vidal, 1998; Szreter & Woolcock, 2004).

Beyond studying the types of social capital, researchers have also examined its content and divided it into three dimensions: structural, relational and cognitive (Adler & Kwon, 2002; Evans & Carson, 2005). The *structural* dimension of social capital is about the relationships and mutual obligations among people. Here it is crucial whether or not a relationship exists.

The *relational dimension* pertains to the nature and quality of the existing relationships. For example, in a collaborating team, the relational dimension involves the trust among its members. Trust can be perceived as what the individual does to society and unity beyond a simple rational behaviour (Kramer & Tyler, 1996). However, trust is more than the belief in what the counterpart in a relationship will do. It is also about reciprocity, which means that a person may be willing to invest in a relationship to strengthen it, but at some point, the person expects to receive something in return (Nahapiet & Ghoshal, 1998). In relation to the structural and the relational dimensions, two other aspects have been investigated. The first comprises the structure of the personal network and the number of *strong ties* (Coleman, 1988). The second consists of more loosely coupled networks with more superficial and *weak ties* (Burt, 1987).

The *cognitive dimension* of social capital is about whether people have common concepts, representations, interpretations and processes. Some examples of questions relevant to online collaboration in relation to the cognitive dimension are as follows: Do we have a common vision? Do we speak the same language? Do we use the same concepts? Do we have the same stories?

The social capital concept has been used to study societies, differences between developing and developed countries and recently, project teams. In their study, Lacity et al. (2008, p. 30) conclude, "Our research found that one of the best ways to transfer knowledge is to invest in social capital. Social capital is simply the idea that knowledge and resources are exchanged, work gets done, and value is created through social relationships."

In relation to teams, Evans and Carson's (2005) interesting study links the performance of distributed and heterogeneous teams (similar to online collaboration as the authors call it in this chapter) to three core processes, namely, communication, social integration and coordination. In this linkage, social capital acts as a moderating structural dimension, meaning that when social capital is low, then distributed teams will be negatively related to group processes and positively related when social capital is high. Figure 2 shows this correlation.

Interviews

Armed with a request for help on online collaboration (project) management and a preliminary understanding of the problem as related to social capital, the authors travelled to India for three weeks. The purpose of our trip was to obtain a better understanding of the problems related to the actual online collaboration and the management of geographically distributed teams in Danske Bank's DCI, where approximately 500 people worked.

Figure 2. Relationships among group processes, social capital and performance (Evans & Carson, 2005)

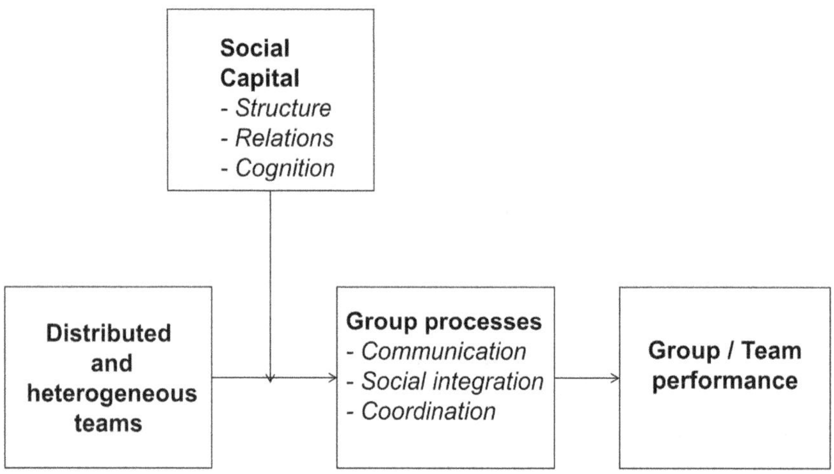

Based on the applicable knowledge, the authors formulated an interview guide as shown in Figure 3.

To pilot test the interview guide, the authors conducted some interviews before leaving for India with a project manager (using a pseudonym called Project Bravo). Some changes and improvements were made after the pilot testing before the interview guide was finalised as shown in Figure 2.

To obtain a broad understanding of the problem domain, the authors asked permission to interview three people with different roles in projects that varied in size, type and scope. Besides interviewing the Indians in DCI in Bangalore, the authors also interviewed five Danish liaison officers and the first vice president, using the same interview guide.

The people whom the authors interviewed worked on the following projects:

- **Bravo:** This medium-sized system management project used Scrum as its development methodology, with one Scrum team in Denmark and another in Bangalore. The project manager, the task manager, the business developer, the tester and two developers were interviewed.
- **Charlie:** In this smaller system management project, the task manager and the developer were interviewed.
- **Golf:** This larger, three-year development project had its main parts developed in India. Three people were interviewed, including the task manager and two developers, of whom one newly arrived at DCI (four months prior to the interviews).
- **Tango:** In this larger system management project, the three people interviewed were the task manager and two developers, of whom one had recently arrived.

All interviews were recorded, with careful and detailed transcriptions jotted down. Furthermore, the authors observed the developments, using technology such as video and telephone conferences. The

Figure 3. The interview guide used in the interviews

1. Your background, Education / Experience
2. The project? Scope? / Organization? / Roles?
3. Denmark – India? People on-site in Denmark? / You? / Use of LOs?
4. Communication: Structure: What? / Daily? / Regularly? / Challenges? / Hindrances & obstacles?
5. Team building: How do you build a team?
6. Networking: Here? / Across to DK? / Community here at ITC? Bangalore?
7. Project Management.
8. Coordination and Control Formal control and follow-up?
9. Common vision and mental model? / Informal?
10. Coordination Mechanisms?
11. Examples
12. Surprises? / Miscommunication / Misalignment / Special or what is different about working here in general?

authors also tried using the telepresence room for a two-hour meeting. For all these observation studies, the authors made careful notes as well.

From Interpretive Coding and Analysis to Design

Our adopted methodology was contextualised and interpretive, and it utilised the case study technique (Pettigrew, 1990; Walsham, 2006). This study was interpretive in the sense that the authors attempted to understand the online collaboration–project management phenomena and the problems therein through the meanings that people assigned to the issues that the authors brought up in the interviews. Thus, reality was accessed through social constructions, such as language, consciousness and shared meanings (Myers & Avison, 2002).

Data analysis also followed the interpretive tradition, using hermeneutics (Myers, 2009). Interview transcriptions and observation documents were coded and analysed. First, the authors found several potential problem categories. A subsequent, more careful qualitative data analysis (Miles & Huberman, 1994) uncovered a number of underlying problem themes, where the authors followed the process inspired by Goetz and LeCompte (1981). First, the authors scanned and coded the data. This step identified several categories of explanations for online collaboration management problems, as well as attributes of both complex and interwoven relationships between problems. Second, the authors compared the data and the categories to uncover all relevant elements. Third, the findings were presented and discussed in a workshop with all the Danish liaison officers in Bangalore and in January the following year, with the whole Indian Management Group in DCI. Finally, in the fourth step, the authors arrived at the conclusion of the study, in the form of a design of a conceptual framework for online collaboration and distributed project management.

INTERVIEW FINDINGS

Generally, the authors note a preponderance of engineers at DCI (compared with Denmark), and most employees have four-year bachelor's degrees rather than master's degrees (the figures show that 35% completed master's degrees). The authors also note that many of the interviewees formerly worked in another Indian company and thereby have experience from an outsourcing arrangement other than the one at DCI. For example, several interviewees previously stayed in the US or the UK before joining DCI.

Problem #1: Social Ties Take Time

The first problem area identified is about social ties. The interviewees' experience is that generally, it takes time to build social ties, which develop over time. In online collaboration projects, social ties take more time, and it is considerably harder to forge social ties with someone whom the interviewee has never met face-to-face. In relation to this problem area, the authors also notice that the Indians have very few professional and technical specialist networks. In contrast, in Denmark, developers often attend "on-the-way-home" meetings and seminars in Dansk IT (Danish Computer Society), *Dansk Projektledelse* (Danish Association for Project Managers) or Tecpoint (Danish association for companies involved in the technical area). One does not find such gatherings in India. However, the equivalent is held internally in DCI by having a quality month, a tech month, and so on.

Problem #2: Not Enough Trust

The second problem area is trust. Again, the interviewees have experienced that it takes time to build trust, and it takes longer to gain confidence at a distance. The authors find that problems #1 and #2 are related in that social ties and trust are mutually reinforcing.

Problem #3: Lack of Shared Vision And Language

The third problem area is about shared vision and shared language for better work. Based on the interviews, the authors find a limited use of a common vision or a roadmap in the projects. This may be rooted in the fact that the authors have mainly studied system management projects. Another aspect of this third problem area is the lack of shared language and concepts. Here, some of the problems are related to national languages (in contrast to professional vocabulary). Thus, if the majority of the project team members had a shared language other than English, they might tend to use this language in meetings and written correspondence, which would exclude the members who had not mastered that language. However, in most projects, English is successfully implemented as the common language. Here, different accents and inadequate language skills occasionally result in miscommunication and discomfort in communicating.

Problem #4: Cultural Distance

Problem area #4 involves issues related to cultural differences, as illustrated in some examples given in the interviews. There are differences in how self-propelled team members are expected to behave when uncertain. Do they ask first or investigate/check for themselves first? The interviewees also cite differences in whether they should follow the specified methods or work out-of-process and in how they perceive performance measurements. There are differences related to the management style, especially the acceptance of formal and hierarchical power and the value of seniority.

All the interviewees agree that Danes take many things for granted, that it is important in India to gain recognition after finishing a task and that measurement is more acceptable in India. One solution that has been tried with some success is explicit training in cultural differences. Another successful initiative is a metrics programme in India that DCI has built. Several metrics programmes have been tried without success in Denmark over the last 10 years – making it clear that metrics and measurements are difficult in Denmark.

Cultural differences especially surface in relation to the management style. In Denmark, it is alright to have a rotating manager (as in Scrum). In India, a team prefers to be configured with a formal leader in order to work. The Danes mention that seniority is surprisingly important in India, illustrating the distance between the two cultures. The Danes have also found that DCI offshore teams do not work as well if everyone is at the same level; employees in India need the hierarchy. However, it is not just seniority that counts. DCI has existed in Bangalore for four years, and it is clear from the interviews that seniority in the Danish Bank counts more than seniority in general (e.g., in companies other than DCI).

Problem #5: Communication Issues

Problem area #5 is about communication issues. Much communication takes place daily, and most of it works quite well. E-meetings work well, and video conferences are used to a great extent. The chat programme is open and used frequently for questions. Nevertheless, cost considerations to some extent limit the diffusion of new and potentially valuable technology. Communication works very well in some teams but not so well in others. Those who communicate have invested a lot of time in it (e.g., meetings every day, chat, etc.).

Some communication issues are again culturally dependent. For example, the authors have been told about the Indians' tendency to answer more based on intention and social structure – not as things really are. In this regard, it seems that Danes and Indians are not good at "reading" each other. This problem may become more severe when they have only online communication.

Problem #6: Lack of Reciprocity

Problem area #6 is about reciprocity. According to the Merriam-Webster Learner's Dictionary (2017), reciprocity can be understood as having/giving the same rights to all or an agreement among different parties to do something similar for one another. In teamwork, neither rights nor exchange of favours may be stated explicitly; nevertheless, a fair balance is expected.

Team members may do tasks for others, with the expectation that the favour will be returned in some way in the future. In the interviews, the cited examples show the interviewees feeling the lack of reciprocity, such as always holding meetings at inconvenient times, individuals taking time by asking questions without paying them back in different ways, and newcomers feeling that they are not given the same opportunities as the seniors have.

Problem #7: Insufficient Team Identification Across Sites

Problem area #7 is about insufficient team identification. The interviewees experience difficulties in identifying themselves with the team or realising the true team spirit, especially with only one or a few members in one location and a larger number of members in another location.

Two solutions make it easier to identify oneself with a team. One is co-location of the team members at a site, as is the case in Project Bravo. Another important vehicle (that the authors have observed) for creating team identification is a video conference attended by all team members.

DESIGNING A FRAMEWORK FOR ONLINE COLLABORATION

Team performance varies significantly. In his book entitled *Software Engineering Economics*, Boehm (1981) finds a factor of 1-to-4 between the best and the worst teams – which on paper are equal. In their book entitled *Peopleware*, DeMarco and Lister (1999) find a factor of 1-to-5, and the 400–500% difference is worth the effort.

Our overall impression from the interviews is that to make online collaboration teams succeed, project managers need to spend much more *time for* and *emphasis on* teamwork. The preliminary focus on social capital is confirmed by the interviews, meaning that the dimensions of social capital seem to play an important role. Thus, the conceptual framework the authors designed focuses on the project managers' ability to facilitate the creation of social capital through all phases of a project.

The authors believe that the interviews and the literature (knowledge base) give strong grounding for the following proposition:

Projects need to build social capital in all its aspects through all phases of building an online collaboration team to ensure a successful project.

In the following section, the authors describe the conceptual framework. First, the theoretical basis for the two dimensions of the framework is explained in the subsections "Phases of the Online Collaboration Team Process" and "Elements of Social Capital". These two dimensions result in a six-by-six matrix, which is then described in detail.

Phases of the Online Collaboration Team Process

In their literature review on teams, Pries-Heje and Commisso (2010) find four primary aspects of interest: (1) the task, (2) team roles, (3) teamwork and (4) the process. The *task* to be undertaken by the team has an influence. For example, the more complex the task is, the more there is a need for a balanced team where all the *team roles* are enacted. *Teamwork* is mainly about two elements – trust and a common vision. Trust (problem #2) is really a prerequisite for an effective team. If the team members do not trust one another, they cannot work well together. They will show their facades instead of their real selves, and facades have never done a good job (Drexler, Sibbet, & Forrester, 1988). In relation to teamwork, the second important point is to have a common vision within the team (problem #3).

Finally, the *process* that the team goes through is crucial. Pries-Heje and Commisso (2010) use the old forming–norming–storming–performing model as a main proponent of the team process. In the Team Software Process - TSP (Humphrey, 2006), the authors also emphasise a good team startup in the form of a distinct launch activity. Commisso and Pries-Heje (2011) have developed a model for building a team in six phases, as described in Table 1.

Table 1. The Six-C model for team building (Commisso & Pries-Heje, 2011)

Constitute	Team and project constituted. Think whether you have the knowledge and competence needed; The team gathers; Feels that "we ARE a team" and wonder " Who am I?"
Clarify	Who are the others, clarify group dynamics; how to communicate; how to decide; rules of conduct; social contract
Commit	Aim and goals, priorities, roles, context, and vision
Carry Out	Working – preferably effectively; continued group dynamics; ongoing communication and coordination
Check	How are we? Do we need to go back and repeat – build more social capital; mid way crisis?
Conclude	We have come to the end; what did we learn

Elements of Social Capital

Social capital has been discussed earlier in this chapter, giving an overview of the existing theory. According to Evants and Carson (2005), social capital has three key elements – structural, relational and cognitive. The *structural* element is the network of ties and relationships formed by group members and reflects the degree to which groups of individuals openly communicate. The *relational* element concerns the nature and quality of the relationships and refers to the trust that exists among the group members. Research findings suggest that network ties that are not strengthened by mutual obligations, trusting relationships, a common language or narratives easily break down (Burt, 2002). The relational element can be decomposed into identification, trust and reciprocity. The *cognitive* element can be described as the shared language and narratives that together form a shared system of meaning. The cognitive dimension of social capital highlights the importance of shared representations, interpretations and systems of meaning among the parties involved, and it can be split into two sub-elements – shared vision, and language and concepts. Hence, social capital can be understood as having six elements, as follows:

1. Structure
    a. Social ties
2. Relation
    a. Identification
    b. Trust
    c. Reciprocity
3. Cognition
    a. Shared vision
    b. Language and concepts

Six-by-Six Framework

In our design science research undertaking, the authors now have a solid and thorough understanding of the need and the problems, as well as the existing knowledge base. After some iterations, the authors have ended up with a design combining the Six-C model (as presented in the preceding subsection) with the six aspects of social capital, allowing the necessary building of enough social capital in all phases of a team building.

In summary, the design is shown in Table 2.

In relation to the analysis of the interview data, the conceptual framework covers the problems that the authors have identified as follows:

- **Problem #1:** Social ties take time. This is addressed with a conceptual framework where projects goes through all the phases of team building, thereby allowing the time it takes to build social ties.
- **Problem #2:** Not enough trust. This is addressed by the strong emphasis on trust building (the third row in Figure 4).
- **Problem #3:** Lack of shared vision and language. This is addressed by rows 5 and 6 in the conceptual framework.
- **Problem #4:** Cultural distance. This is partly addressed in the fields labelled "Hofstede".

*Table 2. The designed conceptual framework consists of a six-by-six matrix combining six dimensions of team building with six dimensions of social capital. The * symbol after some techniques means that the techniques are repeated in other fields.*

	Constitute	Clarify	Commit	Carry out	Check	Conclude
1. Structure / Social ties	Create personal shield	Clarify the human behind What technology?	Hold events that create ties	Celebrate success Create social time	"Light" Retrospective *	Have a retrospective *
2. Relation/ Identification	Make a group portrait	Imagine success Build confidence	Have short feedback loop	Create team pride – talk about success	Process observation * "Light"	Have a retrospective *
3. Relation/ Trust	*Historic trust*	Create team game rules	Explore barriers Define roles	Build performance and knowledge-based trust	Process observation * "Light"	Have a retrospective *
4. Relation/ Reciprocity	*Mindset for online work*	*Mindset for online work*	*Mindset for online work*	Identify trust	Process observation * "Light"	Have a retrospective *
5. Cognition / Shared vision	Create a knowledge map *	Hofstede *	Create a Front page for vision Elicit the Big steps towards the vision Make a Game Plan	Use common vision	"Light" Retrospective *	Have a retrospective *
6. Cognition / Language and concepts	Create a knowledge map *	Team building activities Hofstede *	Carefully plan first meeting Use a common development model or a common architecture		"Light" Retrospective *	Have a retrospective *
7. Others			Map communication needs Create stakeholder matrix		Create a Fishbone (Ishikawa) diagram	

- **Problem #5:** Communication issues. As shown in Figure 2, better communication is a result of the heightened level of social capital, that is, all six rows in all phases in the conceptual framework. Furthermore, row 7, labelled "Others", specifically addresses communication.
- **Problem #6:** Lack of reciprocity. This is addressed by row 4 on reciprocity in the conceptual framework.
- **Problem #7:** Insufficient team identification across sites. This is addressed by row 2 on identification in the conceptual framework.

Our design's overall idea follows from the proposition, "To be successful, a project needs to build all elements of social capital in all phases of team building. To populate the six-by-six matrix, the authors have chosen techniques that a team can use to build a specific part of social capital. Generally, the techniques included come from a number of different sources. The main source is the book entitled *Best Practices for Facilitation* (Sibbet, 2002). The second most important source is the work of Duarte and Snyder (2006). Kerth's book entitled *Project Retrospectives* (Kerth, 2001) is the main source for the Check and the Conclude phases. Furthermore, the authors have been inspired by agile techniques, especially Scrum (Sutherland & Schwaber, 2010). The rest of the techniques are taken from the work of Commisso and Pries-Heje (2011). This choice of techniques is not coincidental. The authors have carefully discussed each of the 36 fields in the six-by-six model. After considering several techniques, the authors have selected those that are especially well suited for both applying in an online collaboration and creating the most social capital. In relation to our research method, this can be regarded as the Build and Justify iteration (middle part of Figure 1).

Space does not permit going through all 36 fields in the matrix, but a few examples are provided here.

Under Clarify and Relation/Identification, there is a technique called *Imagine success*, with the following steps:

- Everyone in the team thinks for himself or herself about the last project that he or she participated in that was implemented successfully.
- Each team member tries to capture the feeling of success and transfer it to an idea of how a successful completion of this project was experienced.
- Each team member tells the team what helped him or her succeed in his or her projects and how success was experienced.
- On a common screen or a large sheet of paper (seen on video), a team member writes "The look of success" and around it, writes or draws each member's perception of success.

Under Clarify and Cognition (both shared vision and language and concepts), Hofstede is cited (cf. Hofstede, 2001), referring to his famous research on national cultures. Using it as a technique, the authors offer the following recommendations:

- With *high individualism* (in Denmark and partly in India), use team building to let individuals speak about themselves first and then talk about cooperation afterwards.
- With *low uncertainty avoidance* (as in Denmark), talk generally about how the team should work and give details later.
- With *high uncertainty avoidance* (as in India), use team building to specify, in a precise and detailed manner, how the members want to work together.
- With *low power distance* (as in Denmark), use team building to let each individual talk about his or her background in relation to the team, and use competition-oriented activities where everyone can win.
- With *high power distance* (as in India), use team building to let individuals talk about themselves in relation to others; do not use activities that could upset the balance of power and hierarchy.

Under Commit and Cognition/Shared language and concepts, the authors urge the teams to use the Danske Bank development model, with the following arguments about its advantages:

- Everyone uses the same terminology in projects.

- It becomes easier to register and understand the data and experiences from earlier projects. The method can become a common framework for communication (e.g., of successes).
- With well-defined phases and documentation for each phase, management is much easier.
- New employees without experience access a well-defined platform to start from.

Carry out and Relation/Identification incorporate a technique called *Create team pride – talk about success*, which includes the following:

- A (project) manager is responsible for telling the members about team success.
- Plan to facilitate an early success.
- Celebrate it.
- Build momentum.
- Create team pride and team identification.
- Make the opponents think; maybe the project actually has the potential to become a success.

DIFFUSION AND ADOPTION IN DANSKE BANK

The authors presented the conceptual framework to the management in Danske Bank on the last week of our visit to India. The authors clearly linked it to the problems that were identified in our analysis of the interviews, as presented in this chapter. The response was very positive. The Danske Bank management was confident that focusing on building social capital had the potential to help them improve their online collaboration (project) management.

It was then decided that the authors should teach task managers, process people and general managers in DCI about building social capital. The authors held a five-day course on the first week of the next year. After the course, the participants should be able to do the following:

- Independently facilitate the startup of and the ongoing work in an online collaboration project team, whose members are based across Denmark and India and have never been together in one physical location.
- Choose appropriate techniques for the six phases of teamwork to facilitate building enough social capital within the team to ensure that the team can work in online collaboration.

The participants' evaluation emphasised the following more general comments:

- Introduce the topics (that some participants considered vague) in a clear and perceivable manner.
- Although the discussed techniques were familiar, put together as a package, it was new and encouraged them to think about what was being done.
- The sensitive subject was handled very well, with good examples/tools.
- The practical tools given to them would really help in day-to-day management.
- Building social capital is the real added value (by our conceptual framework).

At the end of this course, the authors realised that it was necessary to provide a similar course in Denmark. The resulting course took place in February of the same year. Here again, the evaluation was quite positive, and the participants committed themselves to using the conceptual framework.

The conceptual framework was then being used in seven project areas – mainly the so-called system management areas. In all areas, concrete plans were made about how to build social capital for their respective teams. They each performed a cost-benefit analysis, specifically for their own projects, showing that the benefits of building and ensuring enough social capital were much higher than the costs.

Over the summer, the authors evaluated the use of the conceptual framework for online collaboration. A representative sample of interviewees was selected from both India (DCI, Bangalore) and Denmark. The interviews were recorded. For each interview, detailed field notes were made with the help of the recordings. Furthermore, most interviews were conducted by two people, making it possible for one person to concentrate on talking with the interviewee, while the other was taking notes. On the first week of August, the authors then coded and analysed the interview and observation data in more depth, resulting in the following findings:

1. The framework had been used extensively in all seven project management areas.
2. All the people interviewed agreed that the use of the framework was clearly worthwhile and that the outcome was higher in value than the investment in time.
3. The framework had been used more in India than in Denmark. The overall satisfaction with the framework was high in India. Many examples of improved team building and building of social capital were found among the Indian interviewees.
4. In Denmark there were varying levels of satisfaction with the framework.
5. The techniques from the first three C's (Constitute Team, Clarify and Commit) were used considerably more than those from the last three C's (Carry Out, Check and Conclude).
6. Two techniques, the Shield Exercise and the Group Portrait, were especially reported to have been used with very positive results.

Thus, the authors concluded that Danske Bank could make real use of implementing the framework throughout the organisation. In fact, our recommendation had been followed; the management decided to make the framework available to all its projects.

Consequently, the authors trained the trainers in Danske Bank in February and March of the second year after the project began. In the third year, the four Danske Bank trainers were training 60 project managers and team managers in using the framework. On several occasions, the authors have heard (informally) that the people whom the authors trained, as well as those trained afterwards, are still using parts of the framework – if not most of it – to systematically build social capital in online collaboration.

CONCLUSION

In this chapter, the authors have identified seven problems in online collaboration projects with participants from more than one site, in Denmark and in India. The authors have presented a solution to the problems in the form of a framework and an idea of building social capital in all its aspects through all phases of building online collaboration.

To design the framework, the authors have applied design science and used the framework by Hevner et al. (2004). In relation to March and Smith's (1995) work, the conceptual framework is designed as both a *model*, stating that the project – typically with the project manager being responsible - should systematically build social capital throughout the phases of a project, and a *method, contained in the six-by-six fields.* Furthermore, the authors have been using it in several areas and projects, making it multiple *instantiations*.

A natural question is whether the framework will also be beneficial in systematically building social capital for face-to-face collaboration projects. The authors believe that the answer to this question is yes. Increasing social capital improves team performance (Evans & Carson, 2005). However, the framework that is designed and discussed in this paper is aimed at geographically distributed online collaboration. Many techniques that can useful in a classic face-to-face project team are not part of the present framework.

Hence, the contribution of this chapter is the design of a framework that is built for the contingencies of the situation at hand in Danske Bank. All the techniques included in the framework have been selected and adapted to the geographically distributed online collaboration projects of Danske Bank, that is, to fit the cultures of Denmark and India and the technology available to cooperate and collaborate online across both countries.

A natural question towards the end of this chapter is whether the findings would be applicable to another set of circumstances. The authors recently tried the framework in the context of a world-leading chemical engineering company, with online collaboration among countries other than Denmark and India. In this case, the framework had to be somewhat adapted, not in terms of the overall idea of systematically building social capital but how it could be done. Thus, if another organisation wishes to cope with similar problems by systematically building social capital through team building, then it needs to evaluate the techniques in the six-by-six framework in order to clarify the following issues: Will this technique be possible to use in the organisation with its given contingencies? Does the organisation need to adapt the techniques, for example, to the available IT and to the differences in time zones? One of the advantages between Denmark and India is the 3–5-hour overlap in a normal work day, which is not the case between the US and India, for example.

To use the results of this study, an organisation may go through the following steps. First, establish the challenges that it is facing. This can be done through a study involving interviews or a survey using the seven problem areas that the authors identified. If the challenges are similar, the organisation can benefit from adopting the six-by-six framework for systematically building social capital in its projects. Before the organisation can adopt the framework, it needs to take each of the techniques mentioned in Figure 3 and ensure that these are useful in its own context. It should also consider whether other techniques have to be added. In considering this point, the organisation should probably review the experiences from its online collaboration projects and harvest the positive uses of techniques. Fourth, when the framework has been locally consolidated, the authors recommend that the organisation pilot it in 5–7 projects, similar to what the authors did in Danske Bank. Finally, when the pilot project has been evaluated and the results have been taken into account, then the six-by-six framework adapted to the whole organisation can be rolled out.

REFERENCES

Adler, P. S., & Kwon, S.-w. (2002). Social Capital: Prospects for a new concept. *Academy of Management Review*, *27*(1), 17–40.

Benbasat, I., & Zmud, R. (1999). Empirical Research in Information Systems: The Practice of Relevance. *Management Information Systems Quarterly*, *23*(1), 3–16. doi:10.2307/249403

Bhat, J. M., Gupta, M., & Murthy, S. N. (2006). Overcoming Requirements Engineering Challenges: Lessons from Offshore Outsourcing. *IEEE Software*, *23*(5), 38–44. doi:10.1109/MS.2006.137

Boehm, B. (1981). *Software engineering economics. Advances in computing science and technology*. Prentice-Hall.

Bourdieu, P. (Ed.). (1983). *Ökonomisches Kapital, Kulturelles Kapital, Soziales Kapital*. Goettingen, Germany: Otto Schartz & Co.

Burt, R. S. (1987). Social contagion and innovation: Cohesion versus structural equivalence. *American Journal of Sociology*, *92*(6), 1287–1335. doi:10.1086/228667

Coleman, J. S. (1988). Social capital in the creation of human capital. *American Journal of Sociology*, *94*, 95–120. doi:10.1086/228943

Commisso, T. H., & Pries-Heje, J. (2011). *Optimér dit projektteam. Håndbog i teamudvikling*. Samfundslitteratur.

DeMarco, T., & Lister, T. (1999). *Peopleware: Productive projects and teams* (2nd ed.). Dorset House Publishing Company.

Drexler, A. B., Sibbet, D., & Forrester, R. H. (1988). The team performance model. *Team building: Blueprints for productivity and satisfaction*, 45-61.

Duarte, D. L., & Snyder, N. T. (2006). *Mastering virtual teams: Strategies, tools and techniques that succeed* (3rd ed.). San Francisco, CA: Jossey-Bass.

Evans, W. R., & Carson, C. M. (2005). A social capital explanation of the relationship between functional diversity and group performance. *Team Performance Management*, *11*(7/8), 302–315. doi:10.1108/13527590510635170

Friedman, T. L. (2006). The World is Flat: The Globalized World in the Twenty-First century (Updated and expanded ed.). London: Penguin Books.

Gittell, R., & Vidal, A. (1998). *Community Organizing: Building Social Capital as a Development Strategy*. Newbury Park, CA: Sage Publications.

Goetz, J. P., & LeCompte, M. D. (1981). Ethnographic Research and the Problem of Data Reduction. *Anthropology & Education Quarterly*, *12*(1), 51–70. doi:10.1525/aeq.1981.12.1.05x1283i

Hevner, A. R., March, S. T., Park, J., & Ram, S. (2004). Design Science In Information Systems Research. *Management Information Systems Quarterly*, *28*(1), 75–105. doi:10.2307/25148625

Hofstede, G. (2001). *Culture's Consequences: Comparing Values, bahaviors, Institutions, and Organizations Across nations* (2nd ed.). Thousand Oaks, CA: Sage Publications.

Humphrey, W. (2006). *TSP: Coaching Development Teams*. Addison-Wesley Professional.

Kerth, N. L. (2001). *Project Retrospectives: A Handbook for Team Reviews*. Dorset House Publishing Company, Incorporated.

Kramer, R. M., & Tyler, T. R. (1996). *Trust in Organizations: Frontiers of Theory and Research*. Thousand Oaks, CA: SAGE Publications.

Lacity, M. C., Willcocks, L. P., & Rottman, J. W. (2008). Global Outsourcing of Back Office Services: Lessons, Trends and Enduring Challenges. *Strategic Outsourcing*, *1*(1), 13–34. doi:10.1108/17538290810857457

March, S. T., & Smith, G. F. (1995). Design and natural science research on information technology. *Decision Support Systems*, *15*(4), 251–266. doi:10.1016/0167-9236(94)00041-2

Miles, M. B., & Huberman, A. M. (1994). *Qualitative Data Analysis: An Expanded Sourcebook*. Thousand Oaks, CA: Sage Publications Inc.

Myers, M. D. (2009). *Qualitative Research in Business & Management*. London: Sage Publications.

Myers, M. D., & Avison, D. (2002). An Introduction to Qualitative Research in Information Systems. In M. D. Myers & D. Avison (Eds.), *Qualitative Research in Information Systems - A Reader* (pp. 3–12). London: Sage Publications. doi:10.4135/9781849209687.n1

Nahapiet, J., & Ghoshal, S. (1998). Social Capital, Intellectual Capital, and the Organizational Advantage. *Academy of Management Review*, *23*(2), 242–266.

Pettigrew, A. M. (1990). Longitudinal Field Research on Change: Theory and Practice. *Organization Science*, *1*(3), 267–292. doi:10.1287/orsc.1.3.267

Pries-Heje, J., & Commisso, T. (2010). *Improving Team Performance*. Paper presented at the Proceedings of the 33rd IRIS Seminar, Rebild Bakker.

Putnam, R. (1995). Bowling Alone: America's Declining Social Capital. *Journal of Democracy*, *6*(1), 63–78. doi:10.1353/jod.1995.0002

Putnam, R. (2000). *Bowling Alone: The Collapse and Revival of American Community*. Simon and Schuster. doi:10.1145/358916.361990

Sibbet, D. (2002). Best Practices for Facilitation. San Francisco, CA: The Grove Consultants International.

Sutherland, J., & Schwaber, K. (2010). *The Scrum Papers: Nut, Bolts, and Origins of an Agile Framework* (Vol. Draft - 14 May 2010). Retrieved from http://jeffsutherland.com/ScrumPapers.pdf

Szreter, S., & Woolcock, M. M. (2004). Health by association? Social capital, social theory, and the political economy of public health. *International Journal of Epidemiology*, *33*(4), 650–667. doi:10.1093/ije/dyh013 PMID:15282219

Walls, J. G., Widmeyer, G. R., & El Sawy, O. A. (1992). Building an information system design theory for vigilant EIS. *Information Systems Research*, *3*(1), 36–59. doi:10.1287/isre.3.1.36

Walsham, G. (2006). Doing Interpretive Research. *European Journal of Information Systems*, *15*(3), 320–330. doi:10.1057/palgrave.ejis.3000589

KEY TERMS AND DEFINITIONS

Cognition Element: Part of social capital that includes shared values, concepts, and vision.

Process Improvement: A systematic undertaking that typically involves checking or measuring the existing process, acting on the checking/measurement results, planning what to do to improve, and doing it. After some time, plan-do-check-act again.

Reciprocity: Mutual dependence or actions, for example, when one party invests in a relationship, he or she will expect something in return.

Relational Element: Part of social capital involving whom an organisation knows.

Social Capital: The people whom an organisation knows and, by this, the goodwill available. Its source lies in the structure and content of the actor's social relations. Its effects flow from the information, influence, and solidarity that it makes available to the actor. In this chapter, the authors divide social capital into relational, structural, and cognitive elements.

Structural Element: The quality of the relations, including whether they are loosely coupled or strong ties, how much trust exists, and the reciprocity of the relation(s).

Trust: The reliance on the character, ability, strength, or truth of someone or something.

Chapter 10
Knowledge on the Move:
Expansive Learning Among Mobile Workers

Rosemary Francisco
Unisinos University, Brazil

Amarolinda Zanela Klein
Unisinos University, Brazil

Yrjö Engeström
University of Helsinki, Finland

Annalisa Sannino
University of Tampere, Finland

ABSTRACT

Mobile workers are professionals who frequently work on the move, far from a fixed workplace, often performing knowledge-intensive activities. Mobility challenges creation and sharing of knowledge among these professionals, and the existing literature lacks powerful theoretical frameworks conducive to creating supportive learning pathways for them to meet these challenges. This chapter is a theoretical and practical contribution to fill this gap by analyzing a case of expansive learning initiated by mobile workers themselves. Based on longitudinal participant observation, the study traces the steps undertaken by these professionals to create a new artifact that helped them to know what, when, where, and how they needed to perform their work activities.

INTRODUCTION

The knowledge workforce as known in the past is shifting. Global economic factors, increasing professional specialization, improvements in customer relations, and rapid technological advancements, along with the diffusion of information and communication technologies (ICT) and the acceptance of corporate bring-your-own-device programs, has driven a considerable increase in the number of mobile workers in

DOI: 10.4018/978-1-5225-4094-6.ch010

organizations (Chen, 2015; Jarrahi & Thomson, 2017; McDaniel, Reitsma, & Duan, 2016). According to McDaniel, Reitsma, and Duan (2016), "the percentage of information workers working away from the office at least a few times a month—whether it be from home or while traveling or commuting—has increased more than 60% in just three years (2012 to 2015)." In spite of this increase, little is known about the consequences of mobility in organizational practices (Karanasios & Allen, 2014; Reynolds, 2015), or about professional practices outside traditional centralized offices (Jarrahi & Thomson, 2017).

Knowledge work involves activities to create new ideas, new technology or creative content and problem-solving capabilities, and thus regularly requires collaboration due to the complexity to perform it alone (Nelson, Jarrahi, & Thomson, 2017; Vartiainen et al., 2007). However, the context of mobile knowledge workers differs from the work environment of fixed knowledge workers. The term "mobile" means the freedom these workers have to perform their work practices anytime anywhere, mainly where and when the results of the work are better achieved (Harmer & Pauleen, 2012; Vartiainen, 2008). Mobile knowledge workers usually do not have a dedicated desk to work at in their main workplace, and they also work less than 20 hours per week in a fixed place (Bosch-Sijtsema, Ruohomäki, & Vartiainen, 2010); most of their working time is performed alone and on the move. Working alone restricts the opportunity to strengthen relationships (Kietzmann et al., 2013); it increases the lack of support and also motivates the feeling of being "invisible" in the organization (Koroma, Hyrkkänen, & Vartiainen, 2014). As a result, these challenges can reflect directly in knowledge creation and sharing in organizations (Kietzmann et al., 2013).

In this study, knowledge creation and knowledge sharing are seen as collective learning processes embedded in work activities and practices undergoing transformations (Greeno & Engeström, 2014). Among practice-based approaches to learning (Gherardi, 2009; Miettinen, Samra-Fredericks & Yanow, 2009; Nicolini, 2012), activity theory and the theory of expansive learning (Engeström, 2015; Engeström & Sannino, 2010) offer a promising framework for our analysis. Mobile workers often face learning challenges for which there are no ready-made solutions available, having to engage in "learning what is not yet there" (Engeström, 2016).

Learning in the context of mobile work is complex (Lundin & Magnusson, 2003; Muukkonen et al., 2014) because there are fewer opportunities for socialization and informal learning with peers, as in cubicle chats, cafeteria discussions, and impromptu team meetings (Jarrahi & Thomson, 2017; Nelson, Jarrahi, & Thomson, 2017; Vartiainen et al., 2007).

A literature review reveals that studies of mobile workers do not focus on understanding their processes of knowledge creation and sharing; we identified only two studies dealing with this subject (Kietzmann et al., 2013; Lundin & Magnusson, 2003). Most of the research involving mobile workers is related to their acceptance and adoption of mobile technologies (Dal Fiore et al., 2014; Karanasios & Allen, 2014; Mazmanian, Orlikowski, & Yates, 2013; Pauleen et al., 2015; Stieglitz, Lattemann, & Brockmann, 2015). Considering this gap, the research question explored in this chapter is: *how can expansive learning take place among mobile workers?* Expansive learning is a type of learning identified and conceptualized in cultural-historical activity theory (CHAT) to explain qualitative changes in work practices over time (Engeström, 2015; 2016). Through the lens of this theory, the processes of knowledge creation and sharing among mobile workers may be better understood.

The chapter is organized into seven sections. The next section explores the main characteristics of mobile work and mobile workers, and presents a review of the literature on expansive learning. The third section describes the research method, and the fourth presents the research results. The fifth, sixth and seventh sections offer a discussion of the results, the research conclusions and recommendations for further studies, respectively.

THEORETICAL BACKGROUND

This section presents the main characteristics of mobile work and mobile workers. It also introduces expansive learning theory, as a theoretical lens applied in this study to understand the processes of knowledge creation and sharing among mobile workers.

Mobile Work and Mobile Workers

According to Sørensen (2011), flexible work practices can take various forms: telework, home offices, shared offices, multifunctional offices, virtual work teams, and mobile work. Sørensen (2011) additionally points out that these new forms of work require new ways of conducting, coordinating, managing, and measuring work. This is because, in most cases, the decision of when, how, and where to work becomes a primary concern for professionals (Coenen & Kok, 2014).

Mobile work is the most radical form of flexible work (Sørensen, 2011) because it enables workers to carry out their organizational activities outside the confines of the organization, and also to collaborate with other workers regardless of time and location. According to Vartiainen et al. (2007), mobile work involves a change in the definition of the traditional office because of the diversity of workplaces in which professionals can perform their work activities, which in many cases dissolve the boundaries between home and workplace. Moreover, the quality of work done becomes of critical importance, depending on the workplace and the support obtained from the technological infrastructure and tools.

Mäkinen (2012) explains that *mobile* defines the characteristics of a professional who needs to move to different places and perform his/her work practices while traveling, frequently using ICT as a support. Thus, mobile technologies can provide support and extend work beyond the office, independent of the occupational or hierarchical position of professionals, from technicians to executives (Chen & Nath, 2008; Cohen, 2010). Mobile work amplifies the idea that the place where work is done is irrelevant; what is most important is that the work is performed in the best way (Cohen, 2010; Vartiainen et al., 2007).

Regarding the location where work is done, Wiberg & Ljungberg (2001) contend that the importance of where mobile work is performed depends on the tasks to be done, and on the time and space. These authors have elaborated a framework to explain this dependency. According to this framework, there are four categories of mobile work on the move. These categories and their meaning are presented in Figure 1.

The categories shown in Figure 1 reinforce the complexity of the context of mobile workers, who often need to figure out by themselves what, where, when, and how their tasks need to be done. Mobile workers also frequently need to resolve problems, disturbances, and contradictions in their work activity. In this context, the expansive learning theoretical approach may lead to a better understanding of the process of knowledge creation and sharing.

Expansive Learning

According to Engeström (2015), expansive learning is a type of learning in which participants generate transformations and development in their activities. Expansive learning occurs when an individual

Figure 1. The four categories of mobile work on the move (Wiberg & Ljungberg, 2001)

		Place	
		Independent	**Dependent**
Time	**Independent**	**1. Anytime, anywhere:** Tasks that can be done independent of time and place. They can be done anytime, anywhere.	**2. Anytime, particular place:** Tasks that need to be done within on a particular place but can be done whenever.
	Dependent	**3. Particular time, any place:** Tasks that can be done independent of place but at a certain time or in a certain order.	**4. Particular time, particular place:** Tasks that must be done on a particular place and on a particular time.

interacts with his or her community to solve contradictions that permeate their activity. The community re-conceptualizes its motive and object for the collective activity, with a new principle of operation or organization (Engeström & Sannino, 2010). Contradictions manifest themselves as dilemmas and conflicts in the activity (Engeström & Sannino, 2011), and thus can become a dynamic source of transformation and development.

As stated by Engeström & Sannino (2010, p. 2), "In expansive learning, learners learn something that is not yet there. In other words, the learners construct a new object and concept for their collective activity, and implement this new object and concept in practice." Expansive learning promotes the evolution of the activity into a new pattern or structure, and this involves the collective interaction between participants moving across their zone of proximal development (Engeström, 2016).

The definition of the zone of proximal development (ZPD) stems from Vygotsky (1934/1987), who argued that ZPD is the zone in the learning process where a more capable individual helps the less capable individual to solve problems and achieve a new level of development. Engeström (2015) proposed an extended definition of the ZPD according to which the ZPD is the zone in the learning process where participants construct a new, expansive solution to resolve contradictions in their collective activity.

Expansive learning occurs in a learning cycle by undertaking learning actions driven by contradictions and, thus enabling the activity system to evolve into a more advanced form (Engeström, 2015). Figure 2 illustrates the sequence of learning actions in an expansive learning cycle. The typical sequence of these learning actions is described as follows (Engeström, Rantavuori, & Kerosuo, 2013):

1. The first action is questioning, criticizing, or rejecting some aspects of the accepted practice and existing wisdom.
2. The second action is analyzing the situation. Analysis involves mental, discursive, or practical transformation of the activity to find out causes or explanatory mechanisms. One type of analysis is historical-genetic; it seeks to explain the situation by tracing its origins and evolution. Another type of analysis is actual-empirical; it seeks to explain the situation by constructing a picture of its inner systemic relations.

3. The third action is modeling the newly identified explanatory relationship in some publicly observable and transmittable medium. This involves constructing an explicit, simplified model of the new idea that explains and offers a solution to the problematic situation.
4. The fourth action is examining the model, running, operating, and experimenting on it to fully grasp its dynamics, potential, and limitations.
5. The fifth action is implementing the model by means of practical applications, enrichments, and conceptual extensions.
6. The sixth action is reflecting on and evaluating the process.
7. The seventh action is consolidating and generalizing the outcomes into a new and stable form of practice.

The expansive learning cycle presented in Figure 2 refers to the large-scale cycle of local expansive reorganization of an activity (Engeström, Rantavuori, & Kerosuo, 2013). However, a large-scale cycle commonly involves numerous smaller mini-cycles of learning actions (Engeström, 2010). "A smaller cycle may take place within a few days or even hours of intensive collaborative analysis and problem-solving" (Engeström, 2010, p. 82). If the smaller cycles remain isolated events, a larger expansive cycle may not emerge (Engeström, Miettinen, & Punamäki, 1999). Figure 3 illustrates in a schematic way the embeddedness of mini-cycles within larger cycles of expansive learning.

Expansive learning makes possible for those involved in the activity to discuss and identify solutions to their contradictions and problems, generating new ways of working. By resolving contradictions and modifying work practices, knowledge is shared and created and problems are mitigated. For this reason, this theoretical approach was used to analyze how expansive learning may be carried out among mobile workers.

Figure 2. Sequence of learning actions in an expansive learning cycle (Engeström, 2010)

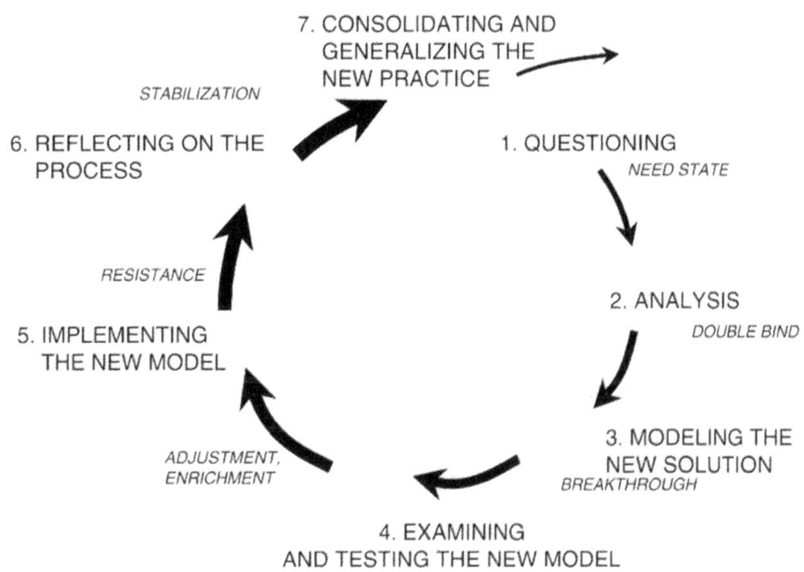

Figure 3. Large and small cycles of expansive learning (Adapted from Engeström, 1996)

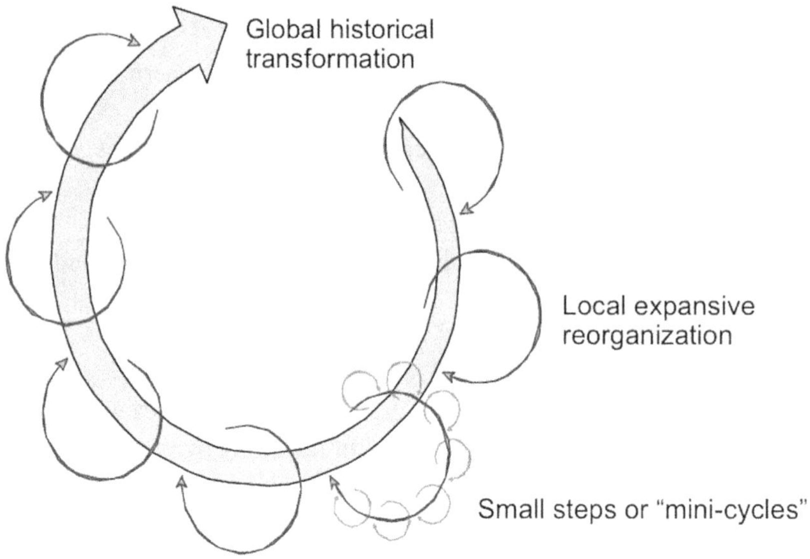

RESEARCH METHOD

The primary research method used in this study was participant observation. Based on this method, it was possible to observe mobile knowledge workers of an IT company. Participant observation was an important step, because, as mentioned by Schultze (2000), to study knowledge workers, it is necessary to focus on what these professionals do (their work practices) rather than what they say they do. The participant observer gathers data by participating in the daily life of the people who he/she studies, thereby coming to an understanding of the issues from their perspective (Becker, 1958; Spradley, 2016).

Participant observation was carried out in an IT company established in southern Brazil. To preserve the anonymity of the company, the name "ITCom" is used in this chapter. This organization has existed for more than 20 years, and is one of the main Brazilian IT groups in the areas of outsourcing and consulting services. During the period of participant observation, ITCom had branches in four Brazilian states, and its activities involved a variety of industries, such as metal fabrication companies and agribusinesses, in Brazil and other countries, such as the United States.

The first author of this chapter worked at ITCom from 2010, first as a systems analyst and later, in 2011, as a systems analyst leader of one of its outsourcing operations. In September 2012, the vice president of ITCom invited the researcher to participate in a research and development initiative called PTD (Program of Technological Development), one of the requirements of this initiative being that the researcher enrolled in the PhD program of a local university. Thus, in January 2013, the researcher started a new position within ITCom, in a research project that had as its primary objective the creation of a virtual space to improve communication and retain organizational knowledge in a single repository. To achieve this goal, a tool was developed, based on Web 2.0 and SharePoint technology aiming at stimulation and support of collaboration between employees. The name "ITCom Social Media" will be used in the remainder of this chapter to refer to this tool.

Although the project and tool were developed for the entire company, the focus of attention was on the managers, because the company wanted to standardize its management best practices and thereby improve the services provided for the customers. For this reason, many interactions took place with managers to identify their main characteristics and work practices. Mapping their processes and responsibilities gave the researcher the chance to ask questions about their practices and the challenges they faced. Three groups of managers were identified as mobile knowledge workers: IT Operation Managers, IT Services Managers, and IT Relationship Managers. The interaction with managers was a motivation for, and a source of understanding of, the research problem of this study. The script used to collect and organize data is presented as follows. For each manager, the following questions were asked to know their practices:

- What activities do you perform?
- What is needed to start the activity (inputs)?
- What is needed to end the activity (outputs/evidence)?
- When does the activity need to be performed (deadline)?
- Who provides the inputs (provider)?
- Who receives the outputs (customer)?
- What are the related activities (predecessors)?
- What are the main resources to perform the activity?

Table 1 shows the profile of participants observed during the fieldwork.

Participant observation requires the researcher to spend considerable time in the field, adopting various roles to gain more understanding of the people and the context studied (Baker, 2006; Becker, 1958). The observation took place from January 2013 to mid-January 2015. The researcher performed the fieldwork from Mondays to Fridays; some exceptions were made during weekends, depending on the work carried out with the participants. The participant observation lasted 541 days. During 2013, a moderate researcher's participation took place to maintain a balance between participation and observation (Spradley, 2016). During 2014, participation with the IT Relationship Managers and the IT Services Managers was more active, because the researcher was more involved with their central activities. In the same year, the researcher became a member of the IT Operation/Project Managers group, assuming the position of IT Operation/Project Manager. This role undertaken by the researcher allowed to test the

Table 1. The 39 participants in the study

Role	IT Relationship Manager (responsible for sales, customer relations and prospects, account management)	14
	IT Services Manager (performs technical pre-sales and has the IT Relationship Manager as his/her internal client)	11
	IT Operation/Project Manager (responsible for delivery of client service contracts; he/she is subordinate to an IT Services Manager and has the IT Relationship Manager as his/her internal client)	14
Gender	Male	33
	Female	6
Type of mobility	Alternating between two fixed locations	14
	Working at three or more places and constantly moving	25

methodology and tool[1] adopted by the company to help new managers, and to empathically understand in practice the problems they faced in their daily tasks.

The observations were made both face-to-face and within the virtual community (as in the study of Nørskov & Rask, 2011), for example by exchanging instant messages and taking part in virtual meetings. This way, the recording of data was ensured for future analysis, and for triangulation with the researcher's written field notes (Angrosino & Pérez, 2001). The participant observation during January 2013 and mid-January 2015 yielded more than 1 GB of digital files and records. In total, 301 files were analyzed: 108 field notes, 75 emails, and 118 instant-message conversations. These files were considered because they contained information about the process of tool development—the central part of expansive learning. Besides online social-interaction data, a variety of other documents were collected, such as PowerPoint presentations, minutes of meetings, reports, and others. Nine participants were also interviewed later, between November 2015 and March 2016, to supplement the data collected via observation and documents.

For a qualitative analysis the NVivo software was used to organize and analyze all the collected data. First, a careful reading of all the records saved inside the database was done. Then, thematic coding (Punch, 2009) was conducted to identify key themes about the profile and the context of the mobile workers. Finally, an analytic coding (Punch, 2009) took place, considering each one of the steps of the process of expansive learning as indicated by Engeström, Rantavuori, & Kerosuo (2013).

CASE: ITCom

The managers were required to work 40 hours a week, but it is very common for them to work more than this. They did not need to register the hours they have worked, and they had considerable flexibility to start and finish their work whenever they want. It was very common to find them working online in virtual spaces such as Skype or Microsoft Lync[2] early in the morning, late at night, or even on weekends. They often worked at home, in the company sites, or at client sites. Some of them, mainly the IT Relationship Managers, needed to travel a lot. They often used their cars or, if the customer was far from the company site, they traveled by air.

The managers generally enjoyed the variety of organizations, situations, people, and challenges to which they were exposed. Common remarks among professionals working most of the time at client sites referred to a feeling of being "invisible" in their own organization. These professionals felt like they were employees of their clients:

I feel better being a mobile worker because usually as you are on the move and working in different places all the time, you end up, as I said before, experiencing different realities, different cultures, different habits, knowing different people, you know. And I like this one, this particularity of being a mobile worker (IT Project Manager).

Another point is that when you go back to the base [...] your relationship with the people is weaker because you're not much there. You spend 50% of your time off, a week in the headquarters and a week on the move, [...] many times the people who spend more time at the headquarters, they have some routines that usually those who are not there all the time do not participate of, because people do not even remember to invite them. For example, the people are going to have lunch every day at the same hour, and then because you always are not there, they do not remember to invite you. So if you do not go there and say, "oh let's have lunch", they do not invite you. You also have this loss of relationship with the headquarters staff (IT Services Manager).

Since there was little standardization in managers' work practices, they needed to deal with the many intricacies of their clients' processes, as well as the processes and methods of their own company. As a result, they often adopted a mix of procedures and methods, but this increased their difficulties in meeting deadlines and the goals of service quality expected by ITCom. In addition, they had to deal with many challenges during interactions with their own community, mainly with support staff at the ITCom headquarters. Hence, discipline, time management, and self-management are some of the required skills for these mobile workers. The interviewees often made remarks along these lines, such as in the following example:

We have some clients with very rigid norms that cause a lot of impact in our lives (e.g., clients that do not let us access their IT infrastructure). So we need to adapt and keep in touch with everything and many times we need help from someone from the headquarters. However, it is common that the people from the headquarters do not understand our "urgency" (IT Relationship Manager).

Regarding the instruments used for work, these mobile workers used a wide variety of tools. Because they needed to be ready for anything, anywhere, they carry an extensive toolset, such as notebooks, tablets, 3G/4G modem, extra batteries, more than one smartphone, and so on. However, it was observed that these workers enjoyed technology and were always aware of technological innovations that could help them in their work. It was common to witness informal chats between mobile workers about new devices that some of them had bought to perform their activities better.

During the participant observation, it was also common to see the commitment of these workers to practices of knowledge sharing with their community. They tended to help their colleagues any time, and especially those colleagues who lacked experience. Therefore, in collaborative problem-solving situations, the more skilled workers helped the less skilled in achieving their task or objectives. An interesting point about these collaborative problem-solving processes among mobile workers is that they occur mainly through the use of ICT tools. The most used tools in these situations were instant messaging applications such as WhatsApp, Microsoft Lync, or Skype. Email and phone calls were also used, but instant messaging were more common because, by comparison, it provided an easier, faster, and cheaper way to contact colleagues who could be of help.

As observed in the literature review and in the empirical data, mobile workers have to operate in a very complex context. They need to deal with the many intricacies related to the different people they need to work with, the many instruments they need to use to support their work, and the rules (from their own company and from the client company) they need to follow to deal with their object and achieve their best results. Therefore, the creation and sharing of knowledge in this context is not a trivial task.

The Process of Expansive Learning Among Mobile Workers

A case of expansive learning among mobile workers was observed during the participant observation. The managers started an informal process of teamwork to use ITCom Social Media for their benefit. The expansive learning steps taken by these workers in the creation of an artifact are presented below. These steps illustrate where, why, what, and how the mobile workers were learning.

First Step: The Need State

The first step in expansive learning is the questioning and identification of needs. This phase started among the mobile workers in February 2013, after the decision to adopt ITCom Social Media was communicated to everyone in the company. The mobile workers had always complained about the many tasks they needed to deal with, and about the lack of formal work standards for the performance of their tasks, which increased in difficulty. Hence, they organized themselves to create a new artifact, based on social media, which could help them define a kind of best-practice guide to perform their work activities.

The first meeting to map needs took place on the 23rd of February 2013. Eight managers participated in this meeting: three IT Services Managers, two IT Project Managers, two IT Relationship Managers and the participant researcher. The researcher was invited by these managers to help them to think about how they could use ITCom Social Media. Thus, they discussed how social media could help them to establish standards and best practices for their work activities.

To track the evolution of this initiative, a meeting was held every two weeks. These meetings were scheduled in advance, for the whole semester, in Microsoft Outlook. The eight managers could participate wherever they were, because the meetings were virtual through Microsoft Lync. Moreover, when someone could not participate, one of the managers was always responsible for the minutes of the meeting to be placed on the company's social media. An example of remote participation is presented in Excerpt 1.

- **Excerpt 1:** Instant messaging conversation (December 30, 2013), follow-up meeting:

IT Service Manager [14:07]: Is the meeting happening?
Participant Researcher [14:07]: We will start right now. Where are you?
IT Service Manager [14:08]: At the airport, call me here.

The main conflict situation articulated by the mobile workers concerned their responsibilities. They complained about the multiple demands and information load they had to deal with every day. According to them, the conflicting demands made it very difficult to achieve the clients' needs: "Many times I need to choose between the client's needs and the procedures of our company … It is tough to perform the procedures as defined and achieve the client's needs" (IT Operation Manager, field notes, February 22, 2013).

Second Step: Analyzing the Situation

The second step is to analyze the historic and actual situation in which the learners are involved. This step started in parallel with the first step and was carried out until April 2013, when the third step started.

It was often observed that the participants had difficulties to grasp the overall situation, or that they did not have enough knowledge to perform their tasks. Thus, they were typically involved in cycles of trial and error until they reached a solution. As one IT Project Manager said, "There are a lot of things to adjust that I have no idea how to do and the time for me is very short, so I end up prioritizing the work with the team" (IT Project Manager, instant message, April 12, 2013). Besides, because no work standards existed to measure to performance, it was common for them to use or adopt different ways to deal with similar situations. This contrasts with the high level of service quality expected by ITCom.

To grasp their situation, the mobile workers created a mind map of all the main activities they needed to perform. One of the IT Services Managers already used mind maps to organize his activities, and he suggested this idea for everyone. A mind map was developed using a mind map tool,[3] and an image of the map was then exported, to be shared via the company's social media. Therefore, for a certain period, the managers needed to consolidate the activities they usually performed, and each was responsible for consolidating their own activities in small groups, according to their roles.

During the first weeks of work, there were some problems with the construction of the mind map. The main issue was how to use the tool in a collaborative way. Some of the managers who had never used the mind map tool before had problems using it. Because only one person could use the mind map file at a time, this task became unproductive. As noted by one IT Relationship Manager in a meeting, "We need to reconsider this tool … it is not working … we need something that is possible for everyone to work on anytime, anywhere, and at the same time" (IT Relationship Manager, field notes, March 9, 2013).

Third Step: Modeling the New Solution

The third step is modeling the new idea for solving the problematic situation observed in the second step. The mobile workers started this step in April 2013.Three participants were elected to find a new way to work in the best-practice guide. These participants were chosen because they were more skilled in the use of SharePoint technology, which was used to develop ITCom Social Media. The chosen participants were one IT Services Manager, one IT Project Manager, and the participant researcher.

To solve the problems that had occurred with the mind map tool, the chosen participants decided to make an experience using the SharePoint resources available on ITCom Social Media. The initial idea was to create a SharePoint custom list[4] that could be updated by any mobile worker at any time and in any place. Due to the format of this list and its content (all the managers' tasks and roles), the participants started to call it the "responsibility matrix." Excerpt 2 represents one of the instant messages about the modeling of the responsibility matrix.

- **Excerpt 2:** Instant messaging conversation (June 6, 2013), modeling the responsibility matrix:

IT Service Manager [09:50]: [Name] Are you there?
Participant researcher [09:50]: Hi, good morning, yes
[…]
IT Service Manager [10:06]: Ok … I'm doing a work with the matrix of activities
Participant researcher [10:08]: Ok, I'm already seeing it here
[…]
IT Service Manager [10:20]: […] We need to put the activities in a logical order of what is done first and last. Also, we discussed this in the last meeting, can you help us?
Participant researcher [10:21]: Wait 1 min … what is the logical order?
IT Service Manager [10:52]: Sorry, the logical order I'm doing here, but I need help because I'm using my "way to do things"
Participant researcher [10:56]: Ok, we need to speak with the others.

Due to the characteristics of the artifact in development – the SharePoint custom list inside ITCom Social Media – an online tool can be modeled, experimented with, and made available quickly. This situation illustrates the occurrence of quick mini-cycles of innovative learning within the large-scale expansive cycle, explained in Figure 3. During the third and fourth step, mini-cycles of innovative learning were taken and allowed the mobile workers to refine and test the model while they were creating it. Excerpt 3 contains one of the instant messages about the responsibility matrix test that occurred before the changes made in the case presented in Excerpt 2.

- **Excerpt 3:** Instant messaging conversation (May 22, 2013), testing the responsibility matrix:

IT Operation/Project Manager [15:11]: [...] Hi, I'm not getting into the responsibility matrix yet
Participant researcher [15:12]: Ok, just a minute
[...]
IT Operation/Project Manager [15:18]: I think it's ok now, the screen appeared
IT Operation/Project Manager [15:18]: How can I know what is my responsibility? There is some filter?

After some months, the SharePoint custom list inside ITCom Social Media became the responsibility matrix artifact. All participants in the creation of the responsibility matrix were already using this name to refer to the artifact. In the matrix, the three groups of mobile workers registered all the activities known and performed by the managers.

Fourth Step: Examining and Testing the New Model

The fourth step is examining and testing the model. Because of the characteristics of the mobile workers, it was observed that this step started when the artifact was still in the process of being modeled in the third step (modeling step). Excerpt 4 illustrates one of the instant messages about the responsibility matrix examination and test.

- **Excerpt 4:** Instant messaging conversation (June 15, 2013), examining and testing the responsibility matrix:

IT Operation/Project Manager [17:21]: [...] Afternoon
Participant researcher [17:26]: Hi
IT Operation/Project Manager [17:29]: Look...We have to think how to 'enumerate' the responsibility matrix... because we see now that it is ordered by alphabetical order
Participant researcher [17:30]: What? I think I didn't follow you...
IT Operation/Project Manager [17:30]: We need something like this:
1. Process
1.1 Sub process
1.1.1 Activity
Participant researcher [17:31]: Ah, ok, understood... but how we will do that? Do you think in input the numbers in a manual or automatic way?
IT Operation/Project Manager [17:32]: So, we need ideas...

In this fourth step, more participants started to experiment and to help in the maintenance of the responsibility matrix. The creators of the matrix invited some of these new participants, while others offered their help in the initiative. By the end of 2013, 15 participants were working on the maintenance of the information in the responsibility matrix.

In this step it is common to adjust the model created, and this happened with the matrix when other mobile workers started to experiment with it. The matrix began with the following information: a) the name and the description of the activity; b) the process name that the activity belongs to; c) the responsible manager's role; d) the deadline of the activity; and e) the name of the support areas (staff who can help the responsible manager). Because the matrix could be improved in real time, other information was added to it. Information such as templates and examples of documents was added to help new managers to perform their activities. Also, because SharePoint technology allows the creation of visions with different information and structure, many visions were also created to help managers find the specific information they needed.

Fifth Step: Implementing the New Model

The fifth step is implementing the new model. In June 2013, the responsibility matrix was available for all managers in ITCom, after the creators of the matrix agreed that the new model could be available for everyone. However, only the group who started the artifact was responsible for its maintenance, with sponsorship from the company. Also, the company provided autonomy for the team and for it to take on additional participants. An interesting situation, namely resistance to the use of the artifact, was observed among certain managers who had not participated in the creation and maintenance of the artifact. Excerpt 5 illustrates this situation.

- **Excerpt 5:** Instant messaging conversation (June 19, 2013), resistance to using the responsibility matrix

IT Operation Manager [14:42]: Hi, how can I export the responsibility matrix?
Participant researcher [14:46]: Export?
IT Operation Manager [14:47]: Yes, put it in excel
Participant researcher [14:47]: Why? Why do you not access it directly?
IT Operation Manager [14:47]: Because I want to do a mind map
[...]
IT Operation Manager [14:52]: There are 94 activities, absorbing them in this way is complicated
Participant researcher [14:52]: Yes ... the idea is that you access directly from the matrix, but we can think in something to help you, based on your suggestions
IT Operation Manager [14:53]: Ok, but in this way I'm limited to a notebook and Internet ... I wanted to do a mind map, to print, to read, and make observations
Participant researcher [14:54]: Yes, I understand, but as I said, we can think on something to help, it helps if you can suggest improvements and I'll send them to the matrix committee
IT Operation Manager [14:56]: Export is my suggestion, with all the items and descriptions
Participant researcher [14:57]: Ok, I will check with the responsible people and I will return to you

In this excerpt, the manager wanted to export the items to an Excel spreadsheet. However, the matrix committee did not allow this feature because they did not want to create a parallel responsibility matrix. According to one of the matrix's creators, during a meeting that anticipated the availability of the matrix for everyone, "If we allow exportation for Excel, the managers will create a parallel matrix. We cannot allow this. We need to have only one matrix, they need to adapt themselves to it" (IT Relationship Manager, field notes, May 22, 2013).

Sixth and Seventh Steps: Reflecting and Consolidating the New Practice

Despite some resistance, the responsibility matrix was consolidated in the company in early 2014. The board of ITCom decided to change the name of the artifact to ITCom Management Methodology. A new vision based on the list was also created that could help new managers. Excerpt 6 shows the email that was sent to all managers from the committee responsible for the matrix.

- **Excerpt 6:** Email (May 14, 2014), new ITCom Management Methodology

Guys, good afternoon!
In this link [...] the new ITCom Management Methodology's vision is available, a preview for you to evaluate. This item has been reprioritized several times due to our demands, but we were able to finish the development this week. (Committee Matrix)

The ITCom Management Methodology continued to be improved in 2014 and, according to the nine interviewees, is still in use. They also mentioned that the instrument continued to be improved by them. As these workers have always complained about the many activities they needed to deal with, this instrument, created by them, helped them to deal with these situations. Figure 4 summarizes the learning steps taken by the mobile workers.

CONCLUSION

Mobile workers are an increasing and understudied population (Jarrahi & Thomson, 2017). The present study aimed to analyze how expansive learning can take place among mobile workers. As already mentioned, only two studies were found (Kietzmann et al., 2013; Lundin & Magnusson, 2003) that addressed, respectively, knowledge sharing and collaborative learning in mobile work. The research goal

Figure 4. Expansive learning among mobile workers

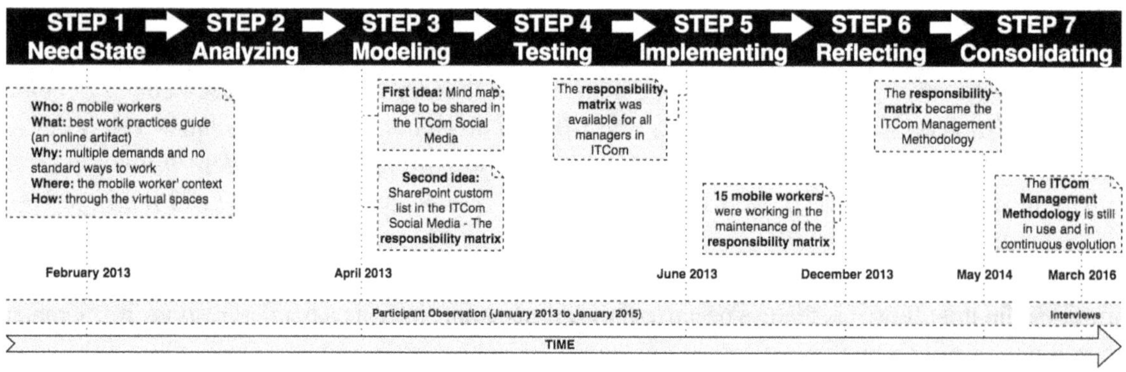

of the first study (Lundin & Magnusson, 2003) was to understand how IT could support learning where work tends to be mobile. In this study, workplace learning is viewed as the development and sharing of knowledge at work, in collaboration with colleagues and peers and the creation of new knowledge is seen as a change in understanding, referred to as process learning. Thus, support for this collaboration is needed. However, in this case, the authors did not specify the type of mobility of workers or how the mobility affects learning.

The research goal of the second study (Kietzmann et al., 2013), in turn, was to analyze how mobility and mobile technologies shape community collaboration and explore the implications for the management of mobile employees. The mobility considered in this study has three dimensions: (1) the dual role of location – location does not matter but at the same time it really does, (2) the physicality of the environment and (3) the formative context of the movement of people, devices and information. This study highlighted the difficulty for mobile workers to participate in communities of practice to share knowledge due to the characteristic that they mainly interact on the move. The authors reinforce that it is time to focus on mobility in a discussion of community-based knowledge sharing, as collaboration across organizational boundaries increasingly occurs through mobile IT.

Our study differs from these two studies, because it seeks a broader exploration and understanding of the context of mobile knowledge workers. Also our study aims at grasping the challenges these workers face to create and share knowledge, especially in problem-solving situations. Finally, our study is the first analysis of expansive learning in the context of mobile work, during processes of knowledge creation and knowledge sharing. With the help of this theoretical approach our study offers insights into who these mobile workers were, what they were learning, why they were learning, where they were learning and how they were learning. Facing an initial problematic situation of multiple tasks, lacking standards and difficulty to meet clients' needs, the mobile workers studied undertook an expansive learning journey. Through this journey, they generated and implemented a new instrument, the responsibility matrix that allowed them to expand their view of the problematic situation and to gain a holistic grasp of it.

As indicated in the literature review, mobile workers perform their work on their own for most of their working time (Di Domenico, Daniel, & Nunan, 2014; Sayah, 2013), and they spend more than 20% of their work hours on the move, away from their main fixed workplace (Bosch-Sijtsema et al., 2010). These were also features observed in the empirical setting of this study, where the mobile workers usually worked on the move and longer than the 40 hours for which they were contracted. It was very common to find them working online after regular working hours.

Since the main duties of these workers are related to working with people (clients, partners, or teams) (Vartiainen, 2008), they participate in diverse situations where they can perform different roles in the projects, in which they are involved, and they usually work on more than one project at a time. In addition, these professionals need to be able to deal quickly with day-to-day impediments (Jarrahi & Thomson, 2017) such as cancelled meetings, cancelled flights, different time zones, etc. These features of mobile work were also observed in our study; for instance, many mobile workers participated in meetings from the space where they were on the move.

Most of the mobile workers observed in this study perform knowledge-intensive work. They are involved in the creation of new ideas, new technology, or creative content. The literature describes these professionals as more independent, self-organized, and devoted to their performance (Yuan & Zheng, 2009). They accordingly have greater concern for their skills and competencies, and are often engaged in improving these (Harmer & Pauleen, 2012).

Also, as observed in both the literature and the analysis, these professionals require new styles of management that encourage and engage them to deliver valuable outcomes without restrictive control and supervision (Harmer & Pauleen, 2012). Besides, the management of these professionals requires an environment based more on trust, commitment, and collaboration rather than on direct control (Kietzmann et al., 2013). ITCom Social Media aspires to provide this kind of environment, though mainly for the mobile workers. When the board of directors of ITCom learned about the artifact in creation, they gave incentives and offered sponsorship to the matrix committee. This helped to consolidate the artifact created by the company's mobile workers.

Considering that mobile workers travel constantly, work in many places (Koroma, Hyrkkänen, & Vartiainen, 2014), and face challenges in meeting face-to-face (Vartiainen, 2008), they mainly use virtual spaces to work and communicate with others. They access these virtual spaces using the instruments or tools to perform their work activity (Dal Fiore et al., 2014). In this research, the instruments used by the mobile workers, especially mobile ICT, helped them to be online, through virtual spaces, and to keep in contact with their co-workers. The main virtual spaces, accessed via mobile devices such as smartphones, notebooks, and tablets, were email, Skype, Microsoft Lync, phone calls, Microsoft Outlook, and ITCom Social Media.

Mobile workers perform their activities not only within their organization but also, frequently, in other workspaces, such as client or partner companies. Therefore, the work processes and rules of their own organization may conflict with the client's needs and the client company's procedures, thereby generating contradictions. To resolve these contradictions, mobile workers adopted a mix of instruments, procedures, and methods to reach their activity goals, which sometimes made it difficult to meet deadlines and achieve the quality expected by the clients or by their own company. Since they worked alone most of the time (Di Domenico et al., 2014; Sayah, 2013), each mobile worker had "his/her" way of doing things. For this reason, in the case studied, they had the idea of constructing a guide to best practice in performing their activities (their primary learning object), in a standard form, that would also help newly hired managers.

The circumstances, infrastructure, and processes of collaboratively developing their work standards represented a clear opportunity for them to create and share knowledge. In line with the expansive learning approach, participants, through their collective ZPDs, provoke transformations and development in their activities (Engeström, 2015). The case presented here clearly shows how this process happened among the mobile workers. It was possible to observe the occurrence of ZPDs through the participation of these workers in problem-solving efforts around the learning object. They engaged in the collective ZPD even though they were physically distant from one another, and even when they were on the move, using a set of tools, such as ITCom Social Media, accessed via mobile devices, to interact and develop a new way of working.

Expansive learning crosses the domains of individual learning and organizational learning, allowing material changes in objects, instruments, rules, and divisions of labor within and between the activities (Engeström, 2015). At the individual level, these workers created new concepts about how they could work better. At the organizational level, they created a new instrument that could guide the managers' way of working in ITCom.

The primary goal of this investigation was to analyze how expansive learning can take place among mobile workers. It was observed that these workers use their ICT tools, especially mobile ICT, most of the time in their collaborative problem-solving tasks and subsequently, to create and share knowledge on the move.

This research makes a theoretical contribution by exploring the concept of expansive learning in mobile work. Since this concept has not been used so far to analyze practices of knowledge creation and sharing in the context of mobile work, this study contributes to the knowledge on this topic.

Furthermore, this study supports recent developments in studies of expansive learning (Sannino, Engeström & Lemos, 2016) demonstrating that this type of learning may be accomplished without deliberate external input from researchers, in the mode of an *intra*-vention. This means that practitioners may conduct successful transformation efforts within their own activities.

SOLUTIONS AND RECOMMENDATIONS

Our initial research question was: How can expansive learning take place among mobile workers? Our analysis indicates that expansive learning processes may emerge among mobile workers when they experience a contradiction between the demands of their complex work tasks and the insufficient shared standards and tools with the help of which those tasks are performed. This kind of tension may trigger an expansive learning process especially when the management introduces novel instruments which require practical adjustment, modification and further development in practice.

In our case, the expansive learning process was initiated and led by the mobile workers themselves, without systematic guidance or planned intervention from above. The entire process was an intravention rather than a traditional intervention (Sannino, Engeström & Lemos, 2016). This testifies to the creative potential of mobile work.

On the other hand, self-initiated and self-directed processes of expansive learning are also vulnerable. They often remain isolated or become interrupted due to lack of awareness and support among the management. Our study points to the need for systematic experimentation with supported formative interventions aimed at expansive learning among mobile workers. Such processes may generate innovative solutions that enhance the quality of the work both for the practitioners and the clients.

FUTURE RESEARCH DIRECTIONS

As possible limitations of this study, its focus on the mobile work of IT professionals, may explain the inclination of these professionals to interact with people through ICT. Besides, due to the diversity of workplaces that these workers can use to perform their work activity, future research may undertake similar studies of mobile work involving other categories of mobile knowledge workers.

The findings of this study also highlight the importance of further research in this topic. It was observed a lack of studies related to mobile knowledge workers and also related to knowledge creation and knowledge sharing in the context of these workers. Furthermore, since there is a forecast (McDaniel et al., 2016) that estimates the increase of mobile knowledge workers in the next years, the use of activity theory and the theory of expansive learning to analyze this context can be seen as an opportunity.

The rise of mobile workers, in turn, is related to the constant evolution and acceptance of the new mobile/ubiquitous ICTs such as biometric readers, wearable, voice control, near-field communications (NFC), augmented reality, etc. Therefore, new technologies and new ways to stimulate collaborative problem-solving and learning can be adopted in the mobile workers' context. The use of augmented reality in the collaborative problem-solving situations is but one such possibility. The adoption of new technologies typically opens up possibilities to expand the object of work and requires reorganization of the work activity. To make optimal use of such opportunities, the theory of expansive learning and formative interventions aimed at supporting expansive learning are powerful resources.

REFERENCES

Allen, D. K., Brown, A., Karanasios, S., & Norman, A. (2013). How Should Technology-Mediated Organizational Change Be Explained? a Comparison of the Contributions of Critical Realism and Activity Theory. *Management Information Systems Quarterly*, *37*(3), 835–854. doi:10.25300/MISQ/2013/37.3.08

Angrosino, M., & de Pérez, K. (2001). Rethinking Observation: From Method to Context. In The American Tradition in Qualitative Research (pp. 673–702). SAGE Publications Ltd. doi:<ALIGNMENT.qj></ALIGNMENT>10.4135/9781446263570

Baker, L. (2006). Observation: A Complex Research Method. *Library Trends*, *55*(1), 171–189. doi:10.1353/lib.2006.0045

Becker, H. S. (1958). Problems of Inference and Proof in Participant Observation. *Source American Sociological Review*, *23*(6), 652–660. doi:10.2307/2089053

Bosch-Sijtsema, P. M., Ruohomäki, V., & Vartiainen, M. (2010). Multi-locational knowledge workers in the office: Navigation, disturbances and effectiveness. *New Technology, Work and Employment*, *25*(3), 183–195. doi:10.1111/j.1468-005X.2010.00247.x

Chen, L. (2015). A Model of Mobile Work Continuance of Knowledge Workers: Evidences from China. In *The Third International Conference on E-Technologies and Business on the Web (EBW2015)* (pp. 43–52). Academic Press.

Chen, L., & Nath, R. (2008). A socio-technical perspective of mobile work. *Information, Knowledge, Systems Management*, *7*, 41–60. Retrieved from http://iospress.metapress.com/index/b412488672201268.pdf

Coenen, M., & Kok, R. W. (2014). Workplace flexibility and new product development performance: The role of telework and flexible work schedules. *European Management Journal*, *32*(4), 564–576. doi:10.1016/j.emj.2013.12.003

Cohen, R. L. (2010). Rethinking "mobile work": Boundaries of space, time and social relation in the working lives of mobile hairstylists. *Work, Employment and Society*, *24*(1), 65–84. doi:10.1177/0950017009353658

Dal Fiore, F., Mokhtarian, P. L., Salomon, I., & Singer, M. E. (2014). "Nomads at last"? A set of perspectives on how mobile technology may affect travel. *Journal of Transport Geography*, *41*, 97–106. doi:10.1016/j.jtrangeo.2014.08.014

Di Domenico, M., Daniel, E., & Nunan, D. (2014). "Mental mobility" in the digital age: Entrepreneurs and the online home-based business. *New Technology, Work and Employment*, *29*(3), 266–281. doi:10.1111/ntwe.12034

Engeström, Y. (1996). Developmental work research as educational research. Nordisk Pedagogik. *Journal of Nordic Educational Research, 16*(5), 131–143.

Engeström, Y. (2001). Expansive Learning at Work: Toward an activity theoretical reconceptualization. *Journal of Education and Work, 14*(1), 133–156. doi:10.1080/13639080020028747

Engeström, Y. (2010). Activity theory and learning at work. In *The Sage handbook of workplace learning* (pp. 74–89). SAGE.

Engeström, Y. (2015). *Learning by expanding: An activity-theoretical approach to developmental research* (2nd ed.). Cambridge, UK: Cambridge University Press.

Engeström, Y. (2016). *Studies in expansive learning: Learning what is not yet there*. Cambridge University Press. doi:10.1017/CBO9781316225363

Engeström, Y., Miettinen, R., & Punamäki, R.-L. (1999). *Perspectives on activity theory*. New York: Cambridge University Press. doi:10.1017/CBO9780511812774

Engeström, Y., Rantavuori, J., & Kerosuo, H. (2013). Expansive Learning in a Library: Actions, Cycles and Deviations from Instructional Intentions. *Vocations and Learning, 6*(1), 81–106. doi:10.1007/s12186-012-9089-6

Engeström, Y., & Sannino, A. (2010). Studies of expansive learning: Foundations, findings and future challenges. *Educational Research Review, 5*(1), 1–24. doi:10.1016/j.edurev.2009.12.002

Engeström, Y., & Sannino, A. (2011). Discursive manifestations of contradictions in organizational change efforts: A methodological framework. *Journal of Organizational Change Management, 24*(3), 368–387. doi:10.1108/09534811111132758

Gherardi, S. (2009). Community of pratice or practices of a community? In S. Armstrong & C. Fukami (Eds.), *The sage handbook of management learning, education, and development* (pp. 514–530). London: Sage. doi:10.4135/9780857021038.n27

Greeno, J. G., & Engeström, Y. (2014). Learning in activity. In R. K. Sawyer (Ed.), *The Cambridge handbook of the learning sciences* (2nd ed.; pp. 128–147). Cambridge: Cambridge University Press. doi:10.1017/CBO9781139519526.009

Harmer, B. M., & Pauleen, D. J. (2012). Attitude, aptitude, ability and autonomy: The emergence of "offroaders", a special class of nomadic worker. *Behaviour & Information Technology, 31*(5), 439–451. doi:10.1080/0144929X.2010.489117

Jarrahi, M. H., & Thomson, L. (2017). The interplay between information practices and information context: The case of mobile knowledge workers. *Journal of the Association for Information Science and Technology, 68*(5), 1073–1089. doi:10.1002/asi.23773

Karanasios, S., & Allen, D. (2014). Mobile technology in mobile work: Contradictions and congruencies in activity systems. *European Journal of Information Systems, 23*(5), 529–542. doi:10.1057/ejis.2014.20

Kietzmann, J., Plangger, K., Eaton, B., Heilgenberg, K., Pitt, L., & Berthon, P. (2013). Mobility at work: A typology of mobile communities of practice and contextual ambidexterity. *The Journal of Strategic Information Systems, 22*(4), 282–297. doi:10.1016/j.jsis.2013.03.003

Koroma, J., Hyrkkänen, U., & Vartiainen, M. (2014). Looking for people, places and connections: Hindrances when working in multiple locations: A review. *New Technology, Work and Employment, 29*(2), 139–159. doi:10.1111/ntwe.12030

Koroma, J., Hyrkkänen, U., & Vartiainen, M. (2014). Looking for people, places and connections: Hindrances when working in multiple locations: A review. *New Technology, Work and Employment, 29*(2), 139–159. doi:10.1111/ntwe.12030

Lundin, J., & Magnusson, M. (2003). Collaborative learning in mobile work. *Journal of Computer Assisted Learning, 19*(May), 273–283. doi:10.1046/j.0266-4909.2003.00029.x

Mäkinen, S. (2012). Mobile work and its challenges to personal and collective information management. *Information Research, 17*(3). Retrieved from http://www.informationr.net/ir/17-3/paper522.html#.VpgtuRHCoUF

Mazmanian, M., Orlikowski, W. J., & Yates, J. (2013). The Autonomy Paradox: The Implications of Mobile Email Devices for Knowledge Professionals. *Organization Science, 24*(5), 1337–1357. doi:10.1287/orsc.1120.0806

McDaniel, T., Reitsma, R., & Duan, X. (2016). *Forrester's Global Business Technographics Telecommunications And Mobility Workforce Survey, 2016: Overview*. Retrieved January 20, 2017, from https://www.forrester.com/report/Forresters+Global+Business+Technographics+Telecommunications+And+Mobility+Workforce+Survey+2016+Overview/-/E-RES136155

Miettinen, R., Samra-Fredericks, D., & Yanow, D. (2009). Re-turn to practice: An introductory essay. *Organization Studies, 30*(12), 1309–1327. doi:10.1177/0170840609349860

Muukkonen, H., Hakkarinen, K., Li, S., & Vartiainen, M. (2014). Tracking mobile workers ' daily activities with the Contextual Activity Sampling System. *Design, User Experience, and Usability. User Experience Design for Diverse Interaction Platforms and Environments*, 289–300.

Nelson, S. B., Jarrahi, M. H., & Thomson, L. (2017). Mobility of knowledge work and affordances of digital technologies. *International Journal of Information Management, 37*(2), 54–62. doi:10.1016/j.ijinfomgt.2016.11.008

Nicolini, D. (2012). *Practice theory, work, and organization: An introduction*. Oxford university press.

Nørskov, S. V., & Rask, M. (2011). Observation of Online Communities: A Discussion of Online and Offline Observer Roles in Studying Development, Cooperation and Coordination in an Open Source Software Environment. *Forum Qualitative Sozialforschung / Forum: Qualitative Social Research, 12*(3).

Pauleen, D., Campbell, J., Harmer, B., & Intezari, A. (2015). Making Sense of Mobile Technology: The Integration of Work and Private Life. *SAGE Open, 5*(2), 1–10. doi:10.1177/2158244015583859

Punch, K. (2009). *Introduction to research methods in education*. Thousand Oaks, CA: Sage.

Reynolds, N. (2015). Making sense of new technology during organisational change. *New Technology, Work and Employment*, *30*(2), 145–157. doi:10.1111/ntwe.12045

Sannino, A., Engeström, Y., & Lemos, M. (2016). Formative interventions for expansive learning and transformative agency. *Journal of the Learning Sciences*, *25*(4), 599–633. doi:10.1080/10508406.2016.1204547

Sayah, S. (2013). Managing work-life boundaries with information and communication technologies: The case of independent contractors. *New Technology, Work and Employment*, *28*(3), 179–196. doi:10.1111/ntwe.12016

Sørensen, C. (2011). Enterprise Mobility: Tiny Technology with Global Impact on Work. In *The Computing Handbook Set – Information Systems and Information Technology* (Vol. 2). Retrieved from http://digitalinfrastructures.org/publications/Sorensen2012-CS-ChapterPreprint.pdf

Spradley, J. P. (2016). Participant Observation (Reissue ed.). Long Grove: Waveland Press.

Stieglitz, S., Lattemann, C., & Brockmann, T. (2015). Mobile Applications for Knowledge Workers and Field Workers. *Mobile Information Systems*, *2015*, 1–8. doi:10.1155/2015/372315

Vartiainen, M. (2008). *Facilitating Mobile and Virtual Work. In 21st Century Management, A Reference Handbook* (Vol. 2, pp. 348–360). Sage. Retrieved from https://www.researchgate.net/profile/Matti_Vartiainen/publication/256498038_Facilitating_Mobile_and_Virtual_Work/links/00b7d52323adc5eef9000000.pdf

Vartiainen, M., Kakonen, M., Koivisto, S., Mannonen, P., Nieminen, M. P., Ruohomäki, V., & Vartola, A. (2007). *Distributed and Mobile Work: Places, People and Technology*. Springer.

Wiberg, M., & Ljungberg, F. (2001). Exploring the Vision of "Anytime, Anywhere" in the context of mobile work. *Knowledge Management and Business Model Innovation*, 157–169.

Yuan, Y., & Zheng, W. (2009). Mobile Task Characteristics and the Needs for Mobile Work Support: A Comparison between Mobile Knowledge Workers and Field Workers. In *2009 Eighth International Conference on Mobile Business* (pp. 7–11). IEEE. doi:10.1109/ICMB.2009.9

KEY TERMS AND DEFINITIONS

Collaborative Problem-Solving: The isolated individual interacts with his community either to get help or to collaborate in problem-solving situations.

Expansive Learning: Participants, through collectives' zones of proximal development, provoke transformations and development in their activities. This result in knowledge creation and sharing both in the individual and in the organisational dimensions.

Knowledge Creation: Creation of something that was not there. In the individual dimension, the creation of new ideas, beliefs, or skills. In the organizational dimension, the creation of new contents, new rules, or new ways of working.

Knowledge Sharing: The knowledge assets (existing knowledge) are shared with others through learning actions (learning in doing) and participation of individuals of the expansive learning cycles (large-scale and small cycles) in collaborative/problem-solving situations.

Mobile Knowledge Worker: Professionals who frequently work on the move (at least 20% of their work time), far from their fixed organization, often performing knowledge-intensive activities.

Mobile Work: The most radical form of flexible work because it enables workers to carry out their organizational activities outside the organization's borders and also collaborate with other workers regardless of time and location.

ZPD (Zone of Proximal Development): Process that involves individuals who are learning, individuals who are teaching and the relationship between them.

ENDNOTES

1 This is a tool created by the managers in this case study of expansive learning.

2 The instant messaging formalized by ITCom and used by these workers to communicate with others.

3 The tool used was the XMind (http://www.xmind.net/).

4 The custom list was created based on the SharePoint tutorial https://support.office.com/en-us/article/Create-a-list-in-SharePoint-0d397414-d95f-41eb-addd-5e6eff41b083.

Chapter 11
Online Leadership and Learning:
How Online Leaders May Learn From Their Working Experience

Ditte Kolbæk
Aalborg University, Denmark

ABSTRACT

Online working environments develop and change continuously, meaning that online leaders and online team members must learn to adapt to change and should utilize emerging possibilities for doing their jobs. The purpose of this chapter is to explore how online leaders learn from experiences developed by and with online teams; the chapter also provides a metaphor by which online leaders' learning is likened to a spiral. The theoretical foundation rests in activity theory and proactive review. The methodological approach involves a case study and netnography, the latter of which is utilized for researching phenomena that occur in the interaction between humans and information and communication technology. Data were gathered online from an online team in a global IT company classified as big business. The following research question was addressed: How do online leaders learn from experiences generated by their online teams by utilizing proactive review as a practice of learning?

INTRODUCTION

Information and communication technology (ICT) is developing rapidly, and thus employees at all levels in organizations must be able to adapt, meaning that they ought to learn accordingly. Ongoing ICT development requires continuous learning in contemporary organizations. In other words, learning is as constant as breathing for organizations. The means and efficiency of organizational learning are dependent on the mindset and skills of the employees and leaders; furthermore, ICT has enabled organizations to create working environments in which employees work in geographically dispersed teams (Hinds & Mortensen, 2002; Schramm & Diederichsen, 2011). These employees rarely or never meet face to face, hence communication and cooperation are mediated by ICT, which may cause problems in aligning

DOI: 10.4018/978-1-5225-4094-6.ch011

work processes, cooperation, and communication. None of these are constant, quite the opposite —they develop and change in accordance with concomitant developments in the technology (Flyverbom, 2012; Hinds & Mortensen, 2002; Schramm & Diederichsen, 2011). Continuous changes in the technology influence ways of leading and ways of participating in team work. The aim of this chapter is to provide an understanding of how online leaders learn from work experiences given that they rarely or never meet face to face with peers, team members, or superiors; rather, leadership and cooperation is mediated by ICT. The research question addressed in this chapter is as follows*: How do online leaders learn from experiences generated by their online teams by utilizing Proactive Review as practice of learning?*

Thus, this chapter explores how online leaders learn from experiences developed by and with online teams. The theoretical foundation lies in activity theory, developed by Engeström (2001, 2007), and Proactive Review (PR), which is a practice of learning from experience in the context of work (Kolbaek, 2011, 2014). Online teams may be described as "Interdependent groups of individuals that work across time, space, and organizational boundaries with communication links that are heavily dependent upon advanced information technologies" (Hambley, O'Neill, & Kline, 2007, p. 1).An online leader may be described as the member of a team who is responsible for setting agendas, moderating interactions, keeping the team on track, and influencing all of its members by interacting online (Huffaker, 2010).

Learning in the context of work can be perceived as mental constructions that may be developed into practices "where knowing how to go on and how to innovate the way of doing are produced, diffused, and kept" (Gherardi, 2015, p. 14). Hence, practices develop continuously, and practitioners must learn accordingly. Practitioners negotiate the practices, and sometimes a practice changes because of contradictions or even conflicts between the practitioners; other times, the practitioners change the practice in harmony. When the latter is the case, the practice is roughly constant; but when modifications are caused by changes outside the practitioners' control, negotiations about new practices may include discrepancies and exude frustration, and the practices may be consequently unstable until the practitioners agree on new ways of doing (Engeström, 2001; Gherardi, 2015).A case study and netnography comprised the methodological approach used for online data gathering and includes observations from a web conference with a chat forum combined with a phone meeting and e-mail interview with the leader. The analysis led to the development of a new model to understand how online leaders learn from their teams and superiors. This chapter is a revised, re-written and extended version of a Danish chapter "Proactive Review som laeringsrum for online-ledere" (Kolbaek, 2017).

BACKGROUND

In order to develop an understanding of how online leaders learn from work experience, the chapter presents a case from a global IT company classified as big business with 25,000 employees throughout EMEA (Europe, the Middle East, and Africa) (Kolbaek, 2014). The company sold hardware, software, and middleware to market leaders in most industries, and it only sold business to business. The company grew through mergers and acquisitions (Kolbaek, 2014), and one of the acquired companies provided a promising product, here named KK. A Top manager Weber was assigned the task of growing this product and its market share by 100% while keeping the contribution margin the same during the fiscal year. This requirement was outside the control of Weber and his team, and they needed negotiating new practices in order to succeed. Weber was aware that this was a learning process for himself and his team – they needed to learn in order to establish and implement new practices, and Weber utilized the learning practice named Proactive Review for fulfilling the requirement. As Weber and his direct reports was a geographical dispersed team, they worked and negotiated online. The learning practice named Proactive Review will be described below. Utilizing ICT for different purposes requires technological knowledge and expertise on the part of the practitioners (Dasgupta & Gupta, 2009; Wenger, White, & Smith, 2009), since technology-mediated communication and cooperation breaks down more often than face-to-face interaction. Online cooperation is based on a common language that is often different from the mother tongue of the online leader and team members. As members of online teams rarely or never meet, the interaction is mediated by IT with the consequence of oral communication becomes more rare, having been mostly replaced by written communication in e-mails or chats with the consequence of declined spontaneity (Flyverbom, 2012; Hinds & Mortensen, 2002; Schramm & Diederichsen, 2011). Online team members do not meet by the coffee machine, and it is rather uncommon that they lunch together, as their lunchtime is dependent on the times zones they live in that may vary. Consequently bodily cues and facial expressions often lack, and the IT mediated communication impacts the manner in which trust is built, and it seems harder to build trust online (Flyverbom, 2012; Hinds & Mortensen, 2002; Schramm & Diederichsen, 2011)-. Working and cooperating online enable the online leaders to follow up on or control work progress without personal contact with the online team member, as the leader may find the data needed online (Flyverbom, 2012; Hinds & Mortensen, 2002; Schramm & Diederichsen, 2011). Mostly, these changes are not spoken about—they are not "an issue," but just happen, and online leaders and team members must struggle to adapt to the changes; in other words, they learn how to cooperate and lead online.

FRAMING ONLINE LEADERS' LEARNING

This section presents the theoretical foundation of the chapter that includes expansive learning (Engeström 2001, 2007, 2010) and Online Proactive Review (Kolbaek 2014). Any leader is situated in a group. In the context of work, the leader is situated in an organization that has a purpose and certain ways of doing things. Engeström (2001, 2007) described the framing of learning in the context of work, which he called the human activity system.

Expansive Learning

Yrjö Engeström (2001) developed a theory for understanding learning in the context of work. He described "the context of work" as a human activity system that includes six interdependent elements which serve as the foundation for learning processes in the work context. He calls these learning processes *expansive learning*. Both the elements and the processes are briefly described below.

The *subject* is a group—here, the team and the online leader—that learns in order to participate in "culturally valued collaborative practices in which something useful is produced" (Engeström, 2001, p. 141). Group activities are directed by the *object,* which may be defined as a general purpose; for example, the vision of the organization or personal objectives like the contributions the online leader and team have committed themselves to. Moreover, "the object of activity is a moving target, not reducible to conscious short-term goals" (2001, p. 136), which means that the object must make sense to the subject and, at the same time, define the direction of all activities as it expresses their meaning within the team (Engeström & Sannino, 2010). The *outcome* refers to the consequences of the object once achieved or, in

Figure 1. Engeström's activity system (Engeström 2001)

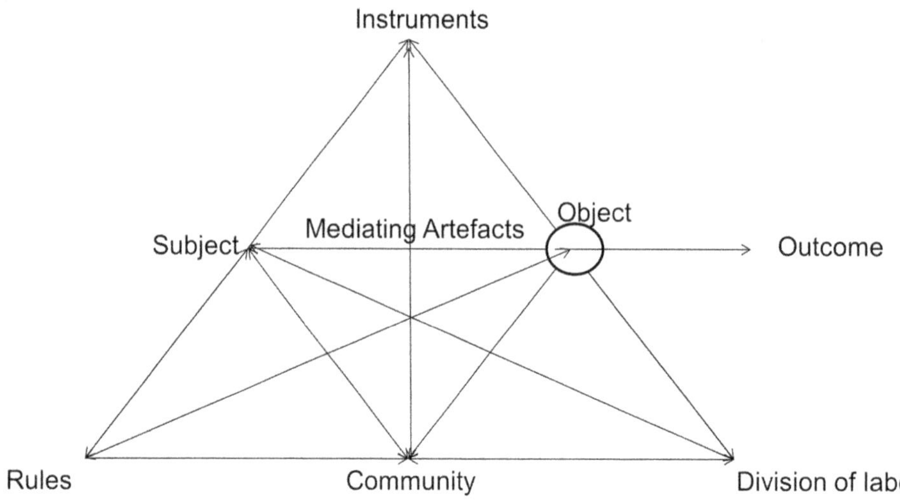

other words, the results achieved by the activities that led to the object. The *instruments* include tools and signs, as understood in their broadest sense, which are involved in achieving the object. Technology, such as ICT, is a good example of such instruments. Other instruments include drawings or manuals, while signs include the language, jargon, and/or logo of the organization, as well as the dress code mandated for the activity system. The *rules* include explicit rules and regulations as well as norms and internalized best practices in the activity system. The *community* is the social group in which the subject is a member; for example, a line of business in a certain company or a working group within a line of business. The *division of labor* describes how tasks are divided within the community and within the subject (that is, a group). The division of labor also points to the division of tasks and power within the activity system (2010). The goal of describing the activity system is to understand the frame of learning in the context of work. The six elements in the human activity system are interconnected such that when one element is "disturbed," the other five elements are affected.

Engeström (2001) called the learning process *expansive learning* whose point of departure lies in changes in the object that, according to the theory, affect all other elements in the activity system. The first step in changing the object is to ask questions about existing practices; in other words, the accepted ways of doing in the activity system. Doing so leads to an analysis of the situation, including considerations of the past, i.e., historical aspects, and inner contradictions in the ways of doing. Having created common ground on issues to learn from, the subject develops a model that describes this foundation and points to possible solutions for the issues in question. The possible solutions are tested in order to understand the pros, cons, and potentials; once the suggested solution has been refined by testing, it will be implemented in the organization. Finally, the subject reflects on the learning process in order to evaluate and consolidate the new practice (Engeström & Sannino, 2010).

Expansive learning describes how an online leader may learn from experiences in the context of work. First, the leader becomes aware of the needed changes in the object. The need for change may be obvious to the team, as its members may have already asked questions or expressed frustration because a practice no longer seems to work efficiently. The need for change may also stem from new organizational requirements coming from the top down; for example, a new strategy that requires new products or services for customers. The team utilizes the tools available for analyzing the issue; for an online team, this would include ICT and a common language. The subject (the leader and the team) work under the conditions set by certain rules applied in their organization, and they must obey the existing division of labor if they wish to avoid power struggles. The team develops and examines solutions that they then implement in the organization. It can be seen that the process of learning starts with changes in the object and the online team (including the leader) needs to change the way they work or the services or products they provide. Ultimately, expansive learning is founded on conversations within the team (Engeström, 2001).

Changes in the object initiate the learning process. As the object is a moving target, there is no specific objective for the learning, and the direction and purpose of the learning activities are unknown. Consequently, it is crucial to clarify the object (Engeström, 2001). The theory of expansive learning lacks a well-described practice for reviewing the learning process, and thus this chapter draws on Proactive Review (PR) in this regard.

Proactive Review

Proactive Review (PR) is a practice for learning from experience in the context of work. PR consists of seven open questions asked in a specific sequence that allows the participants to contribute to the conversation. This conversation may explore unknown territories or identify mistakes, misunderstandings, or contradictions the subject faced in the progress of achieving the object, and this exploration of the past can be tense and emotional (Kolbaek, 2014).

The name presents a contradiction; to review entails looking back whereas proactive entails looking forward, and PR follows both directions: reflections on and analysis of the past (historical aspects), and "proactive" solutions, tests of suggested solutions, and the implementation of these solutions (Kolbaek, 2014).

The starting point of a PR is an experience in the organization that someone wants to learn from, i.e., questioning the object (in the expansive learning terminology) or a change in the online team's objectives. Anyone in the organization may initiate a PR, and the experience may be perceived positively or negatively. A leader accepts the initiative and provides resources for the PR, which may take place face to face or online. This chapter focuses on online PR. The leader formulates the purpose of the PR; consequently, the objective for the learning process is defined, and the participants are informed about the objective before they are invited to participate in the PR (Kolbaek, 2014). The participants are the people who have solved or are going to solve a task together. Since the rules of a PR include honesty on the part of participants, the leader must carefully consider whether he or she should participate, as doing so may influence other participants to be less honest or forthcoming in order to save face in the eyes of the leader (2014). The norms in the organization allow the leader not-to-know and not-to-have-the-answer, as employees at all levels are expected to be capable of finding the best solutions to the identified problems. In a PR, the community includes the participants, their peers, and the stakeholders influenced by the solutions. The division of labor in a PR includes four roles: the participants, the leader, the top management of the organization, and a trained facilitator who conducts the PR (2014). The tangible output from a PR is two-fold: First, an action-and-communication plan that describes how the identified issues are to be solved and implemented in the organization, including suggestions about whom to involve in the process. Second, the identification of obstacles to the solutions cannot be resolved without the attention and involvement of top management. Such obstacles are called *management challenges* and are

Table 1. The seven questions and recommended IT in an online Proactive Review

No #	Proactive Review Questions	The Question Frames	Recommended Technology	The Online Leader's Potential Learning
1	What is the purpose of this PR?	The objective of the learning process	Video conference	Online leader presents the purpose and absorbs participant reactions
2	What was our objective?	Analysis of the past, historical aspects	Chat forum	Not applicable
3	What happened and why?		Discussion forum	Not applicable
4	What should we do next time?	Modeling and testing solutions	Video conference	Not applicable
5	What to report, to whom, when, and how?	Implementation plan	Discussion forum	Important stakeholders suggest changes in work practices, products, or services
6	What management challenge might be addressed from this PR?	Learning bottom-up	Video conference	Changes the manager cannot implement
7	What was your personal highlight in this PR?	Meta-reflection	Video conference	The participants' learning

(Kolbaek, 2017)

transmitted directly from the PR to top management without interference from middle leaders (2014). The outcome of a PR may be changed work practices, products, or services (Brandi & Elkjaer, 2011).

PR consists of seven questions asked in a specific sequence. Each question reveals insights into the past, the future, and the people involved; and when PRs are run online, considerations about the applicable ICT are needed. This study explores an online PR and table 1 gives an overview of these matters, which are explained below.

Question 1: "What is the purpose of this PR?" defines the objective for the learning process. The leader formulates the purpose as an open question, and the participants are supposed to think about this question in preparation for attending the PR. Nevertheless, the leader is supposed to present the purpose as the introduction to the PR, which starts in a video conference wherein the participants, the facilitator, and the leader can see each other. The purpose defines the direction for the conversation that follows. The participants respond to the purpose by asking questions to the online leader, who will in turn learn about the participants' concerns about the case. The leader normally leaves the PR at this point, typically returning when the participants are ready to present the action-and-communication plan.

Question 2: "What was our objective?" focuses on the past, as the participants recall the original target they were intended to achieve. Most often, this target has changed, as Engeström (2001) stressed, and the participants may have worked in different directions because of unnoticed changes. This divergence may be discussed in a chat forum to avoid a prolonged video or web-conference, which has been shown to be too demanding (Kolbaek, 2014).

Question 3: "What happened and why?" allows the participants to look into historical aspects and analyze the past. They identify and discuss the main obstacles to and enablers for obtaining what they actually achieved. The analysis creates a common understanding of the past that becomes the foundation for improving work practices, products, or services in the future (Kolbaek, 2014). The discussion continues in a discussion forum that encourages more taciturn participants to contribute to the same extent as their more outspoken colleagues; this discussion may last for several hours while avoiding the inconvenience of crossing time zones.

Question 4: "What should we do next time?" invites the participants to imagine the ideal future in which they have overcome the obstacles identified in the former question. The participants model suggestions for solutions to the identified problems, discussing the pros and cons of the different suggestions in order to test them out prior to making final suggestions for solutions (Kolbaek, 2014). This is a creative process in which the participants need to risk suggesting "wild ideas;" this is why the interaction continues in a video conference: so that facial expressions and body language are visible.

Question 5: "What to report, to whom, when, and how?" explores how the suggestions for solutions could be implemented in the organization as well as whom to involve in this process. The participants are expected to develop an action-and-communication plan that invites colleagues to contribute to the refinement of the solution as well as its implementation. The interaction takes place in a discussion forum where space and time is allocated for discussing the various parts of the plan; all participants can follow the discussion, and the facilitator is charged with concluding it in order to finalize the plan in due time. The final plan is delivered to the leader, normally in an online meeting where the participants can present the plan and explain its different parts. The leader needs to understand the intentions and suggestions of the plan as he or she is responsible for its implementation. The leader learns the employees' points of view, the new practices, and maybe new products or services suggested in the plan. Lastly, the leader is expected to accept the plan even without having involvement in its development (Kolbaek, 2014).

Question 6: "What management challenge might be addressed from this PR?" enables the participants to identify problems that cannot be solved without the attention of top management; such problems are consequently pushed upward in the organization (Engeström, 2001, 2007). This represents a bottom-up approach to problem identification, and top management may engage with and learn from any hierarchical level without interference from middle managers (Kolbaek, 2014). The facilitator addresses the management challenge directly to the responsible top manager, who then decides what to do and initiates the appropriate activities. This part of the PR should remain oral in order to strengthen confidentiality and build trust between the participants, which would in turn encourage them to be brave enough to formulate suggestions or solutions that their leader cannot handle. Therefore, this question is mediated via video conference. The leader of the participants may know about the management challenge if he or she is invited to the last part of the PR. If not, the leader may learn about the management challenge after top management initiates changes to resolve it.

Question 7: "What was your personal highlight in this PR?" encourages meta-reflection on the learning process among the participants in order to evaluate and consolidate the new practice (Engeström & Sannino, 2010). The participants are invited to share their highlights in the video conference, and the facilitator ends the PR by asking the question "round the table" without comments or further discussion. If the leader is present for the video conference, he or she will learn what the participants have learned. If the leader is absent, he or she will not receive direct insights into what each participant has learned, but may be able to infer the outcome of group learning from the suggestions in the action-and-communication plan.

When the leader realizes there is an issue that must be resolved, a change in the object has occurred, and the PR allows both the leader and the participants (the subject) to question existing practices, i.e., the accepted ways of doing (Engeström & Sannino, 2010). The question "What happened and why?" includes a consideration of the past, historical aspects, and inner contradictions in ways of doing that become the common ground for exploring possible solutions to the issues discussed until final suggestions are put into the action-and-communication plan to ensure their implementation in the organization. The question "What management challenges may be addressed from this PR?" allows the "grass-roots" to get attention from top management on issues that cannot be solved without top-level intervention. Finally, the question "What was your personal highlight from this PR?" enables the participants to reflect on the learning process and evaluate and consolidate the suggested new practices (2010). When the participants have fully discussed the seven questions in the PR, they have learned according to human activity theory (Engeström, 2001).

PRs describe how an online leader may learn from experiences in the context of work. First, the leader becomes aware of the needed changes in the object, which the he or she formulates as the purpose of the PR. The online leader does not participate in the discussion about the past or in the modeling of new solutions; the participants, on the other hand, are allowed to put their experiences into words, meaning that they have the opportunity to express thoughts and experiences that remained mostly unspoken before. It is these experiences that become the common ground for enhancing practices, products, or services (Brandi & Elkjær, 2011). The online leader as well as the community learn from the PR by becoming familiar with the solutions and by becoming involved in further work and the implementation of these solutions. Top management learns from the PR when they receive the management challenge, which expresses certain needs from the "grassroots;" likewise, the online leader and the community learn from the PR when solutions to the management challenge are implemented.

METHODS AND CASE

This study is based on the investigation of an online PR in a global IT company. The intention is to provide a nuanced view of the behavior of the subject (the online leader and the team) who acts in a specific organization; that is, a concrete, context-dependent learning environment (Flyvbjerg, 2004). Hence, this study could be perceived as a case study investigating a pragmatic example of best practices unfolding via online leaders' learning. Studies of learning processes cannot be based on predictive theories, while learning is dependent upon the context (Engerstöm, 2001; Flyvbjerg, 2004; Gherardi, 2015). Consequently, the researcher is allowed to include real-life experiences and to participate in the research area (Flyvbjerg, 2004). A nuanced view should include "the story in its diversity, allowing the story to unfold from the many-sided, complex and sometimes conflicting stories that the actors in the case have told" (2004, p. 430).

The underlying method may be seen as netnography (Kozinets, 2015), as this study is mainly based on data gathered on the intranet of the global IT company. Netnography may be utilized for studying communicative phenomena (2015). Case studies and netnographic studies include "thick descriptions" to describe social worlds that may be "common" for the informants but "strange" to outsiders (2015). As the researcher only meets the informants online, they do not meet in person. Consequently, the researcher is only influenced by the informants, and vice visa, if/when they meet online in video- or web conferences, via email, etc. (2015). The online setting allows the researcher to conduct the investigation from his or her computer independent of time and space; further, the researcher may gather data simultaneous with the ongoing activities or may gather data on activities that have already taken place without incurring travel costs (2015).

The author of this chapter was manager of organizational learning in this global IT company, which enabled her to gather data that external parties would not have access to. Moreover, she had a deep understanding of the organization at which she had been employed for several years prior to the case taking place. This involvement may make her biased, but the bias may be weakened by the fact that she left the company some years ago and is no longer connected to it in any way (Flyvbjerg, 2004).

Case: Leading Online in the Global IT Company

This case study describes learning in the context of a global IT company with 25,000 employees throughout EMEA (Europe, the Middle East, and Africa). This case is chosen, because it seems to reveal *why* an online team needs to learn, namely the requirement of growing product KK and its market share by 100% while keeping the contribution margin the same during the fiscal year, *how* they learn, namely by utilizing the learning practice called Proactive Review, and *what* they learn, namely new practices that enable the online team to achieve the objective. The common language was English, so the majority of the employees spoke a language different from their mother tongue.

Top manager Weber was assigned the task of growing a new product named KK and its market share by 100% while keeping the contribution margin the same during the fiscal year. The data consist of observations from an online PR situated in a web conference but also including a telephone meeting, a chat forum, and slides from the PR. A year after the online PR, the top manager was interviewed, and the internal publication of this interview constitutes part of the data.

In order to achieve the task, Weber needed to learn in the context of the global IT company, which may be described as a human activity system (Engeström, 2001). The *subject* was Weber and his 12 middle managers. The *object* was given and was not a moving target (2001); consequently, it is called the *objective*. The objective was 100% growth while keeping the contribution margin for 12 months. The *tools* included the intranet, the internet, a web conference with a chat forum, and a telephone meeting. Remarkably, video conferencing was not available at this time. All members of Weber's team were accustomed to utilizing IT for communication and collaboration. Weber's team had to obey the *norms and rules* in the global IT company. These rules included the requirement of achieving specific targets measured by year, quarter, and month. If the targets were not achieved, the employee (manager or individual contributor) would leave the company. The norms allowed the online leader not-to-know and not-to-have-the-answer; they also allowed the leader to have confidence in his or her direct reports and capacity to find the best solutions to the identified problems. The *community* was the global IT company, EMEA (as opposed to the US or Pacific). *The division of labor* changed according to the growth of the company. So, local entities in the countries were changed into global lines of business in which the employees worked across national borders with their leader, who was somewhere else. The division of labor in Weber's team was clear for the 12 middle managers with individual responsibilities, making it easy for them to divide tasks between themselves.

The Online Proactive Review

According to Engeström (2001) Weber and his team was the *subject* that learned in order to participate in "culturally valued collaborative practices in which something useful is produced" (Engeström, 2001, p. 141) in this case they participated in the online learning practice of Proactive Review (PR). It was important for Weber to achieve the target in order to keep his job, yet he was aware that he did-not-know what to do and that success was only attainable by involving the middle managers. Weber recognized the need for learning for himself and his team, and he invited to an online PR that would last for 2½ hours.

They all knew of PR as a practice for learning from experience, but utilizing PR to create a strategy was new to them.

Weber defined the purpose of the online PR this way: "How to make 100% growth in revenue and 100% growth in market share of product KK while keeping the contribution margin the same over over the next 12 months?" which was the objective for the learning process. Weber informed the12 middle managers about this objective in the invitation to the PR that asked the participants . to consider both obstacles to and enablers for achieving the PR's purpose. Weber decided to attend the online PR, which was conducted as a web conference with a chat forum and telephone meeting.

Weber welcomed his 12 middle managers to the online PR and presented the purpose again. As the strategy could not be built upon existing experience, the facilitator skipped the question, "What was our objective?" and re-formulated the question, "What happened and why?" as "Think for a few minutes, then please describe individually up to three enablers and/or three obstacles for achieving 100% growth of KK next fiscal year." The facilitator divided the 13 participants into two smaller groups that met in two different phone meetings to present and discuss their individual enablers and obstacles in achieving 100% growth of KK. They received 15 minutes for this conversation, and when they agreed on important obstacles and enablers, they wrote their findings in the chat forum in the shared web conference. The *enablers* included the following: strong value proposition—double performance and half the price; well-defined testing criteria, a strong focus, and dedication from pre-sales, support, and other internal contributors. The *obstacles* were as follows: lack of resource, internal competition, the complexity of the sales process, lack of coherence with other processes, and lack of skills regarding KK.

The question, "What should we do next time" was divided into three themes: What should we stop doing, what should we continue doing, and what should we start to do? Again, the 13 participants were divided into two groups, with one group discussing "What should we stop doing?" while the other discussed "What should we continue doing?" in their phone groups. They worked for 10 minutes and reported their findings in a word document that was saved for use at a later stage. Then, they met in the shared web conference and, within 2-3 minutes, presented their findings to the other group. Activities that should be *stopped* included talking about technical features or products, focusing on the hardware (although important, it is an integrated part of the delivery), internal competition and silo thinking, sub-optimizing in any line of business, lacking ambition, and complexity in terms of the lack of clarity over responsibility for the sales process. Activities that should be *continued* included collaborating with the hardware sales team to present the best possible solution to the customer, improving the innovation of KK, and turning successful customers into future references. Based on these inputs, the groups worked for 15 minutes to decide what to do. Activities to *start* doing included defining a clear sales process (including responsibilities), collaborating closely with hardware sales, developing new applications for KK, defining a partner strategy for selling and supporting KK, and dealing with internal competition.

The question, "What should we report to whom, when, and how?" was discussed for 20 minutes in the web conference, where "stop" and "start" included the themes to be developed and implemented over the next few months; the participants then developed an action-and-communication plan via the web conference, which the facilitator conducted. The facilitator also wrote agreements into the plan. Weber agreed to implement the plan by providing the needed resources and by following up on the monthly team meetings.

The question, "What management challenge might be addressed from this PR?" led to a discussion of two obstacles that the participants could not solve—even though Weber, as a top manager, participated. Within less than 10 minutes, the participants had formulated two management challenges: (a) "a solid process for supporting KK and hardware is missing," and (b) "a proper process for escalations in the sales process of KK is missing." Weber took responsibility for the two management challenges and addressed them to the company's US headquarters.

During the final five minutes, the facilitator asked the participants, one by one, "What was your personal highlight in this PR?" Their one-sentence replies, however, were oral and unrecorded.

The online PR described above does not follow the recommendations provided earlier in this chapter because the case took place years before the recommendations were developed (Kolbaek, 2014).

The E-Mail Interview

The year after, the author of this chapter interviewed Weber about what he had learned from the online PR on product KK. The interview was conducted as an e-mail interview, which was transformed into a story that Weber gave his approval of before it was published in an internal newsletter in the company. Here, the story is presented in its essence, and the numbers in the margin serve as pointers in the analysis.

Weber claimed:

The PR showed that KK has a strong value proposition as it provides double performance for half the price. Now, a year after the PR, we achieved 100% growth of KK and 100% growth in the market share, and we kept the contribution margin the same.

1. *Besides, we have strong and dedicated Tech and Hardware sales teams that proactively collaborate to present the best solution for the customer, and we have succeeded to turn successful customers into reference.*
2. *The partner strategy has proven victorious, as an increasing number of partners have penetrated new markets and the number of sold KK has risen continuously over the last 12 months. On our side, we have developed standard processes for cooperation with external parties.*
3. *The PR led to a new escalation process for KK which has been shown to be efficient in tense situations. Together with the training of employees and partners, we now have the right expectations and behaviors.*
4. *Proactive Review is good for the team because the members get an opportunity for reflection on a certain task or obligation. They become aware of the situation "today, "and they plan future activities that enable them to achieve the objective. The main challenge of a Proactive Review is to implement the action-and-communication plan and do what is agreed to.*

ANALYSIS AND FINDINGS

The analysis of the case is based on Engeström's (2001) human activity theory and expansive learning, and will focus on the role of the tools. Furthermore, the online PR is explored in order to grasp which elements support the online leader's learning. The analysis includes text *in italic style* to refer to the six elements in the human activity system as well as data presented in the last section.

Analysis

Learning in the context of work can be perceived as mental constructions that may be developed into practices. Weber made this learning process conscious by initiating the Proactive Reviews that let to changes in the online team's working practices.

The *object* was string penetration of KK into the market. The objective was 100% growth in revenue and market share of product KK while keeping the contribution margin the same. Weber initiated the online PR as he realized this was a complex task. The online leader may have learned that online PR may lead to the achievement of tough objectives.

The *subject* was Weber and his team who were acquainted with PR and utilized this learning practice in order to participate in *culturally valued collaborative practices in which something useful is produced* (Engeström, 2001, p. 141); namely, increasing the sales of product KK by 100%.

The *instruments* included tools such as ICT that in this case consisted of a web conference with a chat forum combined with a phone meeting. When collaboration is mediated by technology, it is a prerequisite that the subject, i.e., Weber and his team, know how the technology works; otherwise, they may drop out and fail to achieve the object. The IT tools excluded facial expressions, body language, and smells. The case shows that Weber and his team were familiar with the ICT which did not require any learning. In order to solve the issues defined in the Proactive Review, the subject needed to collaborate online with external parties, such as partners or headquarters. Consequently this object was dependent on the tools.

A prerequisite for the online collaboration across geographical borders is a common language, in this case English, which is also an *instrument* according to Engeström (2001). The *norms and rules* in the online PR included the acceptance of not-to-have-the-answer which enabled the leader to initiate the Proactive Review in the first place. Additionally the norms and rules included the common understanding that employees are capable of finding the best solutions to identified problems. During the online PR the tools enabled everyone to attend on equal terms, and the facilitator maintained equal participation in the discussion conducted via phone meetings. The online leader must accept the norm of mutual participation in PR. These norms supported the leader's learning as he could benefit from the discussion and the findings during the PR, and these norms supported the achievement of the objective.

In this case, "finding solutions" occurred online—for example, the chat enabled the participants in the online PR to "talk" simultaneously, whereas no one would notice if someone raised their hand to line up as speakers. It became a norm in the online PR that everyone could chat at the same time, whereas the facilitator asked "round the table" when the conversation was situated in the telephone meeting. These two norms extinguished most of the spontaneity in the discussions. Because the participants in the online PR could not see each other, it was impossible to express agreement, support, or opposition with body language such as nodding, shaking the head, or sighing. Instead, the participants had to express themselves in utterances like "yes," "no," "oh no," or sentences in English, which was a foreign language to all but one of the participants. On the other hand, Weber could not influence his middle managers with his body language, as it was invisible to them; so, in this way, the tools supported mutual participation. The leader may have learned that he participates in a PR on equal terms, as his appearance has the same impact as the other participants. This is different from traditional meetings, where the direct reports often position themselves in relation to the leader. Besides, the leader can ask for clarification of utterances, as they do not disappear as they do in oral communication; which gives him more time to understand, wonder and get back to the participant that delivered the utterance.

PR is founded on mutual conversation, and when the technology avoids important properties of the conversation, such as spontaneity and body language, it raises concerns over whether the online PR is as fruitful as PR conducted in person. This case does not provide data for such a comparison, but according to Weber the PR gave the team members "the opportunity to….plan future initiatives that enabled them to achieve the objective". The results of the PR and this utterance indicate that this online PR was as fruitful as a PR conducted in person, and the online leader may have learned that online PRs are suitable for the implementation of new strategies.

The *community* of this online PR included Weber and his team consisting of 12 direct reports, and the tools did not limit anyone from participating in the PR. The conversations in the online PR made Weber learn the points of view of the 12 middle managers as everybody is supposed to contribute, and the participants have less ability to stay silent on an online PR where the technology makes it visible when somebody is absent.

The division of labor included Weber as the top manager of a line of business, with 12 middle managers who were responsible for 12 different areas in this line of business. *The division of labor* was clear amongst the participants in this online PR. The tools did not influence the division of labor per se, but being ICT-enabled, the tools allowed collaboration across the divisions of labor of the 12 middle managers as well as between Weber's team and other lines of business, i.e., hardware and the company headquarters, as well as external parties such as partners. This PR does not seem to have influenced the initial division of labor in the team, but PR allows suggestions of change in the division of labor.

The online leader may have learned that PR includes the possibility of cooperation across the defined division of labor.

The starting point of this PR was Weber's acknowledgement of not-knowing how to grow the market share and revenue of product KK 100% while keeping the contribution margin the same, and he initiated the PR in order to learn how to achieve this objective that then became the objective for the team learning in the PR. The participants in the PR were Weber's direct reports, who had worked together for some time, and Weber decided to participate in the PR. During the PR the online team learned what was needed in order to achieve the objective, and the online team developed an action and communication plan to ensure the fulfillment of the objective. The PR allowed them to identify obstacles they could not solve in the team, and these obstacles named management challenges were addressed the HQ, hence the PR made them learn some boundaries for decision making.

During the PR, all the participants – including the leader - learned that product KK had a *strong value proposition—double performance and half the price*. The action-and-communication plan included the requirements of *defining a clear sales process, including responsibilities,* and *collaborating closely with hardware sales*. Furthermore, the participants decided to *collaborate with the hardware sales team to present the best solution for the customer, to improve the innovation of KK,* and, if possible, *to turn successful customers into references.* Moreover, the participants agreed to *define a partner strategy for selling and supporting KK and for dealing with internal competition*. Given the online leader did not know this before the online PR he learned main obstacles and suggestions for solutions during the online PR.

The interview with Weber one year later indicated that the participants fulfilled these requirements (1) (2). The new partner strategy was based on the creation of new knowledge about partners that may be important to the new product KK and to collaboration between the middle managers.

Weber addressed the management challenges (*a solid process for supporting KK and hardware is missing*, and *a proper process for escalations in the sales process of KK is missing*) with headquarters, which formulated a new escalation process for product KK in EMEA. Thus, Weber, the 12 middle managers, and their teams needed to learn the new escalation process. Additionally, headquarters provided training for employees that enabled them to support product KK with hardware (3).

The action-and-communication plan and the management challenges initiated learning in the IT company. Twelve months after the online PR, Weber and his team increased collaboration with the hardware line of business and generated cooperation with new partners and new customers. Thus, Weber and his team continued their learning after the online PR, as they changed behaviors and developed working processes that were implemented in the organization. The online PR also led to learning at the company headquarters, which needed to understand that the escalation processes in EMEA should be different from those in US. This development led to learning outside Weber's team; namely, among his superiors, his peers, and all the employees in his line of business who worked with product KK. The IT company proved its capacity to accept changes coming from the bottom up, as headquarters responded positively to the addressed management challenges.

Weber maintained the continuous learning process in regular online team meetings in which the middle managers provided updates and discussed how to overcome new obstacles. However, Weber is aware that he is a bottleneck because *the main challenge of a Proactive Review is to implement the action-and-communication plan and do what is agreed to* (4), which is his obligation as leader of the team.

Findings: A Model of a Learning Process for Online Leaders

This section will present a model showing how online leaders may learn from online PR including implementation of the action-and-communication plan and management challenges.

As a member of an online team the online leader is responsible for setting agendas, moderating interactions, keeping the online team on track, and influencing all of its members. Weber is an example of such an online leader who sat the agenda by defining the purpose of the Proactive Review, and he moderated the interactions in the online team by initiating the PR, while he influenced the online team members to provide obstacles and enablers for achieving the objective, and supported the PR processes in which solutions were identified, discussed and decided. By following up on the action and communication plan the online leader kept the online team members on track.

Weber acknowledged that he did-not-know, and he was confident that his team was capable of achieving the given objective, namely to make 100% growth in revenue and 100% growth in market share of product KK while keeping the contribution margin the same over the next 12 months. Initially Weber learned that he did not know how to achieve this objective, and he consciously started a learning process for himself and his online team by initiating the online PR.

Weber formulated the purpose of the online PR by asking the open question: "How to make 100% growth in revenue and 100% growth in market share of product KK while keeping the contribution margin the same over the next 12 months?"

Weber and his team negotiated a common understanding of enablers for and obstacles to achieving the objective. The interview indicated that he learned that a main obstacle for success was being a "bottleneck" for the implementation of the action-and-communication plan, hence the online PR enabled Weber to learn obstacles and enablers for achieving the objective.

Weber and his online team created solutions that were stated in the action and communication plan and subsequently implemented in the organization. The conversation throughout the online PR made all the participants, including Weber learn what it would take to achieve the objective. The participation in the online PR made them learn that a web conference including a telephone meeting and a chat forum is sufficient for solving complex challenges.

Besides the conversations in the online PR made Weber learn the points of view of the 12 middle managers and the process of finding solutions. Figure 2 illustrates Proactive Review as a learning process for online leaders

The online leader acknowledges that he or she does-not-know and is confident that the team is capable of finding the best solutions for the problem or challenge. The online leader initiates a Proactive Review in order to learn what he/she does-not-know. The online leader formulates an open question that becomes the objective of the learning process. The formulation of the question makes the online leader learn what he or she want to achieve in regard to problem solving, and this objective is at the same time an open question and a specific objective.

Based on mutual conversations and negotiations the participants attempt to answer the initial question. The group identifies and tests solutions and then creates an implementation plan, known as the action-and-communication plan, which is delivered to the online leader. In turn, the online leader learns about potential solutions that may change work practices, products, or services. The online leader then decides which parts of the plan, if any, are not going to be implemented, and informs the group accordingly. The action and communication plan enables the online leader to learn from his direct reports.

Figure 2. The learning process of an online leader
(Kolbaek, 2017)

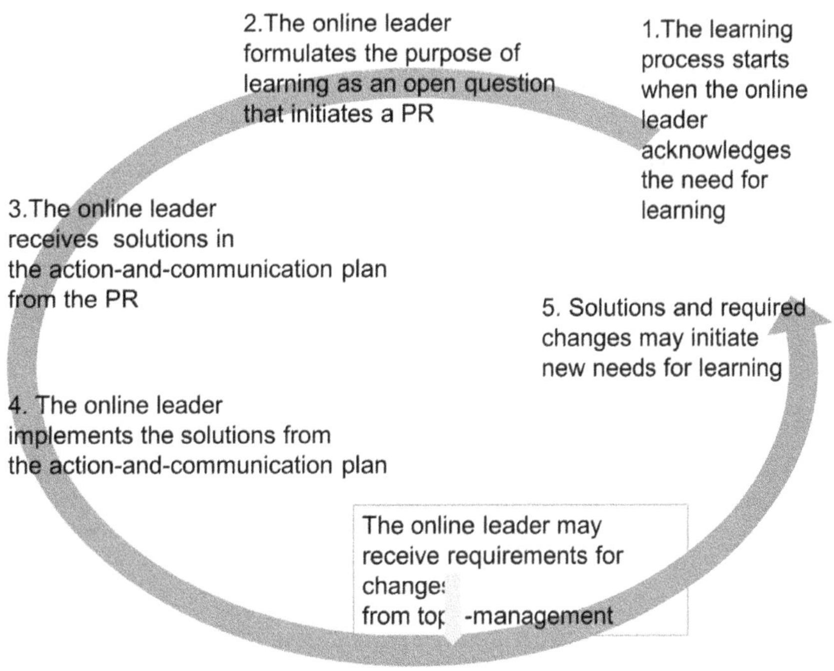

Sometimes the online leader participates in the PR and sometimes he/she doesn't. This is a bottom-up approach to the online leader's learning.

The top management in the organization may change work practices, services, or products as stated by the addressed management challenges, and the online leader needs to learn about these changes and adapt to them. The PR enables the top management to learn from the any organizational level, which is again a bottom-up approach to learning. When the top management implements the changes caused by the addressed management challenges, the online leader needs to learn the new work practices, services or products. This is a top-down approach to the online leader's learning.

By utilizing PR as a process for problem solving, the online leader learns consciously. PR starts the learning process among the participants, their colleagues, the leader, and his or her peers, as well as top management. This case shows that online PRs may be suitable for learning in online teams and for online leaders.

SOLUTIONS AND RECOMMENDATIONS

This case study shows the learning process of a group of online leaders in a global IT company. Online leaders are recommended to utilize Proactive Review as a learning practice that enables the online teams to create *culturally valued collaborative practices in which something useful is produced* (Engeström, 2001, p. 141); in the presented case the team of high-level online leaders achieved a rather complex objective. Online PR is recommended as a practice for learning from experience in online teams, because ICT is developing constantly, making the object a moving target.

In PRs the *subject* of learning is always a group, and the *object* of learning is defined as an open question, quite often the object is concretized into a short-term objective. The *norms and rules* in organization should include the acceptance of leaders that do not-to-have-the-answer. Additionally, the norms and rules should acknowledge that employees are capable of finding the best solutions to identified problems. The online leader must accept the norm of mutual participation in PR. The online leader should utilize *instruments* that are suitable for online PRs being ICT that is well known by the participants, who also need to share a common language. *The division of labor* should be clear amongst the participants in the PR. At the same time the PR may lead to suggestions of change in the division of labor. The participants in a PR have solved a task together, consequently the participants are recruited from the *community*, and the participants ensure different points of views throughout the PR.

Online leaders may learn from experience in the context of work by utilizing PR as a learning practice. Their learning follows a spiral, and the learning process starts when the online leader acknowledges the need for learning, which he/she formulates as an open question to be answered by team members that have solved a task together. The online leader is targeted for a bottom-up approach to learning, when he/she receives and implements the action- and communication- plan. The online leader is targeted a top-down approach to learning, when the top management implements changes caused by the addressed management challenges.

FUTURE RESEARCH DIRECTIONS

Proactive Review is a learning practice very different from learning in courses – being class room learning or online based training, because of the absence of a teacher and the very context related learning. PR invited for future research in both the bottom-up and the top-down approach to online leaders' learning that may uncover what and how online leaders learn in the context of work. It could be interesting to make a comparison in order to uncover which approach is most efficient in regard to achieve pre-defined results. Future research could look into indicators for online leaders' learning, in order to support top management to provide resources for learning practices such as PR. According to Engeström (2001, 2010) the context of learning, being the context of work, influences the why, what and how employees at any hierarchical level learns; hence future research may investigate the influence of the context on online leaders' learning processes, in order to provide recommendations for supportive learning environments in the context of work.

CONCLUSION

This chapter explored the question: *How do online leaders learn from experiences generated by their online teams by utilizing Proactive Review as a practice of learning?*

By evaluating how an online leader from a global IT company learned from experiences developed by and with his team of high level online leaders. The theoretical foundation for this study was activity theory (Engeström, 2001, 2007) together with Proactive Review as the practice for learning from experience in the context of work (Kolbaek, 2011, 2014). The methodological approach was a case study and netnography described the data gathering that included observations from a web conference with a chat forum combined with phone meetings; in addition, an e-mail interview with the online leader took place a year after the PR occurred. The analysis led to the development of a learning spiral as a metaphor for the online leader's learning.

The process of learning starts with changes in the objective that in turn require changes in work practices or in the services or products the online team provides. When someone realizes that a change is needed, he or she asks the leader to perform a Proactive Review, which requires that the online leader acknowledge that he or she does-not-know and be confident that his or her team is capable of finding the best solutions. The leader defines the objective and formulates it as an open question to be answered by the participants in the PR. From the PR, the online leader receives an action-and-communication plan that he or she is obliged to implement.

The findings included modeling the online leader's learning spiral, which started when he/she acknowledged the need for learning. The online leader is not the only one to learn as a consequence of a PR; the online team learns from the experience as well, and their colleagues learn according to the action-and-communication plan. The online leader's learning spiral shows that online leaders in general may learn from experience by utilizing online PRs as a means for problem solving in the online team. When the organization embraces suggested changes coming from any hierarchical level, learning may occur from the bottom up or the top down. Both approaches, however, lead to the online leader's learning.

This chapter contributes to the area of research with the suggestions of online Proactive Review as a learning practice for online leaders, as well as a model to understand the online leaders' learning process being a spiral. The practical implications could be the implementation of Proactive Reviews in more organizations.

ACKNOWLEDGMENT

I would like to thank employees in Oracle for their vast contribution and collaboration in the development and implementation of Proactive Reviews throughout the organization in EMEA. Furthermore I want to thank the reviewers of this chapter for beneficial and thorough feedback.

REFERENCES

Brandi, U., & Elkjaer, B. (2011). Organizational learning viewed from a social learning perspective. In *Handbook of Organizational Learning and Knowledge Management* (pp. 23–41). Chichester, UK: Wiley.

Cole, M., & Engeström, Y. (1993). A cultural-historical approach to distributed cognition. In G. Slaomon (Ed.), *Distributed cognitions–psychological and educational considerations* (pp. 1–46). Melbourne, Australia: Cambridge University Press.

Dasgupta, M., & Gupta, R. (2009). Innovation in organizations A review of the role of organizational learning and knowledge management. *Global Business Review, 10*(2), 203–224. doi:10.1177/097215090901000205

Engeström, Y. (2001). Expansive learning at work: Toward an activity theoretical reconceptualization. *Journal of Education and Work, 14*(1), 133–156. doi:10.1080/13639080020028747

Engeström, Y. (2007). From stabilization knowledge to possibility knowledge in organizational learning. *Management Learning, 38*(3), 271–275. doi:10.1177/1350507607079026

Engeström, Y., & Sannino, A. (2010). Studies of expansive learning: Foundations, findings and future challenges. *Educational Research Review, 5*(1), 1–24. doi:10.1016/j.edurev.2009.12.002

Flyvbjerg, B. (2004). Five misunderstandings about case-study research. In C. Seale, G. Gobo, J. F. Gubrium, & D. Silverman (Eds.), *Qualitative research practice* (pp. 420–434). London: Sage. doi:10.4135/9781848608191.d33

Flyverbom, M. (2012). Globalization as it happens: On globalizing assemblages in tax planning. *Distinktion: Scandinavian Journal of Social Theory, 13*(3), 295–309. doi:10.1080/1600910X.2012.723013

Gherardi, S. (2015). To start practice theorizing anew: The contribution of the concepts of agencement and formativeness. Organization, 1-19.

Hambley, L. A., O'Neill, T. A., & Kline, T. J. (2007). Virtual team leadership: The effects of leadership style and communication medium on team interaction styles and outcomes. *Organizational Behavior and Human Decision Processes, 103*(1), 1–20. doi:10.1016/j.obhdp.2006.09.004

Hinds, P., & Mortensen, M. (2002). Understanding antecedents to conflict in geographically distributed research and development teams. ICIS 2002 Proceedings, 38, 415-424.

Huffaker, D. (2010). Dimensions of leadership and social influence in online communities. *Human Communication Research, 36*(4), 593–617. doi:10.1111/j.1468-2958.2010.01390.x

Hughes, J. (Ed.). (2012). *SAGE internet research methods*. London, UK: Sage. doi:10.4135/9781446263327

Kolbæk, D. (2011). *Proactive review*. Copenhagen, Denmark: DJØF Forlag.

Kolbæk, D. (2014). *Learning from experience in the context of work*. Aalborg, Denmark: Aalborg University Press.

Kolbaek, D. (2017). Proactive Review som laeringsrum for online-ledere. In H. D. Keller & S. Frimann (Eds.), *Lederes laeringsrum* (pp. 149–179). Aalborg, Denmark: Aalborg Universitetsforlag.

Kozinets, R. (2015). *Netnography: Redefined*. London: Sage. doi:10.1002/9781118767771.wbiedcs067

Schramm, M., & Diederichsen, S. (2011). Virtuel ledelse: Skab synlighed, resultater og nærvær på afstand. Viborg, Denmark: Dansk Psykologisk Forlag.

Wenger, E., White, N., & Smith, J. D. (2009). *Digital habitats: Stewarding technology for communities*. Portland, OR: CPsquare.

KEY TERMS AND DEFINITIONS

Action-and-Communication Plan: Includes future actions and the involvement of people in the suggested solutions. The action-and-communication plan is one of two tangible outcomes from a proactive review (PR). The other tangible outcome is the manage challenge. The leader is responsible for implementing the action-and-communication plan even though he or she does not attend the PR.

E-Leader: The leader of an online team. May be used interchangeably with "online leader."

Learning in the Context of Work: Learning in the context of work can be perceived as mental constructions that may be developed into practices. Mostly this learning is unconscious, but it is sometimes beneficial to make this learning conscious, for example by running proactive reviews.

Management Challenge: An issue the participants in a proactive review and their leader cannot solve because it needs attention from higher hierarchical levels in the organization.

Online Leader: The member of an online team who is responsible for setting agendas, moderating interactions, keeping the online team on track, and influencing all of its members.

Online Teams: Interdependent groups of individuals that work across time, space, and organizational boundaries. Online teams communicate and collaborate via ICT. The members rarely or never meet face to face.

Participants: Employees at any level who attend a proactive review.

Proactive Review: A practice to learn from experience in the context of work that could be perceived as an educational design for learning at the workplace.

Chapter 12
Technology and Presence:
Multi-Presence in Online Interactions

Anne-Mette Nortvig
University College Absalon, Denmark

ABSTRACT

This chapter presents and discusses the concept of presence in online dialogues and interactions. The point of departure in this discussion is that of multitasking and social presence. Then, the chapter frames different e-learning scenarios that give participants the impression of being—or not being—present in the same room although geographically dispersed. On the basis of theoretical inspiration from symbolic interactionism and ethnomethodology, educators' new roles in e-learning are discussed in the context of empirical material collected from an e-learning program in a physiotherapy school. In addition, the concept of multi-presence is presented. This chapter argues that multi-presence occurs in both online and offline interactions and that, as a result of technology, is utilized in both professional and learning situations.

INTRODUCTION

In the Western globalized world today, the internet enables people to meet with others when located in different parts of the world and even in different time zones. But, what does it mean to be present in an online situation? What does this presence feel like, and how can it be described? Is online presence different from offline presence? In this book chapter, the concept of presence in both offline and online interactions will be presented and discussed. The theoretical perspective is based on symbolic interactionism, which focuses on interactions as well as interpretations and reflections of interactions in different settings. Thus, the objective of this chapter is to discuss the experience of presence in online settings from the perspectives of time, space, and professional learning. For this purpose, empirical material from interviews and participant observations of an e-learning program at a physiotherapy school in Denmark will be included. 46 students and 12 educators participated in the study that was conducted between 2011 and 2015.

DOI: 10.4018/978-1-5225-4094-6.ch012

The data for this chapter were collected during the author's Ph.D. project and more detailed descriptions are also found in her thesis in Danish (Nortvig, 2015a). This project was concerned with the following research questions: How does e-learning in a professional Bachelor's program influence students' participation and presence during teaching and learning? How does e-learning influence students' opportunities to develop a professional identity during their education? Fieldwork was conducted at a school where e-learning had just been introduced as a way to become physiotherapist. In this e-learning program, students were physically present at campus for three days every second week. The rest of the time, they studied at home or were in internship at hospitals, rehabilitation centers, or clinics. The format of e-learning at the physiotherapy school was specifically geared toward blended learning (Pytash & O'Byrne, 2014) in which the classroom on campus consisted of both traditional on-campus students and of e-learning students at home who accessed the classroom via the internet through Adobe Connect. Thus, e-learning students could see and hear what was going on in the classroom. Moreover, lectures were video-recorded, so students unable to participate in the classroom at a specific time could later watch the lessons.

As it will be shown below, the educators found it difficult to teach using e-learning methods and to meet students online, especially in the beginning. Moreover, as the decision about using e-learning in the physiotherapy school was based more on organizational arguments than on didactical ones, the educators also had doubts about whether their e-learning students would end up having the same physiotherapy competencies as non-e-learning students. Thus, the data presented at following have been generated at a time when e-learning and online education were still new in this specific context and when it was experienced as something very different—especially for educators—from the usual teaching format. Furthermore, the experience of being present both online and in a classroom on campus at the same time was not always felt by educators to be a natural way of being present.

In this setting, the researcher took on the role of participant observer in the e-learning sessions for student physiotherapists. She participated in the e-learning sessions from home and watched online lessons via Adobe Connect. She also interviewed the educators and conducted focus group interviews with two cohorts of first-year e-learning students. One of the foci of this data gathering was the experience of presence in online learning for both physiotherapy educators and students, and a portion of the researcher's Ph.D. thesis will be further elaborated at following. But before addressing the context of the physiotherapy school in order to understand educators' and students' experiences of online interactions, recent research on two concepts related to presence, namely social presence and multitasking, will be presented (see also Nortvig, 2014).

BACKGROUND

Being socially present is not necessarily considered the same as being physically present. Several years ago, Short et al. (1976) defined social presence in the context of telecommunications as "the degree of salience of the other person in the interaction and the consequent salience of interpersonal relationships ..." (p. 65). Here, salience refers to the degree to which a person is perceived as real in mediated communication (Pugsley, 2010), and the authors considered the medium itself to establish more or less social presence according to its ability to transmit nonverbal social cues. However, in the internet age, the concept of social presence became less concerned with the objective qualities of the medium and more about perception (Borup, West, & Graham, 2012; Swan & Shih, 2005). Moreover, the concept of social presence and its relation to teaching and learning has been further developed. Social presence is thus not only about being *perceived* as a real person, but also about *feeling* present as a real person. In current research, social presence has also been defined as the feeling of belonging to a group (Swan & Shih, 2005) and the ability to interact with other students (Dziuban & Moskal, 2001). Much research has been concerned with the finding that students are more satisfied with their online courses if they feel that they belong there and can interact (Hartman & Truman-Davis, 2001). Therefore, the use of technology is often suggested to be balanced with the "human touch of a real person" (Borup et al., 2012). Studies also indicate that social presence is not always related to learning outcome (Beaudoin, 2002) and that increased social cues can be a hindrance for learning, especially to students with low technological efficacy (Lyons, Reysen, & Pierce, 2012). More concepts of presence in relation to students and e-learning have evolved, including telepresence, virtual presence, cognitive presence, and educator presence; most of these are concerned with the presence of educators and learners in face-to-face settings or in online settings and often in comparison. Here, the experience of feeling truly present is often the end goal and considered to be the best or the most authentic means of achieving social togetherness in both offline and online situations.

In our age, however, digital devices and ubiquitous access to the internet are a part of many people's lives. One can easily shift between being present in different spaces or be present in more than one space at the same time. When observing, for instance, a classroom with young people, the following scene is not unusual: The educator presents a lecture while the students give the impression that they are taking notes yet are really viewing Facebook and web shops or instant messaging. Usually, this kind of presence in which one occupies digital and offline spaces at the same time, or constantly shifts between them, is discussed through the concept of *multitasking*. Multitasking can be defined as the performance of several activities at once (Hattie & Yates, 2014). Many studies have been carried out in this area to determine whether students learn more or less while multitasking. At the outset, multitasking students are often considered mentally absent although physically present because they may seem more occupied with interactions on their computer than listening to the professor in the lecture hall. The consequences for learning outcome have been investigated by many researchers, who have found that multitasking is widespread among students with access to computers and the internet (Kraushaar & Novak, 2010). For instance, Fried (2008) finds that the students' use of laptops caused distraction and poorer learning outcome, and other studies (Hembrooke & Gay, 2003; Wood et al., 2012) show that online interactions irrelevant to learning caused a 10% drop in students' grades. However, another study (Golub, 2005) tracked online multitasking without relevance for actual learning activities (e.g., playing computer games, checking e-mail or web pages) and compared these activities with offline multitasking equally irrelevant for the learning situation (scribbling on paper, doing crosswords, etc.): no significant differences were found between the online and offline multitasking activities in relation to learning outcome. In this sense, not only online multitasking activities distract students. Overall, these studies generally focus on multitasking as distracted engagement in online activities wherein the attention of the multitasker is preoccupied with something less suitable—when seen from the educator's perspective. Meanwhile, focused presence and full attention are often understood as the opposite of multitasking (Hattie & Yates, 2014).

However, not all research is in favor of the benefits of so-called *monotasking*. Some researchers (Davidson, 2011; Salvucci & Bogunovich, 2010) argue that instead of multitasking being connoted with distraction, it should be reframed as a *distribution* of attention. Davidson (2011) writes the following:

Some of our moment's agony over multitasking is that we haven't yet figured out which digital fork to use at which times. The Internet is still in its adolescence. We will work out the rules. But until we do, we can expect to feel taxed (p. 282).

This chapter will—among other things—discuss which "digital forks" are used in online interactions, and it will focus on online interactions in the context of learning situations and classroom teaching sessions and compare them to offline interactions.

A THEORETICAL PERSPECTIVE ON INTERACTION AND PRESENCE

Many theorists have studied the phenomenon of presence, and the concept of presence in *offline* interactions will be presented in this section from a theoretical perspective. Mainly, the following sections aim to highlight what happens to interactions if one of the interlocutors starts "multitasking" and stops focusing on the person with whom he or she is talking. The multitasking person might then start to examine the dialogue itself and his or her own role in it.

The religious thinker and educator, Martin Buber (1878–1965), distinguishes between interpersonal relationships that are framed as "I-Thou" or "I-It" (Buber, 1984). He finds that the "I-Thou" relationship is experienced as dialogical and as the "making present" of an event. Moreover, this relationship is experienced as the highest expression of a genuine meeting between two persons. In the "I-It" relationships, the interlocutors objectify each other by categorizing one another and trying to grasp what/who the "Other" is. Thus, the experience of being present is, according to Buber, more clearly found in the "I-Thou" rather than in objectifying "I-It" relations.

However, the objectifying of meetings, situations, persons, and things can also be analyzed and understood from another perspective. Symbolic interactionism is a sociological theory based on the thoughts of George Herbert Mead (1863–1931) and Herbert Blumer (1900–1987), among others. From this theoretical perspective, people always act according to their interpretations of the world and perceived meanings. Therefore, interactions cannot be categorized as less or more genuine per se. Situations are defined and interpreted by the individuals that take part in them; thus, meanings are not objectively true. Only because people are capable of (objectifying and) interpreting situations as well as other people and themselves, they can interact empathically with others (Mead, 1913; 1934). Blumer (1969) writes the following:

... a human being can be an object of his own action... he acts toward himself and guides himself in these actions toward others on the basis of the kind of object he is to himself... Like other objects, the self-object emerges from the process of social interaction in which other people are defining a person to himself (p. 12).

However, to be able to objectify oneself, one must see oneself "from the outside" (Blumer, 1969, p. 13) and place oneself in the other's perspective, looking at oneself through the eyes of the other. Thus, according to symbolic interactionism, perspective change and constant interpretation of the situations are advantages that naturally form people's way of being in the world and interactions with others.

Like Buber, Goffman (1922–1982)—who is often also considered a symbolic interactionist (Scheff, 2005)—views a downside to the opportunity to objectify and constantly interpret the situation in which one participates. According to Goffman, one can possibly lose the feeling of being present in a situation if one starts to interpret more and interact less or, stated in other terms, as follows: if one focuses his or her attention on only interpreting and evaluating a situation, the elements of interaction and presence die out. This is what Goffman calls *interaction-consciousness*: the person is consciously concerned with the way in which the interaction, qua interaction, is proceeding, instead of being involved in the conversation themselves (Goffman, 2005/1967, p. 119). He gives an example of interaction-consciousness in which a hostess at her own party is very concerned about whether the interactions at her party are "going well" (Goffman, 2005, p. 120): "She sometimes becomes so much concerned with the social machinery of the occasion and with how the evening is going as a whole that she finds it impossible to give herself up to her own party" (Goffman, 2005, p. 120). Thus, symbolic interactionists show the need for differentiating between being present in a single spatial setting, which one might call *mono-presence*, and being mentally present or focused in more spaces or with several foci at the same time.

Up until now, the focus in this chapter has been placed mainly on offline settings, but it will now turn its attention toward presence in both online and offline contexts when these exist simultaneously. Thus, examples will be given of how this kind of *multi-presence* was interpreted by the educators at a Danish physiotherapy school.

WANTED: BEING PRESENT TOGETHER

In reference to classroom settings with both on-campus and e-learners present, the experience of not being truly present and being unable to interact with others in specific situations was a frequently expressed theme at the physiotherapy school, especially by educators. Educators frequently said their attention and focus were *split* and divided in a negative way; overall, educators disliked being present in two places, the online and offline classroom, at the same time, as they found this was very demanding. Physically, in this model, educators were present with on-campus students in the classroom and visible, at the same time, to e-learners sitting at home. This setting created the sensation of not feeling truly present in the classroom because specific focus and attention were also placed on the e-learners' space and vice versa. The interviews revealed that many educators would have preferred that the e-learners and the on-campus groups be separated instead of occupying a parallel classroom environment. They had many ideas of how this could be done in order to avoid the experience of "sitting between two chairs" as one educator described. Her colleague agreed, stating the following:

If you did something with the e-learners specifically, then you could be in their sphere . . . because you live in two worlds as a educator [as things are right now]. It's damn hard. You teach the on-campus students and try to make it exciting, 'cause it's not great to look at someone who stays completely still and talks monotonously—and you need it to adjust mikes and video cameras and so on. So, in a way everything ... . . . Really, you live in two worlds. (Catherine, educator interview)

Several educators thus preferred to teach without having to also relate to the e-learners at home too. Moreover, being physically and mentally present with other people in the same room was considered important and central to physiotherapy education because this task is similar to what students themselves must master as physiotherapists.

The students, however, rarely distinguished between the online and offline spaces in such strong terms, and they did not as clearly interpret the online space as another world distinct from the offline world. The students were more likely to see e-learning and online meetings as a use of digital technology and as spaces of collaboration where their "lives take place already." They were used to being physically present in one room while also actively present online or on social media. Among the students, online spaces were associated with the idea of taking a break, forming the space of a "playground," but also with collaboration and spaces for communication alongside the physical learning space on campus. For example, one student in the focus group stated the following: "We have Google Docs open next to Skype, so we can work in the documents while sitting and talking on Skype" (Tina). If one of the technologies does not work as planned, the study groups can utilize other technologies as well: "So, we have Skype running for the sound, and then Adobe Connect for the video, and then Google Docs for writing (laughs). Yeah, pretty crazy ..." (Carol, student in focus group).

Many students stated that, besides using the digital spaces for the group meetings, they also met in Facebook groups to discuss their studies and other practicalities as well as for fun. Moreover, the students used Facebook while on campus, despite the disapproval of their educators. When students got bored during a lesson or wanted to have some fun, they described entering Facebook to see if anybody else was connected. Facebook was also the place to go if they wanted to get in contact with students on campus. Students did not relate this kind of online presence to multitasking but rather saw it as a normal way to be together and interact with people.

Even so, during the empirical field work, the experience of presence was connected to mono-spatial settings, *either* purely offline *or* online, in the context of both face-to-face teaching on campus and online interactions or meetings via Skype or Adobe Connect. The inclusion of technology in such settings essentially expanded the physical space: after computers gave access to the internet, new opportunities arose for doubling one's presence. For instance, the use of video during synchronous and asynchronous activities opened the door to new and multiple ways of being present: the students videotaped activities at one moment in time (i.e., during placement) in order to watch them at another time (i.e., during group work or in class). In doing so, a so-called professional presence could be mirrored and retained on the video, and reflections on interactions with other students or with patients could be discussed among the students later on. Such possibilities to evaluate one's own performance against a benchmark are similarly found by others to have a positive effect on clinical skill acquisition (Maloney, Storr, Morgan, & Ilic, 2013); therefore, the opportunity to double one's presence can be especially rewarding in professional learning situations. Thus, a professional interaction-consciousness (Goffman, 2005/1967) or a conscious focus on and use of multi-presence can represent a learning advantage for student professionals in numerous contexts.

MULTI-PRESENCE: ACROSS TIME AND SPACE

We will now turn to reflections on synchronous meetings via videoconference systems in smaller groups. The experience of mutual presence in small meetings is different from the experience of presence in webinars or videoconferences with large groups; this might be due to the experience of eye contact or the lack of it, among other factors. Mutual gaze is an important part (Moyer, 2016) of body language and social interactions, such as teaching, for example. Both recent and older research has found that eye contact affects students' perceived credibility of the educator (Beebe, 1974). If eye contact is established continuously among students and educators, both engagement and interactions seem encouraged (Andersen, 1979; Korthagen, Noordewier, & Zwart, 2014). Thus, eye contact may support learning (Lin & Creswell, 1989) and help students in the performance of tasks (Dalzel-Job, Oberlander, & Smith, 2011). Similar studies conclude that mutual gaze might form "a multimodal resource in organizing activities" (Mondada, 2006 in Satar, 2013) and that when a educator looks at students while lecturing, students are more likely to listen, pay more attention, and become more interested in learning (Wesson, 2011). Online teaching, however, is more complex because the lecture is mediated via a videoconference system synchronously or asynchronously in a video. Eye contact must be enacted differently online (Develotte, Guichon, & Vincent, 2010), as bodily cues cannot be similarly shown or interpreted via the computer screen as when people are in the same physical space (Pytash & O'Byrne, 2014). In this setting, if participants in a learning or meeting situation are to experience presence via mutual gaze, they have to look into a camera without meeting the gaze of anybody. However, research still suggests that the experience of eye contact is important for the experience of inclusion and not any less important in online settings (Levinsen, Ørngreen, & Buhl, 2013; Orman & Whitaker, 2010).

Despite potential shortcomings, online settings can also take advantage of meetings mediated through video in multiple ways during the teaching and the learning process. Research on online and computer-supported learning often distinguishes between synchronous and asynchronous virtual spaces (Garrison, 2011). Here, the big advantage of synchronous activities is the experience of social presence. Meanwhile, asynchronous collaborations and interactions offer the opportunity to spend time reflecting before interacting in the online space (Giesbers, Rienties, Tempelaar, & Gijselaers, 2014). However, as elaborated at following, these two virtual spaces can also be integrated with advantages for reflections on the specific presence.

Others also discuss the virtual space as different from traditional teaching spaces (if "traditional teaching spaces" even exist anymore). Dourish (2006) highlights that offline and online spaces are now so closely connected that they call for new practices:

The technologically mediated world does not stand apart from the physical world within which it is embedded; rather, it provides a new set of ways for that physical world to be understood and appropriated. Technological mediation supports and conditions the emergence of new cultural practices, not by creating a distinct sphere of practice but by opening up new forms of practice within the everyday world, reflecting and conditioning the emergence of new forms of environmental knowing (p. 304).

Similarly, Wahlstedt, Pekkola, & Niemelä (2008) argue that the design of e-learning should include a focus on creating virtual spaces that enable collaboration and interaction in such a way that would transform the impersonal e-learning space into an intimate and meaningful place. Wahlstedt et al. (2008) state, "A space becomes a place when meanings, constructed through social interaction, cultural identities, and personal involvement are supported and embedded into the environment" (p. 1024). Moreover, pedagogy and participants' critical reflection and own partaking in e-learning activities can be argued to be the elements that transform online spaces into meaningful places, rather than the technology itself, which functions as an artefact in the form of an e-learning system: ". . . it is these kinds of relationships—social interactions— that really help turn a space into a place. This is why placeness is difficult to 'design in' . . ." (Whitworth, 2008, p. 1034).

THE INFLUENCE OF TECHNOLOGY ON PRESENCE

At the physiotherapy school, different strategies were developed for creating a meaningful digital place (see also Nortvig, 2015b). At following, three examples of educators' multi-presence in the blended learning lecture hall will be presented (all names have been anonymized). The first example highlights Joe's ways of being present in the classroom. Joe finds it very important to be physically present with his students in the classroom, and he addresses his questions to them. The students in the classroom answer any questions and help Joe set up the camera to allow the e-learners to also have access to the learning space. However, Joe does not focus his attention on the online students, as his presence is entirely located in the physical classroom among the students present. Interactions are based on verbal and bodily communication between Joe and the present on-campus students. Joe finds that online meetings and teaching are less authentic than traditional face-to-face teaching. This interpretation of technology as a hindrance to the flow of the lesson (Vetere et al., 2012) and the 'the classroom ping-pong', as one of his colleagues puts it, is supported generally by several educators. The technology for online teaching requires attention. For example, a educator must focus on the angle of the camera, the sound, and the lighting.

When the educator is not used to this extra-teaching attention - or teaching-interaction-consciousness - he might try to avoid it. A educator, Philip, described the challenge as follows:

I think it's very difficult to keep the focus on . . . well, if you are to keep the focus on both . . . you really need big reserves of energy to be present both in the classroom with the students and to pay attention to the chat and respond to the e-learning students without making the on-campus students think, "Well, well, now I'm just wasting my time." So, I think it's really difficult to be simultaneously present, virtually and in the classroom—I haven't been able to do it . . . I've ended up just saying "Welcome," [to the online students], and then I've done my teaching. (Philip, educator interview)

Even if Joe has a multi-presence in the blended classroom, he finds this presence too demanding. The right "digital forks" (Davidson, 2011) are impossible to find, so he prefers to act as if the inclusion of digital technology and the online transmission of the lesson do not change anything about his usual lessons or his way of being present.

Another educator at the physiotherapy school, Harold, was found to separate his multi-presence into two places: he had a bodily/verbal presence in the physical classroom and another presence transmitted through technology and written words during direct interaction with e-learners. Questions for the physical class were asked verbally, and questions for the e-learners via chat. Thus, he and the e-learners could interact in the same mode and media and could meet at more equal terms. Harold chose to specifically focus on the e-learners during breaks and before and after the lesson in order to avoid interference with the ping-pong and to ensure smooth interaction with the physical classroom. This separation of focus and presence encouraged the e-learning students to chat with each other during class when they needed help or to ask questions about the subject.

As mentioned, many of the educators at the physiotherapy school wished the two groups of students were separated because they found that different ways of interaction and teaching were required to address both groups. One educator explains this as follows:

I'd prefer a separate e-learning class and a separate on-campus class, so you don't mix apples and oranges. Instead, you could focus on "now, it's e-learning" and focus on the interaction you can have with such a class. Then, it would be on their premises and not on those of the on-campus class—and vice versa." (Catharine, educator interview)

The third example of enactment of multi-presence was found in the performance of Allan. He seemed to be very aware of his multi-presence, as he paid a lot of attention to the students in the classroom and also to the e-learning students via the camera, ensuring that the camera captured him at all times. He spoke directly to the e-learners, saying, for instance, "You, back home, you can just google it." He also shared comments with the off-campus students during the lecture and breaks: "You just take a break, back home, too." During breaks, Allan spoke with the students in the classroom, but just before the breaks would end, he would go to the camera and resume the on-campus chat with the e-learners.

The e-learning students were very active in the chat, and they contributed both socially and in terms of subject matter. Allan kept an eye on the chat, and he verbally integrated the questions asked in the chat as a natural part of the lesson. Procedural information was similarly reported. For example, when he was going to leave the camera view, he said "I'll just go get my coffee." Allan also used a lot of humor when he taught, and this apparently inspired the atmosphere both in the classroom on campus and among the e-learners: the chat was active and included jokes and other content unrelated to the subject matter. One of the students even wrote the following one day at the end of a lesson: "It's so nice to participate here, even without speaking. It's like being there yourself ;o)."

Therefore, in this classroom setting, many questions and answers during lessons involved both groups of students, and e-learning students were more active than on-campus students if interactions were to be counted. Both groups of students were spoken to verbally, and e-learners' comments and answers from the chat were verbally integrated in the classroom discussion as well. The way Allan interacted with students during the lesson seemed to encourage interaction and the strong coupling of the e-learning space with the physical one, thereby expanding and changing the model of the traditional teaching classroom.

During an interview, one student compared Joe's and Allan's teaching strategies, as follows:

Majken: *[Joe] needs someone to turn the camera on and off. Stuff like that ... well, he doesn't show us much engagement and interest. Unlike Allan, you know. He really works hard to give us a good lesson.*

Anne-Mette: *Yeah? How? ... How do you see that?*

Majken: *Umm, I think that the stuff he's made with Wallwhisher, or what's the name ... where you can go in and watch his videos—well, stuff like that ... Then, it's easier to sit at home and relate to ... and I think he does a great job of having the lessons recorded and saying, "Let's take a break" and stuff like that ...*

Clearly, more educators than students found multi-presence to be challenging. This might be due to the fact that multi-presence was required of the former, as educators' lessons were filmed and retained on video, whereas the multi-presence of the latter was optional and retained to a much smaller and voluntary degree. Consequently, the students spoke of multi-presence as something pleasant, convenient, and relaxing. When asked to define e-learning, a student said the following: "E-learning is sitting in the classroom on your private couch" (Student, first year). Empirical data from other research studies find similar expressions related to ease, flexibility, and pleasure (Poellhuber, Racette, & Chirchi, 2012).

In summary, in professional dialogues and interactions, both online and offline, it is important and expected that the professional continuously analyses the process at hand and evaluates whether it progresses as anticipated and intended and if anything should be changed or if new interventions are needed. The professional cannot be completely immersed in interactions or dialogue, as this could take away from his or her professional stance, yet must at the same time be fully present. For instance, a physiotherapist cannot speak of personal matters, such as his or her own illnesses, but instead must focus on conducting anamnesis and treating the patient. However, to be professionally present, or to be multi-present in professional setting, this mode must be learned and practiced during education through reflections on actions and reflections in action (Schön, 1991). In offline situations, multi-presence is not as visible as it is in online ones, therefor well-designed technology can support learning of this professional manner. But, as previously discussed, technology does not act alone but rather influences the learning space that in turn affects those who enter the space to learn.

CONCLUSION

In this chapter, empirical data from a physiotherapy school highlighted both the challenges and advantages that the implementation of e-learning created. E-learning technologies provided educators and students who were in more than one place at the same time to be present, or multi-present, in the same learning space. In the material presented in this chapter, an example of a transformation away from "traditional" teaching activities was evidenced, in which teaching that previously took place in one place and at one time, to a space accessible by many students from many spaces and at different times.

The educators in the physiotherapy school often felt uneasy about being present in both online and offline places at the same time: due to the unpleasant experience of multi-presence, some focused their attention in the offline space and—maybe due to their awareness of the online learners' invisible gaze—they felt only halfway present. In contrast to this, other educators used their multi-presence in strategic and humorous ways so both groups of students experienced the educator's presence in the classrooms.

To understand presence in digital settings, presence across space must be distinguished from presence across time, as both potentially multiply with the use of technology. Examples of multi-presence across space was found in educators' and students' meetings via videoconference, where educators were physically present in both the classroom and online, while the e-learning students were present at home although connected via the Internet and thus present on campus too. Accordingly, examples of multi-presence across time were found in the learning design were videos from the student physiotherapists' placements or ordinary group work on practice and training activities. This physical presence on video could later on be observed from an external perspective. Furthermore, the integration and retention of interactions with fellow students or with patients afforded an opportunity to evaluate and reflect on the (inter)actions and to focus on interaction-consciousness (Goffman, 2005) without losing the experience of being present in the specific situation while it was taking place. Student physiotherapists are to become able to act professionally and to reflect in action (Schön, 1991), and what is more, reflection in action can be practiced by the separation of action and reflection by video. Thus, in this case, video afforded an opportunity to create a didactic focus and a use of multi-presence with reflection on interaction and presence in professional situations as a goal.

As a musician learns to be multi-present while playing actual notes, contemplating those to come, and evaluating the musical expression, so must other professionals learn to exercise multi-presence in professional situations. While they interact with patients or clients, they continuously consider and reflect on the interaction going on. They must balance between being and appearing present in the situation and constantly also seeing the situation from the outside and evaluating and making decisions on that basis. Without the ability to be multi-present, professional reflection in action would not be possible.

FUTURE RESEARCH DIRECTIONS

This chapter briefly touched on the subject of eye contact and the experience of multi-presence in online interactions, which can be enhanced by this bodily cue. However, as health technologies expand, online meetings with health professionals might be more frequent in the future. There is a need for further research in this area on the influence of mutual gaze in online meetings. This research interest might be explored in relation to the experiences of both health professionals and patients as well as in relation to diagnosis and anamnesis when bodily cues such as smell and touch are not accessible.

REFERENCES

Andersen, J. F. (1979). Educator immediacy as a predictor of educator effectiveness. New Brunswick, NJ: Transaction Books. doi:10.1080/23808985.1979.11923782

Beaudoin, M. F. (2002). Learning or lurking? Tracking the "invisible" online student. *The Internet and Higher Education*, *5*(2), 147–155. doi:10.1016/S1096-7516(02)00086-6

Beebe, S. A. (1974). Eye contact: A nonverbal determinant of speaker credibility. *The Speech Educator*, *1*(23), 21–25. doi:10.1080/03634527409378052

Blumer, H. (1969). *Symbolic interactionism: Perspective and method*. Los Angeles, CA: University of California Press.

Borup, J., West, R. E., & Graham, C. R. (2012). Improving online social presence through asynchronous video. *The Internet and Higher Education*, *15*(3), 195–203. doi:10.1016/j.iheduc.2011.11.001

Buber, M. (1984). *Das dialogische Prinzip / Martin Buber* (5th ed.). Heidelberg, Germany: Lambert Schneider.

Dalzel-Job, O., Oberlander, J., & Smith, T. J. (2011). Don't Look Now: The relationship between mutual gaze, task performance and staring in Second Life. In *Proceedings of the 33rd Annual Conference of the Cognitive Science Society* (pp. 832–837). Academic Press.

Davidson, C. N. (2011). *Now you see it: How the brain science of attention will transform the way we live, work, and learn*. New York: Viking New York.

Develotte, C., Guichon, N., & Vincent, C. (2010). The use of the webcam for teaching a foreign language in a desktop videoconferencing environment. *ReCALL*, *22*(3), 293–312. doi:10.1017/S0958344010000170

Dourish, P. (2006). Re-space-ing place: place and space ten years on. In *Proceedings of the 2006 20th anniversary conference on Computer supported cooperative work* (pp. 299–308). ACM. doi:10.1145/1180875.1180921

Dziuban, C., & Moskal, P. (2001). Emerging research issues in distributed learning. In *The 7th Sloan-C International Conference on Asynchronous Learning Networks.* Academic Press.

Fried, C. B. (2008). In-class laptop use and its effects on student learning. *Computers & Education, 50*(3), 906 914. doi:10.1016/j.compedu.2006.09.006

Garrison, D. R. (2011). E-learning in the 21st century : a framework for research and practice / D. Randy Garrison (2nd ed.). New York, NY: Routledge.

Giesbers, B., Rienties, B., Tempelaar, D. T., & Gijselaers, W. (2014). Why Increased Social Presence through Web Videoconferencing Does Not Automatically Lead to Improved Learning. *E-Learning and Digital Media, 11*(1), 31–45. doi:10.2304/elea.2014.11.1.31

Goffman, E. (2005). *Interaction ritual: Essays in face to face behavior.* New Brunswick, NJ: Aldine Transaction.

Golub, E. (2005). On audience activities during presentations. *Journal of Computing Sciences in Colleges, 20*(3), 38–46.

Hattie, J., & Yates, G. C. R. (2014). *Visible learning and the science of how we learn.* New York: Routledge.

Hembrooke, H., & Gay, G. (2003). The laptop and the lecture: The effects of multitasking in learning environments. *Journal of Computing in Higher Education, 15*(1), 46–64. doi:10.1007/BF02940852

Korthagen, F. A. J., Noordewier, S. A., & Zwart, R. C. (2014). Educator–student contact: Exploring a basic but complicated concept. *Teaching and Educator Education, 40,* 22–32. doi:10.1016/j.tate.2014.01.006

Kraushaar, J. M., & Novak, D. C. (2010). Examining the Affects of Student Multitasking with Laptops during the Lecture. *Journal of Information Systems Education, 21*(2), 241–251.

Levinsen, K., Ørngreen, R., & Buhl, M. (2013). Telepresence as Educational Practice in the Third Teaching-Room–a Study in Advanced Music Education. In *Proceedings of the 12th European Conference of E-learning. Sonning Common: Academic Conferences and Publishing International Limited* (pp. 250–257). Academic Press.

Lin, C. A., & Creswell, K. W. (1989). Effects of televised lecture presentation styles on student learning. *Learning, Media and Technology, 15*(1), 37–52.

Lyons, A., Reysen, S., & Pierce, L. (2012). Video lecture format, student technological efficacy, and social presence in online courses. *Computers in Human Behavior, 28*(1), 181–186. doi:10.1016/j.chb.2011.08.025

Maloney, S., Storr, M., Morgan, P., & Ilic, D. (2013). The effect of student self-video of performance on clinical skill competency: A randomised controlled trial. *Advances in Health Sciences Education: Theory and Practice, 18*(1), 81–89. doi:10.1007/s10459-012-9356-1 PMID:22354337

Mead, G. H. (1913). The social self. *The Journal of Philosophy, Psychology, and Scientific Methods, 10*(14), 374–380. doi:10.2307/2012910

Mead, G. H. (1934). *Mind, self, and society: From the standpoint of a social behaviorist*. Chicago: University of Chicago press.

Mondada, L. (2006). Participants' online analysis and multimodal practices: Projecting the end of the turn and the closing of the sequence. *Discourse Studies, 8*(1), 117–129. doi:10.1177/1461445606059561

Moyer, M. W. (2016). Eye Contact: How Long Is Too Long? *Scientific American Mind, 27*(1), 8. doi:10.1038/scientificamericanmind0116-8

Nortvig, A.-M. (2014). E-learning in poly-topic settings. *Electronic Journal of E-Learning, 12*(2), 206–214.

Nortvig, A.-M. (2015a). *"At sidde på skolebænken i egen sofa": En undersøgelse af hvordan e-læring ses at have indflydelse på muligheder for deltagelse og tilstedevær i undervisning og dermed udvikling af professionel identitet i relation til professionsuddannelse*. Aalborg: Aalborg Universitetsforlag.

Nortvig, A.-M. (2015b). "Sometimes, I feel a bit decoupled": Strategies in Videoconference Teaching. In *Proceedings of the 1th D4|Learning international Conference Innovations in Digital Learning for Inclusion* (pp. 81–89). Academic Press.

Orman, E. K., & Whitaker, J. A. (2010). Time usage during face-to-face and synchronous distance music lessons. *American Journal of Distance Education, 24*(2), 92–103. doi:10.1080/08923641003666854

Poellhuber, B., Racette, N., & Chirchi, M. (2012). De la présence dans la distance par la visioconférence Web. *Revue Internationale Des Technologies En Pédagogie universitaire/International Journal of Technologies in Higher Education, 9*(1–2), 63–77.

Pugsley, M. (2010). *An Exploratory Study of Educator Perception of Social Presence: Design and Instructional Practices for New Online Educators*. Academic Press.

Pytash, K. E., & O'Byrne, W. I. (2014). Research on Literacy Instructions and Learning in Virtual, Blended, and Hybrid Environments. ETC Press.

Salvucci, D. D., & Bogunovich, P. (2010). Multitasking and monotasking: the effects of mental workload on deferred task interruptions. In *Proceedings of the SIGCHI conference on human factors in computing systems* (pp. 85–88). ACM. doi:10.1145/1753326.1753340

Satar, M. H. (2013). Multimodal Language Learner Interactions via Desktop Videoconferencing within a Framework of Social Presence: Gaze. *ReCALL: The Journal of EUROCALL, 25*(1), 122–122. doi:10.1017/S0958344012000286

Scheff, T. J. (2005). Looking-Glass Self: Goffman as Symbolic Interactionist. *Symbolic Interaction, 28*(2), 147–166. doi:10.1525/si.2005.28.2.147

Schön, D. A. (1991). *The reflective practitioner: How professionals think in action*. Avebury.

Short, J., Williams, E., & Christie, B. (1976). *The social psychology of telecommunications*. London: John Wiley and Sons Ltd.

Swan, K., & Shih, L. F. (2005). On the nature and development of social presence in online course discussions. *Journal of Asynchronous Learning Networks*, *9*(3), 115–136.

Vetere, F., Green, J., Nisselle, A., Dang, X. T., Zazryn, T., & Deng, P. P. (2012). Inclusion during school absence: Using ambient technology to create a classroom presence for hospitalised children. *Telecommunications Journal of Australia*, *62*(5), 3–27. doi:10.7790/tja.v62i5.353

Wahlstedt, A., Pekkola, S., & Niemelä, M. (2008). From e-learning space to e-learning place. *British Journal of Educational Technology*, *39*(6), 1020–1030. doi:10.1111/j.1467-8535.2008.00821_1.x

Wesson, R. H. (2011). The Impact of Performance Skills on Students' Attitudes towards the Learning Experience in Higher Education. *Issues in Educational Research*, *21*(1), 22–22.

Wood, E., Zivcakova, L., Gentile, P., Archer, K., De Pasquale, D., & Nosko, A. (2012). Examining the impact of off-task multi-tasking with technology on real-time classroom learning. *Computers & Education*, *58*(1), 365–374. doi:10.1016/j.compedu.2011.08.029

KEY TERMS AND DEFINITIONS

E-Learning: The use of digital technology for teaching and learning. Often in a blended learning format.

Mono-Presence: The experience of being visible and present in a single space for instance in a classroom or in a web conference (while alone in the physical room).

Monotasking: The opposite of multitasking, the performance of a single activity.

Multi-Presence: The experience of being visible to other people in more (digital) spaces at the same time (e.g., online in a video/web conference and analogously in the classroom at the same time).

Multitasking: The performance of several activities at once.

Online Space: A place where people can meet online (e.g., discussion, learning, and collaboration).

Social Presence: The experience of being perceived as a real person, and the feeling of being present as a real person in an online setting.

Section 3
Cases and Good Practices

Section 3 provides rich cases and recommendations for overcoming challenges in online collaboration and leadership. Practitioners share their experience and insights in online leadership, and they provide hands-on recommendations for overcoming challenges in the online work environment. In addition, an interview with a digital nomad is presented for further research and inspiration. Practitioners may benefit from Section 3 as a guidebook in handling challenges in online collaboration and communication in contemporary organizations. These practitioners could be global managers, people who work online, or HR staff that want to improve the organization's online working practices.

Chapter 13
Developing the Practice of Online Leadership:
Lessons From the Field

Bonnie Cheuk
Euroclear, UK

Jane McKenzie
University of Reading, UK

ABSTRACT

Online leadership largely develops through practice: taking the first steps into what seems a relatively uncontrollable environment is scary. Learning to adapt one's leadership style to suit conditions in the online world is an experimental process that benefits from coaching and guidance from experienced online leaders. This chapter distils 10 years of recent experience the first author gained helping senior managers develop an influential online presence at the same time as implementing digital strategies in three organizations. It starts with a brief review of face-to-face leadership theory, which most leaders encounter in traditional development programs, identifies how online practice differs, explores the phases of a typical leader's journey from conscious incompetence to conscious competence online, and highlights practical interventions that both develop capability and diffuse change organization wide. It concludes with a summary of useful characteristics for change agents supporting the enterprise transformation, which is usually the aim of a move to online working.

INTRODUCTION

Ubiquitous online connectivity has significant implications for enterprise leadership: globalization, democratization of influence (Shirky, 2010; Tapscott & Williams, 2006), cyber-security risks, access to big data, radical discontinuities fuelled by innovation, crowd sentiment, or market upheaval combine to produce a degree of complexity and unpredictability that undermine the efficacy of conventional conceptualisations of leadership, largely exercised face to face. Planning, direction and control of a reasonably contained enterprise in a relatively stable environment is no longer the norm. In what McAfee

DOI: 10.4018/978-1-5225-4094-6.ch013

dubbed Enterprise 2.0 (2006), direction comes through lateral interactions rather than detailed planning: networks of knowledgeable individuals across cultures combine their collective intelligence to interpret directional intent in complex conditions (Bonabeau, 2009), using the 'cognitive surplus' (Shirky, 2010) in the network and the so-called 'wisdom of crowds' (Surowiecki, 2005). Critically, then, formal power in the upper echelons of the organization no longer affords control. Leadership becomes a collective practice both amongst the top management team (TMT) and distributed more widely amongst those with insights relevant to the situation. It emerges when people forge collaborative relationships that become influential within an organizational system and share responsibility for outcomes (Bolden, 2011). Often those relationships are mediated by technology. The most influential forces shape the adaptation of the organizational system. Such disaggregation of power, reliance on meritocracy and community concern, *potentially* allows an enterprise to make better use of available knowledge in the network, better decisions, learn faster and become more agile and adaptable (McKenzie & Van Winkelen, 2004). In the long term, these are crucial factors for dynamic capabilities (Teece, 2007, 2014; Teece, Pisano, & Shuen, 1997) that allow organizations to 'survive and thrive in turbulent times' (Economist, 2009), because they heighten collective capacity to sense signals, seize opportunities and respond innovatively in what has been called a VUCA world - volatile, uncertain, complex and ambiguous (Pasmore & O'Shea, 2010).

Distributed leadership supports collective sensemaking (Maitlis & Christianson, 2014; Weick, Sutcliffe, & Obstfeld, 2005) in VUCA contexts (Lüscher & Lewis, 2008; McKenzie, Woolf, van Winkelen, & Morgan, 2009) and co-creates results through shared, collaborative activity (Edwards, 2011). But the context for exerting influence affects how leadership is distributed (Avolio, Kahai, & Dodge, 2000; Thorpe, Gold, & Lawler, 2011). The possibility of multi-media, many to many interactions through advanced information technology (AIT) collectively known as Web 2.0 has moved more work online; leadership practice is now enacted more frequently in virtual team settings (Hambley, 2007; Lipnack & Stamps, 1999; Ziek & Smulowitz, 2014). Hence the context for leadership as 'a social influence process' (Parry, 1998) increasingly depends on the quality and availability of electronic channels between highly interconnected network nodes who may be individuals or groups (Kolb, Prussia, & Francoeur, 2009). Avolio et al (2000) dubbed this intersection of technological infrastructure and distributed leadership influence 'e-leadership'. Online engagement distributes leadership influence further across time and space (location and the physical constraints upon leadership time become much less relevant). Leaders influence is more durable, because messages are searchable and their presence is enduring. These 'online breadcrumbs' leave a trail of evidence that re-enforce leaders' reputation and value systems for better or worse (Avolio & Kahai, 2003). Yet, as this chapter argues, e-leadership development cannot be separated from face to face leadership practice. An appropriate leadership style should be manifested *consistently* across all communication channels, whether these involve face-to-face interaction, traditional one-way broadcast (e.g. press, newsletters, videos, digital information etc.) or online dialogue (e.g. blogs, wikis, and virtual community spaces): unexplained inconsistencies in espoused views and values create impressions of inauthenticity, promote misinterpretation of messages, reduce trust and damage leaders' reputation, thus weakening their capacity to influence and guide the organizational ship. By implication, leaders must be mindful of how they manifest their presence and values both off-line and online. Developing organizational leadership that can work effectively in virtual and physical interactions across a complex, networked system (McKenzie & Aitken, 2012; Uhl-Bien, Marion, & McKelvey, 2007) is clearly a high priority.

This chapter is an inductive exploration of the first author's learning about how to alter an entire leadership system. It draws from more than 10 years research and practical experience developing leaders at multiple levels in three different enterprises and helping them to adapt to the full implications of leading in a Web 2.0 enabled enterprise. Data comes from her personal reflections, educational presentations, peer reviewed publications and internal documents. The primary aim of the chapter is to offer practical guidelines for others working to integrate e-leadership and face to face practice, so that readers may appreciate the implications, considerations and consequences (both intended and unintended) of exhibiting what Cheuk encapsulates in the term Leadership 2.0 (Cheuk & Dervin, 2011). This combination of developing leaders and a moving the organization to a more collaborative networked model transforms the whole enterprise. Cheuk's insider research (Bartunek & Louis, 1996) on changing the enterprise ecosystem, whilst modelling on line leadership for others, provides a unique perspective and a wealth of examples of real, although necessarily anonymised, practical experiences of CEO's and Executives learning to 'work out loud' (Semple, 2012) in multinational companies.

There is no one-size fits all approach for developing Leadership 2.0 capability. The principles elicited are the essence of what is common in real world leaders' learning, but should be used with an awareness that organizational context, its history, trajectory, culture and narrative all play a role in the development pathway. The most basic tenets are to encourage leaders to be mindful of their moment to moment responses to distributed dynamics, to surface the entangled implications so they actively yet consistently influence the prevailing sentiment through appropriate channels and develop a leadership presence that remains authentic yet adaptive both online and offline.

Although it is important that leaders remain true to themselves in both contexts, their modus operandi and engagement style will change. As they venture into unfamiliar online territory, they are learning in public how to create an influential presence. This can only be done by trial and error. Unfortunately, mistakes can be amplified in the online space. Yet, in Cheuk's experience, the best trigger for unlearning ineffective leadership behaviours comes from successfully transforming mistakes into opportunities. In practice, that prospect generates a level of fear that needs active measures to be overcome. Most senior leaders benefit from ongoing coaching.

Encouraging regular mindful review of their behaviour and communication style is useful. However, the art lies in creating safe opportunities for them to experiment for real, listen to employees' responses, reflect on their practice and adapt. This represents at least 70% of senior leaders' development process. At the same time as practicing online leadership, they are modelling the intended leadership ethos for others, hence shaping what the organization becomes: ideally more collaborative, and better able to use valuable knowledge in the organizational network. Beyond the senior team, e-leadership champions and online community facilitators are powerful supports for encouraging reflexivity, overcoming fear and raising mindfulness of the link between action and consequences. Training provides a framework and foundation for future leaders' development often using lessons learned by respected e-leaders. In many ways, it is a classic action learning process (Coghlan & Coughlan, 2015; Coghlan, Shani, Roth, & Sloyan, 2014) in which organizational change and personal change co-evolve through cycles of practice and reflection, informed by sound theoretical principles. Both bottom up interventions and top down development are required to change the whole ecosystem.

WHERE ARE WE NOW?: THE BACKGROUND TO LEADERSHIP 2.0

Arguably, leadership development is at a crossroads. Adult learning theory emphasizes the importance of experiential and situated learning (Kayes, 2002; Lave, 1996) and the role of conversation and community for developing effective practice (Lave & Wenger, 1991; Wenger, 1998). Consequently leadership development in many organizations follows a 70:20:10 principle (Rabin, 2013), proportionately using experiential learning in challenging real life assignments (McCall Jr, 2004); supportive coaching and mentoring from those with credible leadership experience; and content input derived from prior research. To date, little research has identified what challenging 'virtual life' assignments support learning, there are few role models with credible online experience and the content of most leadership literature came from research in different conditions.

Research offers little consensus about what leadership is and only limited advice about which style works best in any given context. Famously, in crisis a more controlling transactional leadership style is considered more effective than open, transformational leadership approaches, simply because delivering a fast response, *democratically*, in turbulent highly uncertain and ambiguous conditions is hard: unequivocal direction gives followers safety and clarity to act coherently (Snowden & Boone, 2007). However generally, once the chaos surrounding crisis recedes, research proposes leadership for a knowledge economy is about creating the conditions for others to perform; by listening to and valuing the plurality of followers' knowledge, expertise and capabilities, working with their needs, beliefs and values, developing their capabilities, and inspiring their voluntary and discretionary effort in the service of something they consider worthwhile.

Past research, generated by investigating face to face leadership, has been the source of content for development activities, guidance for coaching conversations and the design of experiential learning assignments. It will have also shaped senior leaders' views of good leadership practice. Research variously emphasized the personal traits, individual characteristics, behaviours, authentic, values-based styles, and meaningful relationship exchanges between leaders and followers as vital attributes (Northouse, 2016). Few traditional leadership theories comprehensively suggest suitable attributes for situations where power is more equally distributed. The somewhat paradoxically styled conception of Servant Leadership, first identified by Greenleaf (1977), identifies some factors relevant for leading in networks of equals, because it downplays leadership authority and direction, focusing more on developing others, being authentic, inspiring, valuing followers, exhibiting humility and stewarding the organization (Van Dierendonck, 2011). Yet, much theory and experience absorbed by senior leaders relied on assumptions about leadership being invested with superior authority from position power, which naturally endowed a responsibility to direct the organization through command and control of followers. How deeply embedded these assumptions are in leaders' current style affects how comfortable they will be operating in a networked online context.

Conventional leadership theory for face to face relationships assumes time, distance and localised cultural norms naturally bound leadership practice. In a virtual context, temporal, physical and cultural boundaries do not constrain leadership influence. This extends social reach and undermines the hierarchical nature of followership, allowing the values of distributed individuals and groups to gain leadership influence (Avolio et al., 2000).

Perhaps the only way we can presently define Leadership 2.0 is simply and generically: as "a social influence process" (Parry, 1998 p85) in a networked setting, where individuals and groups influence each other "to achieve a common goal" (Northouse, 2016 p6). Ultimately, distributed leaders are expected to exert influence responsibly to co-ordinate the diverse factors shaping organizational outcomes, so that groups collectively realize organizational intent and satisfy critical local priorities. Thus, leadership involves helping others "understand why certain goals and activities need to be accomplished: As such it constitutes a process of facilitating individual and collective efforts to learn and accomplish shared goals in an organization" (Berson, Nemanich, Waldman, Galvin, & Keller, 2006 p579). Together, these definitions position leadership as an *ongoing* process rooted in interpersonal influence, allow for its enactment at multiple levels, but neglect the implications of translating the activities into an online context.

Early research examining leadership in an Enterprise 2.0 businesses distinguishes distribut*ive* leadership from distributed leadership, by virtue of the fact that there is "something special about leadership associated with the upper echelons, particularly in conditions of rapid and discontinuous change" (Brown & Gioia, 2002 p409). The TMT shapes the cognitive and affective context for others (Avolio, Sosik, Kahai, & Baker, 2014), and hence both the identity of the organization and how others identify with it. The TMT share responsibility for different aspects of engagement and communication; their choices and presence collectively co-ordinate the undulating flow of social influence in virtual communities, teams and distributed leaders, which otherwise could become unchanneled forces overwhelming the organizational system. Further, since the TMT make strategic decisions about which AIT to invest in, they continuously shape how leadership and new technology become entangled and mutually affect each other in forming the system. Brown et al. contend that the speed, ambiguity and complexity of such activities requires the TMT to focus on creating a 'holographic learning organization' (2014 p416) in which leading and learning are widely distributed and technology enables every element to be aware of the knowledge of the whole. Thus, the TMT need a strategic perspective on which technological advances could improve distributed leaders' information access and careful consideration of the ethical implications of those choices.

Further, being embedded in complex, interdependent and international value networks (Allee, 2000) means the views of knowledgeable stakeholders - customers, suppliers, competitors and multi-national legal, market and environmental interests also influence organizational performance. Leaders are nurturing, developing and adapting collective capabilities (Teece, 2007) within a broader ecosystem to ensure the enterprise thrives when external conditions change (Harreld, O'Reilly, & Tushman, 2007; Kleinbaum & Stuart, 2014; Teece et al., 1997). . Technology cannot discriminate between stakeholders' access, without TMT reviewing the risks and establishing appropriate boundaries for external online engagement. Thus, the TMT must judge the degree of virtual separation between organization and external business ecosystem, in conjunction with the ability of the internal leadership system to respond appropriately.

Research into shared and distributed leadership is in its infancy (Northouse, 2016 p365). When applied to what is called 'online leadership' most research is in the context of facilitating learning communities at middle manager level (Huffaker, 2010; Kim, 2000). Some has examined virtual team leadership in the context of distributed project groups, considering how leaders operate virtually across cultures to co-ordinate distributed and team relationships (Hambley, 2007; Ziek & Smulowitz, 2014). The more inclusive concept of e-leadership emerged in 2000 (Avolio et al., 2000) and emphasized the interdependence of technology and leadership as mutually informing elements at multiple levels as a system. Yet, there is limited work on e-leadership development and "how AIT can completely transform the way leaders are developed is hardly mentioned in the leadership literature at all" (Avolio et al., 2014 p126).

In the Leadership 2.0 development process, senior leaders and leaders of groups, projects and teams are all enacting adaptive and complex forms of leadership and learning (deRue, 2011; Uhl-Bien & Marion, 2009; Uhl-Bien et al., 2007) that combine face to face and virtual activities in the very midst of effecting a technological transformation of the organization. Adapting to contingent contextual forces like valued heritage, local culture, turbulent sentiments, rapidly changing goal and task requirements and peoples' motivational and affiliation needs (Bass & Bass, 2008) requires versatility. Building collaborative relationships, capable of leveraging the value of heterogeneous knowledge domains that contribute to the evolution of the complex organizational system requires crossing functional boundaries (Carlile, 2004; Rau, Möslein, & Neyer, 2016). Dispersed functional groups have different expertise, norms, values and priorities that lead to different interpretations of leadership communications. Such conditions suggest a leadership development that encourages listening, sensemaking, empathy, negotiation and dialogue to evolve a shared understanding of what organizational goals mean at local level. This is renowned for being a difficult process (Bruns, 2013; Tabrizi, 2015). Of necessity, it involves mix of formal and informal communication across levels but between equals. Adapting that approach to the online context additionally requires understanding what is different about that environment.

WHERE NEXT?: IDENTIFYING THE CHALLENGES OF LEADERSHIP 2.0

The previous arguments raise many questions about what good leadership practice looks like now and in the future. In line with the literature, Cheuk highlights how online context fundamentally alters the dynamics of leadership practices in five fundamental ways, all of which are experienced as risk to leaders' reputation and credibility. She suggests that venturing into the unfamiliar context outlined in the literature creates anxiety, tension and fear for leaders who learned their craft in more conventional settings. Kahneman's work (2011) suggests this is a significant hurdle, because people naturally overestimate the risk of unlikely events occurring, and overweight risks that are salient for them in their decision making.

1. A Web 2.0 infrastructure allows for multimedia user content generation and real time, many-to-many networked interaction, thereby 'democratizing' the power in communications (Tapscott & Williams, 2006). Channels that demand attention could include anything from blogs to wikis, on line collaborative discussion spaces, webinars, phone conferences, micro messaging systems, to large scale organizational 'jams' - day or week long ideation sessions that bring together people from around the world into creative conversations as a stimulus for strategic innovation (Bjelland & Wood, 2008). Consequently, existing leaders suddenly become immersed in multiple conversations with many more staff than they would have previously. Whilst it offers opportunities to extend their communication reach well beyond the confines of physical proximity, it challenges them to develop ways to handle the abundance of content through multiple channels and they fear being overwhelmed (Jones, Ravid, & Rafaeli, 2004).
2. Despite its multi-media capabilities, most digital channels alter and often reduce the rich combination of visual, auditory, kinesthetic signals (Slater, Usoh, & Steed, 1994) naturally arising in face to face communications. This alters essential sensemaking cues (Weick, 2002; Weick et al., 2005). Manifesting leadership online and 'working out loud' (Semple, 2012) should prompt careful consideration of how to compensate for the characteristic limitations of the different online channels in order to exert influence. Thinking about when to be open and when to be more circumspect about sharing messages becomes a bigger issue. Leaders, who fear saying too much or interacting inappropriately, hesitate about engaging online.

3. Digital communications are generally recorded for posterity. Hence, whatever leaders say endures well beyond the immediate moment (Avolio & Kahai, 2003). Content becomes searchable, thereby prolonging its reach and range. The style and content of all communications continue to present leadership presence to strangers with unknown concerns in diverse contexts. Leaders give a powerful impression of themselves, conditioning perceptions of the organization and the TMT for new recruits and existing staff in different time zones and locations, long after their words were written or spoken or videoed. Further, once anyone has an equal opportunity to broadcast divergent responses across the leader's network, a leader's point of view unavoidably becomes work in progress. Many find this uncomfortable.
4. Historically, technology offered little more than an accessible and searchable storage medium, which systematically controlled how information was consumed and managed. People could ask questions of the content, but the information stored was moderated and managed and access routes controlled. Once technology supports multi-directional, *informal* information access, suddenly the basis of power changes and the risks increase. The same technology that gives everyone equal opportunity to voice their points of view also gives everyone more choices about who to follow. Anyone with a compelling message, presented in an authentic way, can gain a following. This increases the chance that someone with convincing information will emerge as a leader with a competitive message.
5. Moving beyond the carefully controlled environment of crafted and polished communications at planned events, leaders suddenly need to focus more carefully on how the content of their message and the process of communicating conveys who they appear to be and what they want to become known for. Inviting greater involvement in forming their thoughts and ideas about critical business issues, demands different judgements about when to speak and when to listen and how to handle difference and division constructively. When leaders and followers honestly 'work out loud' (Semple, 2012) together, it opens the door to unconstrained positive *and* negative reactions that requires skill to manage authentically on an ongoing basis.

Cheuk has researched and worked with these challenges over the past 10 years and publicly articulates the whole as Leadership 2.0 (Cheuk & Dervin, 2011). Clearly the physical face-to-face dimension of leadership will not disappear, so leaders are not asked to abandon previously effective practices. Followers are unlikely to be inspired by two different experiences of the same person. Leadership should be leadership, in whatever context. Arguably, in future, it makes little sense to distinguish between online and offline leadership. In the long term, the big question is how can leaders learn to harness the whole portfolio of digital and physical channels to engage with people, read the signals of forthcoming opportunities and threats to better inform strategic judgements? In the short term, that means learning how differences in digital channels shape peoples' experience of leadership. As capability advances, the next question is how to integrate effective online working with what leaders already do in face-to-face encounters, and make better use of time and energy by leveraging the technology portfolio so that leadership activities in the two spaces become complementary and mutually supportive.

These big questions prompt many practical concerns, which leaders need help with: What is involved in this process of influencing others on an ongoing basis, through multiple online channels? How can I project my leadership presence, when I won't physically know my followers or understand their world in any meaningful way? How can I manage the direction of the conversation, the abundance of content and its interpretation without physical presence? How could my online leadership activities complement physical events?

To suggest some answers, this chapter examines the key phases of leaders' experience as they develop their online practice, considers some effective forms of support for their development and how on leadership practice can be diffused more widely throughout the ecosystem. All the quotations, in italics below, are taken from an interview (by the second author) with Cheuk, (hereinafter referred to as Bonnie to distinguish the practical foundations from theory), as she reflected on her extensive experience, working with leadership teams in three separate organizations to support their transition to confident participation in digital and physical environments. The themes that emerge are strongly informed by theories of communication and sensemaking in uncertain and ambiguous situations (Dervin, 1998), by Bonnie's own blog, her own PhD research, as well as presentation and written material that she shared freely. The subsequent sections are the result of an inductive analysis by the second researcher within the context of the relevant leadership theory outlined above. A research project run collaboratively with Bonnie during 2014 provided qualitative evidence of how Bonnie actively models the process of working out loud which she encourages others to follow. Together the theory and practice suggest some general practical principles for leadership development in the light of the challenges that working online represent.

HOW LEADERS ACTUALLY DEVELOP THEIR ONLINE PRACTICE

This section explores some of the pivotal transitions points, where leaders need support if they are going to move forward, and suggest practical interventions that facilitate their capability development.

Step 1: Overcoming the Fear of Moving Into a Radically Different Space

Most leaders immediately recognise the risks involved in moving into the online space. Few naturally see its potential benefits. Learning happens best in a safe environment. Fear of things that threaten one's role or personal situation tend to disengage people from development opportunities (Lund Dean & Jolly, 2012). So, in the first phase of developing e-leadership, it is crucial to take steps to diffuse the many sources of fear that can prevent leaders both from taking their first tentative steps online and then sustaining involvement. Initially, established senior leaders benefit from some personalised support and coaching to demonstrate the benefits and help get them past their concerns.

The big, open unpredictable arena in which conventional communication control mechanisms are suddenly absent and the smallest communication error can be amplified many times seems threatening. Conversational etiquette is two-way, informal and frequent, *'so you need to realise that you cannot control what you want people to hear'* even though it is recorded for posterity. Followers will make sense of the message, interpreting it in the light of their issues and context. Naturally, leaders become anxious about the risk of posting something that could be misinterpreted or become a hostage to fortune later. In truth, the same risk is present in a physical environment, it just gets less attention because the scale seems more controllable. Bonnie deliberately raises this very point in coaching conversations with leaders. At times, personalised support has extended to ghost writing some of the early content to illustrate good practice; more often, it has been as simple as providing emotional backing. This takes the form of sitting with senior leaders, encouraging them to press the "send" button, showing them how the posts appear online. She then re-enforces positive behaviour by giving regular feedback on how often the post has been viewed, surfacing the positive value from comments received and creating a memorable story around the impact. Although this seems time consuming, the investment has multiplicative future value.

Positive stories become powerful 'use cases' that can convince others of the benefits of venturing on line, as well as providing evidence of leadership by example that serves to confirm the direction of the organizational intent.

Problems accessing and navigating around the technology system can become convenient excuses for non-participation. Time can be another big reason for avoiding the opportunity. *'I've got my P & L to run and if this doesn't work it is taking my time and taking away my resources.'* So, one of Bonnie's first priorities is *'to think through the user experience: from the first moment you send them a welcome email, to when they click through on different devices you have to make the online tools frictionless.'* Creating an environment where new online leaders find what they need and immediately appreciate the value to them requires advanced preparation. She develops a community of champions, those who are naturally positive about new ways of working, who 'seed' the online environment with key information and interesting conversations, so that senior leaders find relevant and exciting material on arrival. The aim is *'for them to see the value right away and think wow this will really save me time!'* That both quickly invalidates early avoidance strategies, and ensures people are less likely to go back to old ways of working.

Cyber security is high on many TMT agendas. Fears over leaks from online conversations are likely to be greater in organizations where external social media use has historically been blocked and where even email access is quite locked down. Working with the risk and compliance team to establish a simple guide to online etiquette that contains enough reminders about what it is safe to share and how to consider what should remain confidential is vital to pre-empt resistance. However, it is a delicate balance. Warnings to be cautious create tension and if they are too strong people interpret them as a mandate to stop sharing. Guidelines should be strong enough to steer appropriately open behaviour, but not so restrictive that they re-enforce fears. Following through on breaches is important too, so that people recognise there are real consequences of non-compliance.

Authority in an online world comes from the quality of postings. Content is king for gaining influence; role and position power have much less impact. However, when everyone is equal in the process of content generation, leaders can find it hard to know where to start the conversation. They are also aware that staff *'can talk with you, but they can also by pass you too. So, the network out there could leave you the lonely guy in the middle that no-one connects too!'* For any learner, being the odd one out is naturally a scary prospect (Lave & Wenger, 1991). For an established leader, who is still required to deliver on mainline business goals, working online can seem both disruptive of tried and tested routines and threatening to the security of their authority and subsequent ability to deliver results. To stimulate thoughts about suitable content, Bonnie starts her conversations with senior leaders by tapping into their most pressing concerns, *'What is the stay awake business issue? What are you trying to achieve?'* Immediately that removes the focus on digital, yet opens possibilities for discussing how online working could help them reach their goals. She draws on her wealth of positive stories to illustrate how other leaders with similar issues have benefited from extending the reach of their engagement activities. She cites examples that demonstrate the power of developing a following of people who don't report to you but aspire to learn from you, and *'the advantage of not getting a filtered view of the world'* The answer to the 'what is in it for me?' question also seems to be the answer to the 'why should I bother?' question!

Bonnie makes two further points in counterpoint to senior leaders' fears and resistance, *'By not engaging, you are sending a powerful message about your leadership. The time spent is a good way to understand your people.'* In addition you can *'leverage your investment'* of time by taking advantage of the digital long tail (Andersen & Nissley, 2007). *'No matter how much money you put into an event, and how good it is, when you say goodbye, that's it! If you leverage online it becomes searchable, the emotion can come back in 6 months' time or be felt by a new joiner in a year.'* The message that leadership passion remains accessible to many more people long after the words are written or spoken can be a powerful convincer, when combined with help in composing messages that communicate what someone stands for as a leader, and the potential to save time, money and personal energy. As Bonnie points out, *'what once took twenty town halls around the world to engage people with the company direction of travel, now can be done with maybe three or four town halls combined with online engagement in between, allowing the leader to achieve more with less'.*

The organizational context tends to colour initial levels of concern over any move to Enterprise 2.0. Bonnie's first major experience of transforming an organization by introducing leaders to the benefits of networked Web 2.0 culture was at the Environmental Resource Management (ERM) consultancy (Cheuk & Li, 2013). Here people were more familiar with a less hierarchical and more collegial way of working. Leaders were less inherently wary of transferring established community knowledge sharing to an online platform. Whereas, when she moved to the banking industry, the essential conservatism and concern for information security that are part of the norms of the industry presented a bigger mental barrier to work through. Adjusting the leadership development process to honour the most fundamental norms and demonstrating how online working does not compromise them is critical to success.

Step 2: Developing a Distinctive Communication Style to Re-Enforce and Amplify Reputation

Once leaders have taken their first active steps into this new environment, concerns over style, timing and how to manage conversations in a convincing way herald a second stage of leadership development activity. Often this is the hardest, because it all comes down to judgement in context. There is no right or wrong answer about how often to post, how chatty and personal to be, the extent to which you share personal views beyond the business issues, when to respond, how to turn small mistakes into big opportunities, how to handle controversial topics or some of the surprise and unintended consequences of what has been said. Much depends on the situation and the individual leader.

Obviously, individual leaders come into the online world with various styles that have worked well for them. Caution is required. The online world magnifies poor offline leadership behaviour like aggression or excessive ego. In that instance, Bonnie advises that, before ever stepping into the online space, focused development and reflection on how to achieve results differently is the only safe response. Most leaders with a well-established face-to-face persona have unconscious habits, traits and values that characterise them for others. Some people are naturally more cautious, conservative and risk averse; some more adventurous; some serious and reflective, some naturally like to talk and are more outgoing and *'bubbly'*. Generally, the message is don't try to change their inherent nature but raise conscious awareness of how they manifest it, by encouraging leaders to be more mindful of and reflect more on their how communications and actions could be interpreted. As Bonnie tells them, *'when people meet you in the lift, they want you to be the same person they know from your online posts.'*

For some leaders, that means openly blogging about events outside work, sharing what is on their mind, reflecting on everything, whether that is a holiday experience, a child's wedding or big societal issues like whether Great Britain should leave the EU (Brexit). Others feel uncomfortable opening up so much, preferring to project a view of themselves as *'a serious business person'*. So, they decline to share personal experiences or seemingly trivial reflections. Style may also affect the frequency of posting to either online conversations or blog posts. Some leaders post less frequently, but have such an engaging style and tone that it still connects. Others post very regularly in a more reflective mode. Both are equally valid stances. There is no right or wrong answer about style. Consistency is the important factor. In Bonnie's view, what is important is to *'be yourself'* and *'have a view'* then *'articulate how and why you have come to that view and invite others to join in, not just to support your view but so you can learn their views as well.'* That usually requires extra effort to be explicit about what people can expect. She encourages leaders to invest considerable, conscious effort to make it crystal clear that they are open to and want to hear alternative views, however surprising or contentious they may be and to reflect carefully on what it takes *'to create the chemistry between you and your people'*. For example, when starting to blog, rather than just say welcome to my blog, it's worth expressing, in their own words, sentiments like, *'I really want to hear from you and I don't mind if you are honest, blunt and direct. I welcome surprises and constructive comments, because that is how I get to learn from all of you.'* It is worthwhile to set expectations explicitly rather than leaving it to subjective interpretation, by saying something like, *'I am very busy, so that means I may only post every 4-6 weeks and I can't actively engage in all the conversations, but I promise I am watching and listening.'* Then people are less inclined to feel their valuable comments are being ignored.

Sharing deeply held views is likely to come across more authentically, because emotion will infuse the messages, making it easier to *'create a chemistry with people you have not met'*. However, by encouraging leaders to generate content about what they really want to change in their world, what they are they passionate about, what they want to be known for as a leader, it is almost inevitable that the content of their blogs will create controversy. Perhaps the safest topics to start with are topics that most people will enjoy: but, in Bonnie's experience, leaders are lucky if they avoid hitting up against something controversial quite early on. The challenge then, is to manage the ensuing conversation constructively, listen and show respect for others' contributions and engage in dialogue that generates some deeper understanding of one another.

Online leadership activity leaves a trail of 'breadcrumbs' from previous digital material leaders created. Over time this creates an impression of who leaders are, what beliefs and values they hold and how they will address issues and concerns in any engagement. Online reputation and trust is still fragile if controversy arises before a leader has established an online presence that manifests a pattern of open, authentic and balanced behaviour: people have no reference point against which to make sense of a specific response. For example, if a leader shuts down a conversation once in 500 times, it is easier to attribute this to the nature of the issue rather that a lack of openness. That makes a far less negative statement about open leadership than if the occurrence represents only one in ten instances, as can happen early in the development of leaders' online presence. Nevertheless, dealing with controversy and different points of view offers the best opportunities for coaching, reflection and impactful learning. The traditional role of the communications department vetting messages for possible faux-pas and stepping in with scripted plans is not the answer. The key is to encourage leaders to think creatively about how to turn an apparently divisive or controversial topic into something positive in the long term; turning small mistakes into bigger opportunities. Bonnie recalls a situation where a senior leader of a new environmental consultancy was keen to promote engagement with the new online collaboration platform. He publicly offered a prize of two round the world tickets for the best profile on the new platform. This created an outcry, because the high carbon footprint of flights was anathema to organizational values. However, the leader turned it round by acknowledging the point and using it to engage in a discussion about what sustainability really means, what were the factors affecting it and how could the organization really make a difference. It became a real win/win solution, because the leader gained credibility and a bigger following through the ensuing conversation and the organization developed deeper understanding of its collective values for a sustainability policy. In the end the two flight tickets were not awarded, but no-one was disappointed (Tan, 2008).

A more experienced leader's recent blog post about the problems with Brexit also became quite divisive. The leader responded to the heated discussion by making explicit his acceptance that everyone is entitled to their views, and so long as everyone respected each other it was a good opportunity collectively to build understanding of where the two sides were coming from. This example models an authentic and genuine desire to engage with the big issues, without trying to impose one's views on other people.

Bonnie neatly summarises both these examples of good leadership in a networked communication model, *'It is not rule by democracy, but rule by better understanding of where decisions and judgements come from.'* That doesn't require censorship, but dialogue so that each side builds a better understanding of the other. Managing that process can be challenging. For example, it may be tempting to slip into corporate speak when someone contradicts you, which undermines perceptions of authenticity. The best leaders start to explore the source of the different views, using what they learn to build a bridge to a deeper understanding of the differences between both sides of the argument and then explore what similarities might unite both sides. Public spaces can leave some people feeling more reticent to comment, particularly if the issue is sensitive or divisive. When people are physically present, it's easier for leaders to identify those who are quiet and draw them out. Online, some senior leaders choose to acknowledge the hidden voices and give them the option to comment anonymously via separate mail box. Actions like these say a lot about a leader's respect for others, their recognition of how power can distort debate and their desire to treat people as equals.

There are obviously times when a topic is confidential or the full story is too sensitive to share and would unnecessarily unsettle a wider audience. Listening to online sentiment and thinking about the audience is key to judging what should or should not be shared beyond the bounds of physical conversations. However, it is often possible to identify a wider angle on the topic that could be shared to gain a deeper insight into the feeling on the ground.

Teaching judgement in these situations is hard. But as Bonnie argues, judgement is a part of leadership, wherever it is exercised. And traditionally judgement is developed through active, practical learning: sensing, trying something, reflecting, modifying and revising the approach. Hence supporting practice with mechanisms like coaching conversations to formalise reflection, in a safe environment is vital for senior leaders with an existing reputation to manage.

Step 3: Spreading Leadership 2.0 Across the Whole Organization

Developing on line leadership throughout the organizational system requires a multi-dimensional approach that addresses change from both top down and bottom up. Senior leaders, who do it well, can become role models for others. Coaching is one approach to raise awareness of how leadership can influence outcomes, but it can only be part of a solution for changing a whole ecology. The more personal one-to-one solutions are resource intensive. Further, in Bonnie's view, there is a danger in recruiting a whole raft of so called 'digital coaches'. The label evokes the wrong image, because the change is not just about *'going digital'* but about leadership as a comprehensive practice. In addition, coaching seems a fluffy, woolly, unmeasurable process, so value for money becomes a big consideration in scaling it. Nevertheless, using coaching strategically is a necessary part of the overall transformation process. In

Bonnie's organization, digital coaching is deliberately embedded in the induction process for new hires to welcome them into the online environment. This tangibly demonstrates that online leadership is considered central to the organizational way of working right from the start of employment.

Bonnie often involves senior leadership in sharing experiences during leadership development programmes to focus attention on live business topics like leading for innovation and growth in a complex world. Online techniques are deliberately positioned as part of a portfolio of communications tools for achieving these business goals, rather than something outside of the core business focus.

In formal development programmes, staff are educated to rethink their basic communication principles by learning how to host better conversations. This includes time to think about how to make the communication more personal, show vulnerabilities and uncertainties without undermining people's confidence in the leader, how to put difficult topics on the table, confront people in a constructive way, open dialogue up, listen and invite people to share. Another simple but expedient technique Bonnie uses what Dervin (1998) calls 'verbings'. By increasing the use of verbs rather than nouns - e.g. thinking, acting, existing, reflecting rather than thought, action, existence, reflection - different points of view become less concrete, less normative and hence less finished and more open for discussion. This is a subtle way of inviting others into the conversation to share their perspectives and contribute to more informed sensemaking on route to a final decision. Leadership development emphasizes the value of listening, showing respect for other peoples' point of view manifesting a degree of humility alongside ideas for developing a capacity to deal effectively with surprises and things that may be hard to hear. Much of the learning comes from the examples and stories of how others have achieved the results, which is more convincing than instruction.

In her own leadership of three organization's digital transformations, Bonnie also runs regular awareness session to show people what is happening, share stories about the positive benefits others accrue from participation and listen to feedback which informs revisions to the transformational approach. She also facilitates a community of champions' network, who become positive advocates across the business and amplify the effects of all the other activities.

Bonnie sums the impact of the whole process up as 'knowledging' (Cheuk & Dervin, 2011), blurring the lines between sender and receiver, top and bottom of the organization. In this way, the TMT can find the connections between the authentic, distributed voices and strategic goals, spot threats and opportunities more quickly and incorporate a broader spread of knowledge and expertise into a distinctive way forward for the organization. And because everyone has been involved in the conversation, more people actively engage with the final strategy. It becomes a win/win situation.

Step 4: Ascending to the Next Level

The final stage of leadership development is one that most organizations have yet to achieve. Ultimately the goal would be to reach a point where leadership is a seamless experience of online and offline activities that become an integral part of how people making sense of what is happening inside and outside the organization. Once people fully grasp the potential of online, they can weigh up which is the most appropriate and effective channel to engage with relevant people in the organization and balance the mix. They can also think about how to blend the different channels, link and re-enforce content from one to another. Then Bonnie argues, they can really leverage the time and effort they invest in communication activities to optimise the reach and range of influence and maximise opportunities to hear from different groups.

With repeated experience of reflecting on their online communication styles, leaders learn to be more informal and mindful of how to host good dialogue. Bonnie is now beginning to see signs that online working also improves face-to-face interactions, increasing leaders' sensitivity to the impact of their communication practices.

Even so, Bonnie finds most organizations are cautious about extending leadership online presence externally, except for prepared statements on a web site. Engaging with a wider stakeholder community amplifies the risk of incautious, informal communication being interpreted in ways that harm organizational reputational and create collateral damage. Clearly greater external engagement has benefits in terms of access to more insights, opportunities to influence market sentiment, and organizational reputation. However, the blend of content and channels should be part a very careful communications plan. To date, few organizations have made great strides with this. In Bonnie's current organization they are just beginning to consider the possibilities now that they have a mature online leadership team with several years of experience of good practice.

Emerging technology for big data analysis of external information will become part of the portfolio for engaging with the outside world. Bonnie sees the potential for leaders to better gauge sentiment for themselves, identify emerging issues and patterns from unstructured online conversations more effectively. This would give leaders faster access to insight and intelligence and deliver real strategic benefits.

WHAT THIS MEANS FOR LEADING ORGANIZATIONAL TRANSFORMATION

When organizations invest in implementing collaborative, social, networked technologies, they are embarking on a journey that can transform both the leadership and the whole enterprise, because it changes how decisions are made, how strategy is developed and how innovation arises. Hierarchical relationships become networks of knowledgeable equals all contributing differently to the overall performance of the business. Inevitably that creates tension for those who are learning to operate effectively in such different conditions. Current familiar ways of working, relating to others and leading will at times seem contradictory to the new requirements to work, relate and build virtual presence. That tension must be worked through for the organization to get full value from their investment. Leadership support and development is a critical strand in that process.

Enterprise transformation is a long-term change programme. Its impact can, however, be tangible. At ERM, the intent to move to a more collaborative culture took four years and won awards. The fact that greater engagement and collaboration in the service of a common purpose produced a measurable increase in company value delighted the CEO.

However, to sustain the momentum, simultaneous top-down and bottom-up movement was required, to foster a new style of egalitarian and distributed leadership throughout the ecosystem. Such a big challenge requires a certain type of a change agent to join things up in the middle. That person and their small team are leaders in their own right. This requires some quite distinctive, additional attributes, because as change agents they are both the earliest movers and first among equals in that process. Analysing Bonnie's approach suggests some important attributes,

- A multi-disciplinary knowledge and the capacity to understand how to translate the complex business issues which keep leaders in different functions awake at night into digital communication.
- Deep appreciation of the potential and risks of digital technology to outline a vision of what is possible and recognize how technology could help leaders address their business issues.
- A capacity to thrive amidst uncertainty and tension, and the practical know-how to work through difference and use it constructively and creatively.
- An ability to coach, teach, inspire and develop others, which benefits from,
    - A capacity to listen and empathize with the concerns and fears of groups of people across the organizations, use a range of positive stories and creative solutions to allay those concerns and encourage learning. Without sensitivity to potential sources potential resistance, change can grind to a halt (Ford, Ford, & D'Amelio, 2008).
    - Practical knowledge of the art of what good communication and dialogue entails. The challenge of engaging others and helping groups bridge the differences between deeply held points of view through real and honest dialogue is often woefully underestimated (Isaacs, 1999; Kolbaek, 2012). Yet it is a critical skill for change agents and a vital part of the organizational learning process.
    - A passion for the benefits of working collaboratively.
    - An evidence base of past online breadcrumbs that model the sort of online behavior others can learn from.

FUTURE OPPORTUNITIES FOR LEADERSHIP RESEARCH

The process of developing unconscious competence in online leadership remains work in progress. Concern to distinguish online and offline leadership are, in practice, a false dichotomy. It seems important now, because rapid technology advances have allowed complex global business infrastructures to emerge faster than leadership research can blend offline and online practice into a seamless theoretical position. Once leaders become unconsciously competent in leveraging the technology investment to increase the reach and range of their engagement and communication activities being mindful of how to translate their leadership style to suit the different conditions becomes unnecessary. Research to understand how recipients experience a leadership 'persona' manifested across physical and virtual spaces and what that means for sustaining influence would be useful. Over time, leadership practice in both spaces will evolve together, as learning that develops new styles of communicating and relating online alters face-to-face leadership approaches.

In a recent review of the e-leadership literature, Avolio et al emphasize the co-evolutionary nature of the relationship between leadership and technology. Rapid ongoing advances in AIT capability, using robotics the semantic web and big data for decision making will continue to affect the leadership dynamic, just as 'the leadership dynamic affects the faithful or unfaithful appropriation of AIT' (Avolio et al., 2014 p105). Further, research will be required to identify how people learn to integrate these different modes of interaction into a total leadership system, particularly as technology continues to evolve faster than human behavior. This raises many ethical and practical considerations about the locus and mechanisms of leadership for future organizational practice, which may benefit from sensing and tracking positive and negative sentiments and distributed and distributive leadership responses.

CONCLUSION

Transforming the Means for Organizing

Introducing collaborative technology and developing online leadership capability are two strands of a major change program that transforms where and how work is organized. Initially the process depends on moving leaders from a predominantly command, control and competition mind-set to one in which co-ordinate, converse and collaborate play a bigger role in decision making, strategy development, change management and innovation. Like most change programmes, it starts with bold ideas but emerges out of the accumulation of many small steps to address the fears of those who are instrumental in realising the change - namely those with leadership influence. The chapter outlines how to develop leadership practice that engages and influences others online, but argues that leadership style should be manifested consistently both online and offline.

Each organization starts the journey from a different set of norms, values and degree of confidence with technology. Such contextual variables are integral considerations in the design of a process to evolve a complex ecosystem of people, processes and technology that constitute what we think of as 'an organization'. In addition, each leader starts with a unique and distinctive style of leadership. This chapter has outlined some principles for developing leadership judgement to intervene effectively within a complex ecosystem. It may start with senior leaders developing online presence, but at the same time, it only becomes sustainable if the process of developing leadership throughout the network, occurs simultaneously. The development process requires a mix of developmental approaches to educate both ends of the leadership spectrum. By considering leadership theory and research through the lens of 10 years of successful practical experience supporting digital transformation, this chapter outlines a systemic approach to developing what the first author has called Leadership 2.0 that is better adapted to a Web 2.0 world.

The insights offered are designed both to support senior leaders' learning, through coaching and emotional support so they become more willing to enter what appears to be an environment fraught with risk, and, at the same time, re-enforce commitment to a new leadership style from the time people join the organization and throughout their ongoing development. In this way, the change gets embedded across the whole ecosystem and becomes self-sustaining. Such a process depends on the capabilities of a small team of digital change agents with the vision, knowledge and leadership skills to model the way forward for others, so the chapter concludes with some ideas about what those capabilities might look like.

REFERENCES

Allee, V. (2000). Reconfiguring the value network. *The Journal of Business Strategy*, *21*(4), 36–40. doi:10.1108/eb040103

Andersen, C., & Nissley, C. (2007). *The long tail: How endless choice is creating unlimited demand.* New York, NY: Random House Business Books.

Avolio, B. J., Kahai, S., & Dodge, G. E. (2000). E-leadership. *The Leadership Quarterly*, *11*(4), 615–668. doi:10.1016/S1048-9843(00)00062-X

Avolio, B. J., & Kahai, S. S. (2003). Adding the "e" to e-leadership. *Organizational Dynamics, 31*(4), 325–338. doi:10.1016/S0090-2616(02)00133-X

Avolio, B. J., Sosik, J. J., Kahai, S. S., & Baker, B. (2014). E-leadership: Re-examining transformations in leadership source and transmission. *The Leadership Quarterly, 25*(1), 105–131. doi:10.1016/j.leaqua.2013.11.003

Bartunek, J. M., & Louis, M. R. (1996). *Insider / outsider team research* (Vol. 40). Thousand Oaks, CA: Sage Publications.

Bass, B. M., & Bass, R. (2008). *The Bass handbook of leadership theory: Research and managerial applications* (3rd ed.). New York: Free Press: Simon and Schuster.

Berson, Y., Nemanich, L. A., Waldman, D. A., Galvin, B. M., & Keller, R. T. (2006). Leadership and organizational learning: A multiple levels perspective. *The Leadership Quarterly, 17*(6), 577–594. doi:10.1016/j.leaqua.2006.10.003

Bjelland, O. M., & Wood. (2008). An inside view of IBM's 'innovation jam'. *MIT Sloan Management Review, 50*(1), 32.

Bolden, R. (2011). Distributed leadership in organizations: A review of theory and research. *International Journal of Management Reviews, 13*(3), 251–269. doi:10.1111/j.1468-2370.2011.00306.x

Bonabeau, E. (2009). Decisions 2.0: The power of collective intelligence. *Sloan Management Review, 50*(2), 45–52.

Brown, M. E., & Gioia, D. A. (2002). Making things click: Distributive leadership in an online division of an offline organization. *The Leadership Quarterly, 13*(4), 397–419. doi:10.1016/S1048-9843(02)00123-6

Bruns, H. C. (2013). Working alone together: Coordination in collaboration across domains of expertise. *Academy of Management Journal, 56*(1), 62–83. doi:10.5465/amj.2010.0756

Carlile, P. R. (2004). Transferring, translating and transforming: An integrative framework for managing knowledge across boundaries. *Organization Science, 15*(5), 555–568. doi:10.1287/orsc.1040.0094

Cheuk, B., & Dervin, B. (2011). Leadership 2.0 in action: A journey from knowledge management to "knowledging". *Knowledge Management & E-Learning: An International Journal, 3*(2), 119–138.

Cheuk, B., & Li, H. (2013). Using social media to co-create new business strategies at environmental resources management. *Global Business and Organizational Excellence, 32*(4), 49–58. doi:10.1002/joe.21494

Coghlan, D., & Coughlan, P. (2015). Effecting change and learning in networks through network action learning. *The Journal of Applied Behavioral Science, 51*(3), 375–400. doi:10.1177/0021886314540210

Coghlan, D., Shani, A. B., Roth, J. M., & Sloyan, R. (2014). Executive development through insider action research: Voices of insider action researchers. *Journal of Management Development, 33*(10), 991–1003. doi:10.1108/JMD-06-2012-0072

deRue, D. S. (2011). Adaptive leadership theory: Leading and following as a complex adaptive process. *Research in Organizational Behavior, 31*, 125–150. doi:10.1016/j.riob.2011.09.007

Dervin, B. (1998). Sensemaking theory and practice: An overview of user interests in knowledge seeking and use. *Journal of Knowledge Management*, *2*(2), 36–46. doi:10.1108/13673279810249369

Economist, I. U. (2009). *Organizational agility. How business can survive and thrive in turbulent times*. Retrieved from London: http://www.emc.com/collateral/leadership/organisational-agility-230309.pdf

Edwards, G. (2011). Concepts of community: A framework for contextualizing distributed leadership. *International Journal of Management Reviews*, *13*(3), 301–312. doi:10.1111/j.1468-2370.2011.00309.x

Ford, J. D., Ford, L. W., & D'Amelio, A. (2008). Resistance to change: The rest of the story. *Academy of Management Review*, *33*(2), 362–377. doi:10.5465/AMR.2008.31193235

Greenleaf, R. K. (1977). *Servant leadership*. New York: Paulist Press.

Hambley, L. A., O'Neill, T. A., & Kline, T. J. B. (2007). Virtual team leadership: Perspectives from the field. *International Journal of e-Collaboration*, *3*(1), 40–45, 47–48, 50–64. doi:10.4018/jec.2007010103

Harreld, J. B., O'Reilly, C. A. III, & Tushman, M. L. (2007). Dynamic capabilities at IBM: Driving strategy into action. *California Management Review*, *49*(4), 21–43. doi:10.2307/41166404

Huffaker, D. (2010). Dimensions of leadership and social influence in online communities. *Human Communication Research*, *36*(4), 593–617. doi:10.1111/j.1468-2958.2010.01390.x

Isaacs, W. (1999). *Dialogue and the art of thinking together*. New York: Currency Doubleday.

Jones, Q., Ravid, G., & Rafaeli, S. (2004). Information overload and the message dynamics of online interaction spaces: A theoretical model and empirical exploration. *Information Systems Research*, *15*(2), 194–210. doi:10.1287/isre.1040.0023

Kahneman, D. (2011). *Thinking fast and slow*. London: Allen Lane an imprint of Penguin Books.

Kayes, D. C. (2002). Experiential learning and its critics: Preserving the role of experience in management learning and education. *Academy of Management Learning & Education*, *1*(2), 137–149. doi:10.5465/AMLE.2002.8509336

Kim, A. J. (2000). *Community building on the web: Secret strategies for successful online communities*. Berkley, CA: Addison-Wesley Longman Publishing.

Kleinbaum, A. M., & Stuart, T. E. (2014). Network responsiveness: The social structural microfoundations of dynamic capabilities. *The Academy of Management Perspectives*, *28*(4), 353–367. doi:10.5465/amp.2013.0096

Kolb, D. G., Prussia, G., & Francoeur, J. (2009). Connectivity and leadership: The influence of online activity on closeness and effectiveness. *Journal of Leadership & Organizational Studies, 15*(4), 342-352.

Kolbaek, D. (2012). *Proactive reviews: How to make your organization learn from experience*. Copenhagen: Books on Demand GmbH.

Lave, J. (1996). Situated learning in communities of practice. In L. B. Resnick, J. M. Levine, & S. D. Teasley (Eds.), *Perspectives on Socially Shared Cognition* (pp. 63–81). Washington, DC: American Psychological Association. Retrieved from http://www.ecologyofdesigninhumansystems.com/wp-content/uploads/2012/12/Lave-Situating-learning-in-communities-of-practice.pdf

Lave, J., & Wenger, E. (1991). *Situated learning. Legitimate peripheral participation (learning in doing: Social, cognitive and computational perspectives)*. Cambridge, UK: Cambridge University Press. doi:10.1017/CBO9780511815355

Lipnack, J., & Stamps, J. (1999). Virtual teams: The new way to work. *Strategy and Leadership*, *27*(1), 14–19. doi:10.1108/eb054625

Lund Dean, K., & Jolly, J. P. (2012). Student identity, disengagement, and learning. *Academy of Management Learning & Education*, *11*(2), 228–243. doi:10.5465/amle.2009.0081

Lüscher, L. S., & Lewis, M. W. (2008). Organizational change and managerial sensemaking: Working through paradox. *Academy of Management Journal*, *51*(2), 221–240. doi:10.5465/AMJ.2008.31767217

Maitlis, S., & Christianson, M. (2014). Sensemaking in organizations: Taking stock and moving forward. *The Academy of Management Annals*, *8*(1), 57–125. doi:10.1080/19416520.2014.873177

McAfee, A. (2006). Enterprise 2.0: The dawn of emergent collaboration. *Sloan Management Review*, *47*(3), 21–28.

McCall, M. W. Jr. (2004). Leadership development through experience. *The Academy of Management Executive*, *18*(3), 127–130. doi:10.5465/AME.2004.14776183

McKenzie, J., & Aitken, P. (2012). Learning to lead the knowledgeable organization. *Strategic HR Review*, *11*(6), 329–334. doi:10.1108/14754391211264794

McKenzie, J., & Van Winkelen, C. (2004). *Understanding the knowledgeable organization: Nurturing knowledge competence*. London: Cengage Learning EMEA.

McKenzie, J., Woolf, N., van Winkelen, C., & Morgan, C. (2009). Cognition in strategic decision-making: A model of nonconventional thinking capacities for complex situations. *Management Decision*, *47*(2), 209–232. doi:10.1108/00251740910938885

Northouse, P. G. (2016). *Leadership: Theory and practice*. Thousand Oaks, CA: Sage Publications.

Parry, K. W. (1998). Grounded theory and social process: A new direction for leadership research. *The Leadership Quarterly*, *9*(1), 85–105. doi:10.1016/S1048-9843(98)90043-1

Pasmore, B., & O'Shea, T. (2010). Leadership agility: A business imperative for a VUCA world. *People and Strategy*, *33*(4), 32.

Rabin, R. (2013). *Blended learning for leadership: The ccl approach*. Retrieved from https://www.ccl.org/wp-content/uploads/2015/04/BlendedLearningLeadership.pdf

Rau, C., Möslein, K. M., & Neyer, A.-K. (2016). Playing possum, hide-and-seek, and other behavioral patterns: Knowledge boundaries at newly emerging interfaces. *R & D Management*, *46*(2), 341–353. doi:10.1111/radm.12185

Semple, E. (2012). *Organizations don't tweet, people do: A manager's guide to the social web*. Chichester, UK: John Wiley and Sons.

Shirky, C. (2010). *Cognitive surplus. Creativity and generosity in a connected age*. London: Allen Lane - The Penguin Group.

Slater, M., Usoh, M., & Steed, A. (1994). Depth of presence in virtual environments. *Presence (Cambridge, Mass.)*, *3*(2), 130–144. doi:10.1162/pres.1994.3.2.130

Snowden, D. J., & Boone, M. E. (2007). A leader's framework for decision making. *Harvard Business Review*, *85*(11), 69–76. PMID:18159787

Surowiecki, J. (2005). *The wisdom of crowds*. Garden City, NY: Anchor Books.

Tabrizi, B. (2015). 75% of cross-functional teams are dysfunctional. *Harvard Business Review*, (June): 2–4.

Tan, E. (2008). *Leadership 2.0*. Retrieved from http://greenchameleon.com/gc/blog_detail/leadership_20/

Tapscott, D., & Williams, A. D. (2006). *Wikinomics. How mass collaboration changes everything*. New York: Portfolio.

Teece, D. J. (2007). Explicating dynamic capabilities: The nature and microfoundations of (sustainable) enterprise performance. *Strategic Management Journal*, *28*(13), 1319–1350. doi:10.1002/smj.640

Teece, D. J. (2014). The foundations of enterprise performance: Dynamic and ordinary capabilities in an (economic) theory of firms. *The Academy of Management Perspectives*, *28*(4), 328–352. doi:10.5465/amp.2013.0116

Teece, D. J., Pisano, G., & Shuen, A. (1997). Dynamic capabilities and strategic management. *Strategic Management Journal*, *18*(7), 509–533. doi:10.1002/(SICI)1097-0266(199708)18:7<509::AID-SMJ882>3.0.CO;2-Z

Thorpe, R., Gold, J., & Lawler, J. (2011). Locating distributed leadership. *International Journal of Management Reviews*, *13*(3), 239–250. doi:10.1111/j.1468-2370.2011.00303.x

Uhl-Bien, M., & Marion, R. (2009). Complexity leadership in bureaucratic forms of organizing: A meso model. *The Leadership Quarterly*, *20*(4), 631–650. doi:10.1016/j.leaqua.2009.04.007

Uhl-Bien, M., Marion, R., & McKelvey, B. (2007). Complexity leadership theory: Shifting leadership from the industrial age to the knowledge era. *The Leadership Quarterly*, *18*(4), 298–318. doi:10.1016/j.leaqua.2007.04.002

Van Dierendonck, D. (2011). Servant leadership: A review and synthesis. *Journal of Management*, *37*(4), 1228–1261. doi:10.1177/0149206310380462

Weick, K. E. (2002). Leadership when events don't play by the rules. *Reflections: The SoL Journal*, *4*(1), 30–32. doi:10.1162/152417302320467535

Weick, K. E., Sutcliffe, K. M., & Obstfeld, D. (2005). Organizing and the process of sensemaking. *Organization Science*, *16*(4), 409–421. doi:10.1287/orsc.1050.0133

Wenger, E. (1998). *Communities of practice: Learning, meaning and identity*. Cambridge, UK: Cambridge University Press. doi:10.1017/CBO9780511803932

Ziek, P., & Smulowitz, S. (2014). The impact of emergent virtual leadership competencies on team effectiveness. *Leadership and Organization Development Journal*, *35*(2), 106–120. doi:10.1108/LODJ-03-2012-0043

KEY TERMS AND DEFINITIONS

Change Agents: The key individuals who are motivated and tasked with promoting and encouraging change. In the context of this chapter, the change is around transforming business practice to increasing collaboration and knowledge sharing through effective use of internal social technology platforms.

Ecosystem: The complex interactive system of people, processes, and technology that constitute what is understood as organisation. In this chapter, whilst we acknowledge that generally a business ecosystem would include a range of organisations in the external business environment, we are largely focusing on the relationships between elements of the internal ecosystem that generate what is locally understood by organisation.

Leadership 2.0: A set of alternative management values and practices driven by a set of coherent assumptions about the nature of human communication which enable an organisation to take advantage of Web 2.0 technologies. Leadership 2.0 challenges the standard information transmission model, empahsising 'knowledging' as an evolving process of meaningful knowledge exchange through dialogue. Employees are encouraged actively to seeking information to meet their unique needs rather than being spoon fed.

Leadership Development: The ongoing process of educating and developing leadership thinking.

Mindfulness: Paying attention to what is happening from moment to moment without being biased by presuppositions; becoming more acutely aware of one's own role in framing interpretation of sensory input, possibilities, and constraints.

Networked Organizations: A non-hierarchical perspective on organizations that assumes that the collective evolves as a result of social interactions facilitated by technology networks. The shape of business outcomes is affected more by key influencers in the network who gain authority from their knowledge and connections rather than through their hierarchical position.

Reflection: A process of (re)consideration and (re)evaluation of past (re)action or future behaviour and action which relies on questioning. It may be internally generated by a mindful individual or elicited in developmental dialogues with peers or coaches.

Chapter 14
The Virtual Coffee Break:
Virtual Leadership – How to Create Trust and Relations Over Long Distances

Mads Schramm
Netpsykologer, Denmark

ABSTRACT

A large number of studies show that trust is crucial to success as a virtual leader. The purpose of this chapter is to provide a theoretical and practical perspective on the challenges of creating trust and strong relationships to the virtual leader. The chapter presents the consequences of virtual collaborations if trust is low or high. Next, it points out how the virtual leader can work to create trust. The chapter presents "The Circle of Trust," which consists of several important behavioural attitudes for the virtual leader. The meaning of metacommunication in the form of clear agreements on the use of communication is outlined. Several communication forms—e-mail, videoconferencing, and online informal communication—are examined, focusing on how these forms of communication can consciously promote trust in virtual collaborations. The chapter provides a collection of the most important solutions and recommendations for the virtual leader.

INTRODUCTION

Virtual leadership can be defined as a form of leadership in which "individuals or groups are geographically dispersed, and interactions are mediated by technology" (Avolio et al, 2009, p. 440). Studies on how virtual leaders work show that most of them focus primarily on clarifying goals, results, milestones, and frameworks. These items are essential, but the main challenges of virtual leadership lie elsewhere. Researchers Rosen, Furst and Blackburn (2006); Weisband (2008); Mitchell and Zigur (2009); and Nyström, and Asproth (2013) suggest that traditional leadership tools rarely have a decisive influence on whether virtual leadership succeeds, and that organizations should focus on the ability to:

DOI: 10.4018/978-1-5225-4094-6.ch014

1. Establish trust and ensure good relationships
2. Ensure communication quality
3. Ensure common values and goals on the virtual team
4. Work with different cultures and personalities

These are essential qualities in all effective leaders--virtual or not. When these qualities are highlighted as specific challenges for the virtual leader, it is because physical distance requires virtual leaders to avail themselves of alternative methods to ensure that these basic qualities for successful cooperation are present.

A virtual leader uses several forms of communication and technologies, including e-mails, video conferencing, phones, and other collaborative software. The leader consciously chooses the right communication forms for different messages and tasks, based on the situation. This chapter will explore how online leaders in general can work to create trust and good relationships in an online collaborative context for which the leader is responsible.

An important part of building trust concerns creating spaces for informal communications that enable knowledge and understanding to increase. It is one of the challenges the virtual leader faces, as virtual communication often is characterized by efficiency and formality (Bradley & Vozikis 2004). Therefore, this chapter focuses on how the virtual leader can use various communication technologies -- especially e-mail and online meetings – to support building trust and good relationships.

WHY TRUST IS SO IMPORTANT IN VIRTUAL COLLABORATIONS

Trust in online teams and collaborations is crucial to the success of virtual teams, according to Jarvenpaa and Leidner (1999); Greenberg, Greenberg and Antonucci (2007); Weisband (2008); and Hoch and Kozlowski (2014).

There are several different definitions of trust. For example, Gibb (1991) believes "trust is freedom from fear." Butler (1991) defines trust as "a willingness to risk being exposed to another person whose behavior is beyond control" (Butler, 1991, p. 650). The last definition is perhaps the one that best fits into a working context because both leader and employee lack control over the other person, and at the same time, the other person is acting in a way that is important to the team. This is particularly true when talking about virtual leadership, in which the leader has little power to control people concerning anything other than results.

Stephen Covey has researched trust, and he says, in his book *Speed of Trust* (2006), that trust is important for two parameters in organizations -- speed and price:

- When trust decreases, speed also will decrease, and costs will increase.
- When trust increases, speed also will increase, and costs will decrease.

Thus, trust is fundamental to leaders' ability to create and nurture growth, innovation, collaboration, and better execution by a team or the organization as a whole, Covey (2006) says.

Trust becomes so central in virtual collaborations because human imaginations, or what could be called "self-created images," more easily come into play in virtual collaborations (Schramm & Lønsted,

2001). Employees spend a large amount of time interpreting and guessing each other's intentions and behaviors, with the least amount of time spent on virtual leaders' intentions and behaviors.

When people in a virtual team rarely see each other, the "space for interpretation" is bigger than in a normal leadership environment, e.g.," I have not heard from my manager in a long time. Is it because everything is as it should be, and my manager has full confidence in me, or is it because my manager is not interested in me and my work?" Also the virtual leader must pay attention to his or her own self-created images, e.g., "He (employee X) does not respond so quickly, so (he is probably a little dull), or she (employee Y) constantly asks questions, so (she is critical)." Based on very little information, leaders and employees try to make sense of situations based on daily collaborations, and they are often quick to categorize each other. Trust is important in this context. If there is a high level of trust, the interpretation of a given leadership behavior will be positive, e.g., "I have not heard from my manager in a long time – so he trusts me." However, if trust is low, the same behavior may lead to a negative interpretation: "My leader is not interested in me" (Covey, 2006).

These perception processes take place every day, at all levels, and greatly influence employee experiences working in a virtual-based organization. How is the information sent around, read, and understood? How are the stories and rumors that are parts of all organizations' DNA interpreted? There are fewer options for corrections, clarifications, and explanations of expectations, as there are no breaks at the coffee machine or lunches in the canteen when working remotely. Trust is also important in terms of whether employees dare to report doubts, questions, or problems with reaching deadlines, problems with well-being, or home problems. Today, one of the major challenges is how to handle stress and ensure a good work-life balance. It's hard for the virtual leader to "feel" and perceive whether employees are thriving. It is important for the virtual leader to know what is going on with individual employees when they're not seen in person on a regular basis (Brake, 2006).

Informal communication has an important role to play when it comes to building trust in general. Research shows that when we do not meet face to face, informal conversations and other communications are difficult to have, but still they are an important part of the trust-building process (Bradley & Vozikis, 2004; Chaulat, 2012). Communication in online work tends to gravitate toward more formal communication styles (Bradley & Vozikis, 2004; Chaulat, 2012), so it takes a conscious effort and development of virtual collaborative skills to create trust and informal communication, despite long distances.

"THE CIRCLE OF TRUST": TRUST-BUILDING BEHAVIOR

The following section will examine "The Circle of Trust" (Schramm & Diederichsen, 2011), which contains nine key behaviors for the virtual leader to create trust with employees. Over the past 10 years, the author of this chapter has worked with over 500 leaders through training focused on virtual leadership. The theoretical part of the training has been based on literature reviews on virtual leadership as presented in the book *Virtual Leadership* (Schramm & Diederichsen (2011). "The Circle of Trust" was inspired by Covey (2006) and his description of key leadership behavior to build trust, and Greenberg, Greenberg and Antonucci (2007), who have examined trust in virtual leaders and virtual teams. For the practical part of the training, the leaders have been asked to share their best practices when it comes to building trust in an online environment and to lead virtually. "The Circle of Trust" has been developed based on the large volume of input from literature studies, best practices, and discussions from virtual leadership training.

Figure 1. The Circle of Trust (Schramm & Diederichsen, 2011)

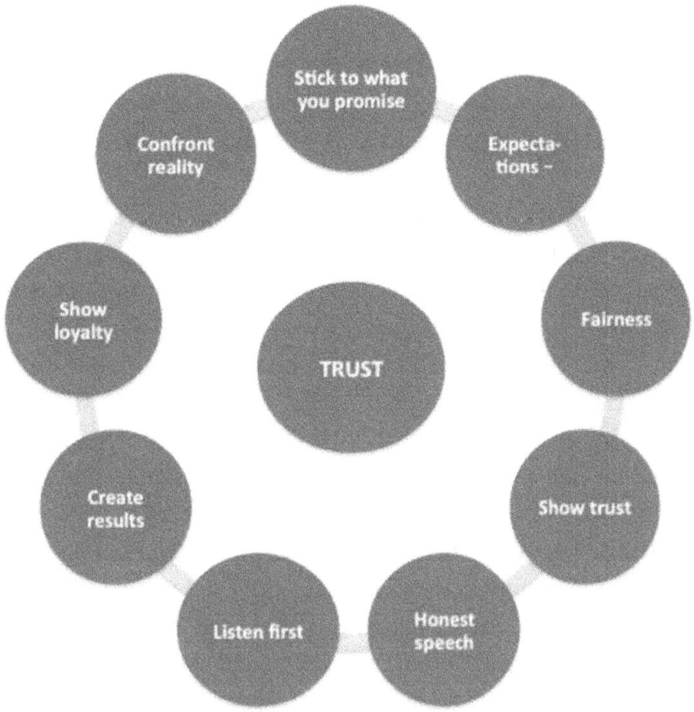

The challenge as a virtual leader during a busy day is to practice these behavioral patterns and at least be aware of how employees actually experience the leader's conduct. It is possible that the virtual leader thinks that he or she already does what is described in the circle, but do the employees agree with that assessment? Does a leader's conduct have the intended effect, or is it read and interpreted differently? In all circumstances, what virtual leaders *do* is far more important than what they *say*. Words, signals, behaviors, and stated intentions can create conditions for trust. When these words are followed by appropriate behaviors, trust will increase, sometimes dramatically, Covey says (Covey, 2006).

Behavior 1: Keep What You Promise - Do What You Say

Keep what you promise and do what you say: This is the fastest way to build trust in any context. Breaking obligations or promises is the fastest way in which trust can be destroyed. It's about being open and real, and telling the truth in a way that can be controlled. "False behavior" is making commitments that are so vague or evasive that no one can hold you to them.

The challenge for the virtual leader may be that generally, when employees are "out of sight," they are "out of mind." Thus, it can be hard for the virtual leader to remember whether he or she got around to everyone, returning with answers to questions, when personnel do not meet at random in the cafeteria or in the hallway. The virtual leader may need to compensate for this via calendars, lists, schedules, or other methods to get an overview of agreements and things he or she promised employees.

Behavior 2: Clarify Expectations - Use Metacommunication

Matching expectations is based on the principles of clarity, responsibility, and accountability. *False behavior* is pretending to be clear about expectations, but not enough to identify the particular circumstances that promote meaningful accountability. In every interaction, there are expectations, and to what extent these expectations are met or violated, trust is affected. Undeclared expectations are often the cause of broken trust.

Within communication research, "metacommunication" often is talked about (Bateson, 1972). Metacommunication concerns communicating about how we communicate and what we communicate. In a virtual context, metacommunication becomes a key concept. Clearly defined agreements must be reached on how to communicate.

Joint team agreements, clear goals, and precise task and job descriptions are always important in a team. However, in a virtual team, they are even more important (Lindholm, 2012). Likewise, expectations are a crucial element. The interpretation of what is happening in virtual collaborations, as mentioned earlier, is greater than in a normal team and organizational cooperation, often causing problems, conflicts, unclear task solving, mistakes, and misunderstandings (Chaulat, 2012). In addition to common goals for tasks and clarifications of roles, there are two areas that stand out in accordance with common agreements, known as Team Operation Agreements (TOAs): trust and communication.

A virtual team is largely an autonomous team. The leader is not there for good reasons, and team members must largely cooperate with each other and know what to do in everyday work situations. After agreements are facilitated by the virtual leader, it is a good idea to answer questions from the team and put the agreements in writing, making them available in a common area online that is available to everyone on the team. The virtual team must decide: How often and in what medium will we have online meetings? How fast will you answer each other? What if you do not understand an e-mail? How do we ensure that informal communication takes place? What communication technologies do we use? How do we create trust in the team? What actions do we consider to be trustworthy? What can the individual employee do and what can the virtual leader do to ensure a high level of trust in the virtual team?

It's all about anticipating the issues and areas that can be challenging to make the team work optimally: The mapping, metacommunication, and creative thinking about how we work remotely with relationship-based communications and trust-building are some of the pieces in the virtual leader's puzzle (Schramm & Diderichsen, 2011).

Behavior 3: Fairness

There are two important dimensions of fairness in a leader. The first is to hold oneself accountable, and the second is to hold others accountable. When things go wrong, and the leader begins to blame or accuse others, the leader must retreat and ask, "What is my own role in it?" Likewise, when employees know that everyone will be held accountable to certain standards, it creates consistency and fairness. When the virtual leader does not hold all employees accountable, it will create a sense of inequality and uncertainty. Fairness is part of virtual cooperation, according to Hunsaker and Hunsaker (2008). A virtual leader should be aware of how he or she allocates leadership time and focus. Employees working remotely are often concerned about whether their leader is aware of how much they work, what they do, and whether they miss out on exciting tasks and career opportunities. Employees at different locations may often find, or suspect, that their leader spends more time communicating with other locations. It's about being transparent, so employees can see what the leader spends time on and why the leader spends extra time on specific teams or tasks. This way, there is no gossip and speculation about the virtual leadership's distribution of time and tasks.

The virtual leader can be visible and accessible at a distance by being an effective communicator of his or her messages (Shriberg, 2009). One way is to create a "communication plan," including scheduled online and face-to-face meetings, as well as provide contact data on how to reach the leader. Employees can easily live with a virtual leader whom they rarely see if the leader is perceived as being accessible. Accessibility is about responding promptly to employee inquiries and having an "open door" policy in the form of an open calendar, open inbox, and open phone. By being an accessible virtual leader, it becomes almost synonymous with being a "visible" leader. If the leader is accessible to all employees, trust increases, and the leader is perceived as fair to all employees (Shriberg, 2009).

Behavior 4: Show Trust and Focus in the Informal Dimension

This behavior is based on the principles of reciprocity and a basic belief that most people can be trusted, want to be trustworthy, and will apply their own standards when they're asked to trust someone. The opposite of this behavior is to withhold trust. Knowledge of each other - including knowledge about competencies, but also some knowledge about personal preferences and backgrounds – makes it easier to trust each other (Caulat, 2012). One of the virtual leader's tasks when taking the helm of a new team is to use time and resources to create knowledge of each other. Virtual teams that perform best have social relationships beyond the virtual workplace (Caulat, 2012).

One of the common challenges in virtual collaboration is communicating beyond solving problems or addressing difficult cases. Informal, slightly easier communications rarely, or perhaps never, take place as a counterweight. If the employee only talks with his leader when there are problems, there may be an imbalance that hinders the creation of trust and relationships remotely. Virtual collaboration is often very effective, formal, and action-oriented, and the virtual leader must be aware of informal approaches, which rarely happen naturally (Scannell, Abrams, & Mulvihill, 2012). Therefore, the leader should ensure that there is room for informal communication on a regular basis. It may be informal phone calls, in which the virtual leader casually asks about tasks and more personal/private matters (see more in the next section on e-mail and online meetings).

Behavior 5: Honest and Direct Speech

Open and honest communication: The opposite behavior is lying or deceiving. "Fake" behavior is either withholding information or saying one thing to some employees and something different to others. The opposite of "fake" behavior is to hide, cover things up, or keep details unclear. Such behavior can mask hidden agendas, meanings, and goals. Fake behavior is making things look different from what they really are.

It is important that the virtual leader communicate in a clear, precise, and open manner, especially the "why" part, i.e., reasons and explanations for decisions and reasons for managerial choices. When employees are far-flung, it's often difficult for them to figure out how things are connected. They do not hear any informal talk in the canteen and do not have the chance to run into the leader in the hallway and ask clarifying questions. With honest speech, there is also honest feedback. If employees know that the leader gives them feedback as soon as there is something the leader is in doubt about, or is not satisfied with, it creates trust. The virtual leader should ensure that he or she gets notified about his or her performance and how he or she is perceived in virtual collaborations. The virtual employee does not know how his or her communication is decoded, read, and understood when collaboration is remote. It is not possible to read each other's body language and immediate reactions. Therefore, honest talk must be a part of metacommunication. An honest-feedback culture is a key element of virtual collaboration. Virtual leaders and employees must discuss how they will communicate, including the importance of providing honest feedback.

Behavior 6: Listen First

The virtual leader must listen and understand before making significant decisions. Otherwise, he or she risks acting on assumptions that are completely wrong, in ways that turns out to be counterproductive. The opposite is speaking first and listening last -- or not listening to anyone. Such a leader forces his or her agenda through without considering whether others have different perspectives that can benefit the organization.

Working from a distance, the virtual leader and employees may not know what's going on at other remote locations. Therefore, it is important to be curious and listen to different perspectives. The practice of conducting good clarifying meetings and soliciting clarifying questions is important. To do so, the virtual leader should engage in tasks and issues, and try to understand employees' perspectives, creating trust and showing interest in the individual employee or team. Collecting such information can easily be done by landline phone or various online meeting forms.

Behavior 7: Create Results

This challenge is related to the principles of accountability and performance. The opposite behavior is about not delivering and instead participating in activities *pro forma* rather than delivering results. If employees perceive the virtual leader as a person who works toward learning, growth, and renewal, they develop confidence in the leader's ability to behave in a rapidly changing environment. The results are always evaluated in relation to expectations. It is always important to deliver results that will build trust consistently - on time and within budget.

To improve matters, there are two strategies that are especially useful: getting feedback and learning from errors. When the virtual leader makes a mistake, the leader must identify the error and learn how to change to get better results next time. The leader also should encourage others to take relevant risks and learn from their mistakes. It is not always possible for employees to know where a virtual leader is, or what he or she is spending the most time on, but it is important to inform and communicate about the results the leader wants. Also, the virtual leader should be clear about how leadership time is used to work upward and onward for the team and the department.

Behavior 8: Show Loyalty

The virtual leader must acknowledge others for their role in producing results. The opposite of giving recognition is to take credit for what others do. *False recognition* is to recognize someone positively when he or she is present, but downplay that worker's contributions and take all or most of the credit when he or she is absent. The second dimension is to talk about others as if they are present. *False loyalty* is to speak nicely about and praise people who are present, then trash them behind their backs. This is based on the principles of loyalty and recognition.

Recognition is important for virtual employees. The risk of feeling outside, isolated, or unseen is potentially greater on a virtual team (Chaulat, 2012). When the virtual leader is not there to make contact, it is important to have goals and a leader with whom to celebrate successes and praise positive results to show that the company appreciates employees' efforts.

Employees working away from the head office can feel like they are getting short shrift as far as attention and need to know that they are included and recognized. Recognition consists of more than praise. It is equally important to get involved in decision making, such as getting asked about decisions and being taken seriously. Here, online meetings become an important focal point.

It is important for the virtual leader to be a role model by showing loyalty to different groups and focusing on "us" and "we," i.e., the team, instead of collective groupings, such as "Danes," "Germans," or the impersonal "those people in the head office."

Behavior 9: Confront Reality

This behavior is about bringing difficult issues to the forefront and tackling both good and bad news. Confronting reality implies demonstrating courage, accountability, attention, and respect. The opposite of this behavior is to ignore reality and act as if it does not exist. Confronting reality openly will promote open interactions and quick results.

Informal leadership may occur when the virtual leader is not present at the location. It can be difficult from a distance to discover whether there are conflicts or mistakes being made by a team. There may be a new signal that the leader must pay attention to, or it may be that the leader is in the "cc" field of e-mails about signs of disagreement among team members. It may be said during online meetings that there is a tense mood that needs to be addressed. In any event, the most important thing is that team members know that they can contact the leader if they have any doubts, disagreements, or conflicts arising within the team.

Summary

"The Circle of Trust" points out important concrete leadership behaviors that can help create a starting point for trust-based virtual leadership. If the virtual leader adheres to these behavioral patterns and checks with his or her team members to find out whether, in fact, this is also the way they experience the leader's behavior, then this is a good platform for exercising virtual leadership and creating good relationships with online employees.

The virtual leader usually has a whole range of communication modes to choose from, including face-to-face meetings, phone calls, online chat functions, different collaboration software, blogs, different types of online teleconferencing options, and e-mail. The virtual leader chooses communication forms and technologies for different types of communication and achievement of leadership tasks. Examples of how to use online meetings and e-mails to build strong, trusting relationships are provided below.

ONLINE MEETINGS

Online meetings are important platforms for a virtual leader to push through his or her messages, step out on stage as a visible leader, and gather a scattered team (Ferrazzi, 2015). This section of the chapter raises awareness of the advantages and disadvantages of various meeting types to create the right dynamics. Used in the right way, online meetings are crucial for creating a level of trust and strengthening relationships. The following section will present several important points related to online meetings, with inspiration from Ferrazi (2015), Caulat (2012), and Ferran and Watts (2008).

Pros and Cons of Online Meetings

Online meetings are often very effective and task-oriented because the distance and medium make the participants focus on the task at hand, with small talk and casual conversations usually not in the mix. Consequently, online meetings are often shorter than regular meetings. The "pros" of online meetings include saving time and money on transportation, making it easier to bring people together wherever they may be, keeping track of projects and tasks, and sharing knowledge. The "cons" include the lack of informal dialogue and the lack of body language and other non-verbal cues. Long online meetings are often more strenuous because we still, more or less consciously, try to read participants' body language through the screen.

It's easier to interrupt each other and disrupt the flow of an online meeting. There often will be "meetings in the meeting," in which a few participants may talk together in small groups in one corner of the video link. This can reinforce the feeling that it is a dispersed team. If the employees do not use the information and communication technology that often, it may prevent them from attending or holding online meetings, or it may make them uncertain of their roles in these meetings.

Types of Meetings

Online meetings are not suitable for all types of meetings; thus, a virtual leader should consider the purpose of the meeting and the meeting format before the organization spends time and resources to bring people together. The most relevant online meeting is the fixed, short meeting, in which the participants know each other, such as weekly meetings between the same people, in which the procedures are the same, e.g., informational meetings, briefings, status meetings, or update meetings, in which the need arises quickly, and decisions must be made promptly. Other forms include meetings to exchange professional experiences, assessments, and coordinated work without having to make difficult decisions.

Meeting types that are a little more challenging in the virtual setup include leadership meetings with final negotiations, contract meetings, meetings on budget challenges, and leadership meetings in which complicated or difficult themes must be addressed, such as exit interviews and issues related to employee performance, when decoding body language is more relevant to gauge reactions.

Brainstorming meetings and those in which someone wants to show participants graphic elements on whiteboards or other media can be challenging. Technology has been evolving rapidly to accommodate these media needs, but it requires some experience and solid meeting skills to pull it off well online. Generally, online meetings with more than eight participants are challenging if the purpose of the meeting is to engage participants in discussions and maintain a high level of energy (Ferran & Watts, 2008).

Consider the Symmetry in Online Meetings

When only one person is attending the meeting remotely, and the rest of the team is assembled somewhere else in a video conference, this is an asymmetrical relationship.

Ferran and Watts (2008) point out that employees' feelings of fairness and justice, as well as the use of technology, are aspects that the virtual leader must be aware of. Any sign that the virtual leader treats people working remotely differently from the people who are physically present will be interpreted as meaning remote workers are less important. This managerial attention to justice and fairness, outlined in "The Circle of Trust," is significant and comes very much into play with online meetings.

There are differences in how communication unfolds, depending on how participants are organized in relation to each other. Symmetry, or lack thereof, can be an important consideration in how status is established and organized in online meetings. With each participant sitting at his or her own computer, equal-participant status seems to be created. The participants all have the same status, sitting in their own space and participating in the joint meeting. When participants are sitting in the same place in a group, they seem to have more power in the conversation than someone in a different location, even if that remote person is the boss. This is because group face-to-face interaction strengthens the group's position. The dynamic is very natural, creating an outsider vibe. It requires special attention and discipline not to create a "meeting within the meeting," in which the remote person is left out.

If two relatively equal groups are sitting in different places while participating in a video conference, the participants are equal in status. However, a power imbalance may occur if the manager is present at one of the two group locations. One solution would be for the manager to alternate between group locations when attending these regular online meetings to achieve some balance in the power dynamic. This can positively affect the more subtle and unconscious asymmetrical dynamic of the meetings.

Another online meeting form is the one-on-one meeting that supports relationship building and feedback, in which foundations of trust can be established. Many of the positive behaviors from "The Circle of Trust" are possible to practice in one-on-one interviews. Roughly 15-20 minutes of one-on-one talks every one or two weeks can make a big difference in the trust factor (Knight, 2016). However, some of the same considerations in online meetings with multiple participants come into play, e.g., how does a virtual leader create a trusting environment in front of a screen? Consider lighting and sound conditions that enable the leader and employees to see each other properly. Try to ensure that eye contact is possible. Send the participants the meeting agenda before the meeting, in case special preparations are necessary.

Recommendations for Good Online Meetings

A first step in having good online meetings is to establish meeting rules, which should be discussed and agreed upon jointly among all participants.

If participants cannot physically meet, invite as few participants as possible to the online meeting. It is important to send out the agenda beforehand, so that participants can prepare comments and questions in advance. This also can reduce wasted time on explanations of background data.

The moderator ensures that the Internet connections are checked and ready when the meeting starts. It can be an advantage to have two moderators -- one handling the agenda and the other handling the flow, i.e., ensuring everyone is heard, etc. The participants can learn how to conduct online meetings by actually doing so, which is easily achieved by taking turns as moderator. Thus, everyone in the team gets experience with online meetings and the technology around it; everyone learns what is difficult and easy about conducting virtual meetings. The disadvantage may be that the meetings in the beginning might be characterized by having less-experienced meeting moderators.

In online meetings, participants should be able to "zoom" in on each other, depending on the videoconferencing service. How to sit, how far away to be from the screen, how the light is set, etc., can send a message. Using clear body language, such as a nod, makes it clear that you can hear and see each other. "You will work together body-language wise"; thus, the participants can better preserve the flow of the online meeting (Ferrazi 2015). There is a need to check the energy level in online meetings; therefore, the participants should use more explicit communication, and the moderator should check and ask whether all can see and hear to contribute to the conversation. The moderator should provide enough time and a place for people to talk freely. Like face-to-face meetings, just a few participants can take control of conversations online. For some participants, it is too easy to sit silent for an entire online meeting, and it is unproductive because silent participants do not contribute to results. Therefore, it is recommended that the moderator organize talk-arounds, in which everyone can express his or her views/concerns in the group.

The moderator should follow the agenda and summarize what has been decided before ending the meeting to align participants' perceptions of the outcome and expectations with the future. The moderator may finish a little earlier than planned, so there is time to check out and get dismissed properly from the meeting (Caulat, 2102). Butler (1999) points to the importance of continuously evaluating meetings and talking about how the team ensures meeting continuity.

Team Building Through Informal Virtual Events

In addition to the more formal, structural, and procedural aspects of online meetings, it is also important to work with the informal communication on the different types of online meeting formats, which helps tie everything together (Caulat 2012).

The author of this chapter has asked the members of six virtual teams from several international companies to describe initiatives they have worked with to build trust and relationships using different types of online meetings. The initiatives have been particularly targeted at getting the informal dimension into their virtual team collaborations. The following summarizes several initiatives in which team members assess how they have had a positive effect on trust and relationship formation:

- ***Virtual Coffee Breaks*:** The virtual leader frequently asks virtual team members to drink a cup of virtual coffee, i.e., retrieve coffee at the coffee machine, and sit down in front of their webcams, or meet in a chat room, for idle talk about weekend experiences, vacation plans, football, or family birthdays. The informal coffee and canteen talk has moved into cyberspace, or rather into more informal chat rooms.
- ***Virtual Handshake*:** The virtual leader checks into the meeting, holds brief introductions, and the participants greet each other. The moderator provides five minutes at the beginning and at the end of the meeting for informal chats about what people are doing or what they are occupied with or concerned about.
- ***Virtual Team Building*:** A team used Google StreetView to show each other the streets they lived on, their houses, and their neighborhoods, while at the same time telling each other about their cities and what characterizes the area where they live. It is described as another way to visit each other, even though they work in different parts of the world.

- ***Virtual Friday*:** Team members set aside time for playing games, such as a local-countries quiz, in which members learn about each other's cultures through quiz questions about home countries. Or they play the Two Truths and a Lie Quiz, in which each person prepares two things about themselves that are true and one that's false. Others must figure out which item is false. The virtual team members also can play standard videogames online together. Prizes can be awarded for best team performance or contributing to a team, or a prize can be awarded for creating good communications within a team.
- ***Virtual Celebrations*:** After completing a project, the team could meet online to celebrate, across multiple locations, at a virtual party. They may drink champagne, eat chips, and share a toast together, and the team leader could facilitate various competitions in which people are divided into small teams and compete against each other.

These examples have been tested in virtual-meeting setups in which the participants rated them as having a positive effect on team cohesion. The effect of introducing and testing these initiatives throughout the organization contributed, according to the respondents, to creating greater attention to each other in everyday life, even though they are all dispersed. The colleagues said these initiatives made them feel "less out of sight, out of mind" and allowed them to learn more about each other. Technology can be used to build relationships and trust in the virtual workplace, but the informal aspect should, over time, overtake the traditional, more formal framework, which technology creates easily.

Summary

A virtual leader should approach online meetings as an instrument to strengthen the team, especially in terms of trust and relationships. It requires that the virtual leader prioritize these aspects and emphasize the value of spending time and effort on informal contact, as well as work time in meetings, by implementing various technological options in the best way possible (Schramm & Diederichsen, 2011). It is important for a virtual leader to strengthen the ability to facilitate effective online meetings. He or she must be familiar with the technology and online meeting dynamics, and should take the opportunity to establish relationships with employees in the best possible ways.

E-MAIL: AN IMPORTANT LEADERSHIP TOOL

Employees and surroundings can be used to assess a virtual leader. A large chunk of communication time takes place through e-mail. Many virtual leaders use more than half of their leadership time handling e-mails. It even could be called "leadership by mail" (Schramm & Diederichsen, 2011). The way the leader manages and writes his or her e-mails is a very central leadership realm in which conscientious practices can build trust and strong relationships (Schramm & Diederichsen, 2011).

E-mail leadership can help gauge leaders' virtual identity using metrics such as how quickly the leader responds to employee e-mails and the way the leader responds. Does the leader write short messages or long instructions? Does the leader write e-mails on the weekend and late at night, then expect a quick answer from employees? Does the leader use emoticons and other personalization tools in e-mails to consciously build relationships? Overall communication context helps create the image of a leader, and online employees pay close attention to virtual leaders' e-mails.

Important Aspects of E-Mail Communication

Research by Byron (2008) sheds light on virtual leaders' e-mail practices, as well as e-mail users in general. The first finding Byron (2008) calls "The Negative Effect," in which e-mail recipients are generally inclined to perceive the content of e-mails as being more negative than intended by the sender. At the same time, managers and online employees with some knowledge of each other interpreted e-mails more accurately.

Additionally, Byron (2008) found that employees who received e-mail from higher-ranking leaders in the organization's hierarchy were more likely to perceive e-mails as more negative than intended. Thus, Byron (2008) said, employees generally have an increased tendency to read a leader's e-mail more negatively than the manager intended.

Furthermore, Byron (2008) found that known and shared rules or agreements regarding the use of emotional content/expressions in e-mails moderated senders and recipients' inaccurate perceptions of emotions in e-mails (Byron, 2008). Thus, emotional cues such as emoticons can weaken inaccuracies or perceptual bias. The more explicit a person's message, through more precise emotional clarity, the greater the chance the receiver will interpret the e-mail accurately. The use of different ways to clarify, soothe, and emphasize the message may be a good idea, but it must be done with care.

This knowledge from Byron's research can be used by virtual leaders to write e-mails more clearly, with messages tailored to the recipient, and the leader's ultimate intent kept in mind.

Consider Whether E-Mail Is the Right Medium

Another important consideration for leaders concerns whether e-mail is an appropriate medium for conveying messages (Byron, 2008; Pedersen & Pedersen, 2006). If the e-mail involves a complicated message, entails many emotions, or requires a quick response, or if potential conflicts are tied to the e-mail's content, e-mail likely is not the correct medium.

It is recommended that the virtual leader consider whether an important e-mail might be misinterpreted by those who receive it. A "second opinion" enables the sender to make the e-mail clearer, avoid formulations that are open to interpretation, and, in general, increase the likelihood of getting the message through clearly, with a minimum negativity effect. Therefore, it is recommended that you get an additional perspective from a colleague before sending a sensitive e-mail.

E-mail is a communication form in which the leader can show trust and bring in elements from "The Circle of Trust" by providing positive feedback, making informal contacts, and using written language in a creative way.

Reflect the Other

A general rule of thumb is to try to "mirror" the other person in e-mails. If the virtual leader has an employee who writes detailed, formal e-mails, it is a good idea to reflect that in the e-mail response. If a team member uses emoticons and sends generally emotional e-mails, the virtual leader may try to reflect this tone of voice in the e-mail without compromising who he or she is as a leader (Bishop, Riopelle, Gluesing, Danowski, & Eaton, 2010).

Just as human beings use body language and tone of voice to communicate with each other in a face-to-face situation, the virtual leader should use that same approach in e-mail communications. It creates trust when we experience, understand, and reflect each other in our communications.

Considerations for E-Mail Writing

When writing important e-mails, the following questions should be considered (Byron, 2008; Schramm & Diederichsen, 2011):

- Who is the recipient, and do cultural or personality differences require any discretion?
- What is your mutual power and position? What is your mutual understanding?
- What should the e-mail achieve? Is an e-mail suitable for this message?
- Consider whether the e-mail creates trust and good relations. Use the same language/tone as if the person were present in the room.
- Will the receiver understand the background (the why) in the message/request for action?
- Are you clear in your communication, i.e., do you need a fresh pair of eyes to read your e-mail before you send it?

E-Mail Codes of Conduct

A common approach and guidelines for e-mails are important for virtual leaders and virtual teams. Although a leader may be reflective, thorough, and skilled at handling e-mails, it requires that employees also have clear expectations and guidelines on how to read, write, and relate to e-mails and how to be responsible if there is anything they do not understand or are uncertain about.

A virtual leader should deal with and set down guidelines for questions such as: What is good e-mail culture on our team? Do we use e-mail in a positive way? How do we reduce the number of e-mails? How soon should we answer e-mail? How do we answer e-mail? How do we use the "cc" function in e-mail messages?

Joint agreements and an e-mail code of conduct can be a big help for organizations and virtual teams. They balance expectations, help reduce misunderstandings, and even reduce overall e-mail volume (Bishop, Riopelle, Gluesing, Danowski, & Eaton, 2010; Byron, 2008).

Summary

A virtual leader is largely assessed on his/her e-mail communication skills. When managers and employees rarely see each other, it is the leader's responsibility to communicate effectively in e-mails (friendly, formal, patient etc.), depending on the receiver. An extra dimension is how others perceive the virtual leader. Good e-mail communication and consciously focusing on using e-mail to build relationships and trust are important skills in a virtual leader's repertoire.

SOLUTIONS AND RECOMMENDATIONS

This section will describe and sum up important aforementioned points that are needed for the virtual leader to build trust successfully.

Trust is built online much like it is offline. It begins with knowing each other and proceeds to trusting. However, technical constraints mean that connecting and building trust online must be a deliberate pursuit. The virtual leader needs to adapt routines in handling the online medium, and he or she should be aware of both pitfalls and possibilities in online media.

1. The virtual leader should pay attention to his or her own "self-created" images. Based on very little information, managers and employees try to make sense of situations based on daily collaborations and are often quick to categorize each other.
2. A common challenge of leading virtually is team members' perception of being "out of sight, out of mind." The virtual leader needs to compensate for this by utilizing calendars, lists, schedules, or other methods to get an overview of agreements and other items that he or she promised virtual team members.
3. A virtual leader should be aware of how he or she allocates leadership time and focus. Employees working remotely are often concerned about whether their leader is aware of how much they work and whether they miss out on exciting tasks and career opportunities. It's about being transparent, so employees can see what the leader spends time on and why the leader spends extra time on specific teams or tasks.
4. Virtual collaboration is often very effective, formal, and action-oriented, and the virtual leader must be aware of informal approaches, which rarely happen naturally. Therefore, he or she should ensure that there is room for informal communication on a regular basis. It may be informal phone calls, in which the virtual leader casually asks about tasks and more personal/private matters.
5. It is important that the virtual leader communicate in a clear, precise, and open manner, especially the "why" part, i.e., reasons and explanations for decisions and reasons for managerial choices. When employees are far-flung, it's often difficult for them to figure out how things are connected.
6. Working from a distance, virtual leaders and employees may not know what's going on at other remote locations. Therefore, it is extra important to be curious and listen to different perspectives. Good clarifying meetings and soliciting clarifying questions are important practices.
7. Recognition is important for virtual team members. The risk of feeling outside, isolated, or unseen is potentially greater in a virtual team. When the virtual leader is not there to make contact, it is important to have goals and to celebrate successes and positive results, showing that the company appreciates the efforts.
8. E-mail leadership can help gauge leaders' "virtual identity" using metrics such as how quickly the leader responds to team members' e-mails and the way the leader responds. A general rule of thumb is to try to "mirror" the other person in e-mails. Just as human beings use body language and tone of voice to communicate with each other in a face-to-face situation, the virtual leader should use that same approach in e-mail communication. It might be a good idea to get an additional perspective from a colleague before sending a sensitive e-mail.

9. Joint agreements and e-mail codes of conduct can be a big help for organizations and teams, as they balance expectations, help reduce misunderstandings, and even reduce overall e-mail volume.
10. Used in the right way, online meetings are crucial for creating a level of trust and strengthening relationships. A first step to having good online meetings is to establish videoconferencing rules, which should be discussed and agreed upon jointly among all participants. It is important to send out the agenda before the meeting, so that participants can be prepared in advance.
11. It can be an advantage to have two moderators so that one handles the agenda and the other handles the flow, i.e., ensuring everyone is heard, etc. Besides, it is recommended to evaluate meetings continuously and talk about how the team ensures meeting continuity.
12. Furthermore, it is important to work with the informal social kit on the different types of virtual meeting formats that help tie everything together, e.g., *Virtual Coffee Breaks*, in which informal coffee and canteen talk moves into cyberspace, or rather into more informal chat rooms. With a *Virtual Handshake,* participants check into the meeting by holding brief introductions or greeting each other. *Virtual Team Building* assigns virtual teams to work regularly in different kinds of virtual team-building exercises. With *Virtual Celebrations*, the virtual leader and virtual team members meet online across multiple locations to celebrate the completion of a project at a virtual party.
13. The concept of "metacommunication," as described in "The Circle of Trust," largely can be seen as the key to ensuring good communication and a high degree of trust in online collaborations. The virtual team and virtual leadership must be able to communicate about communication. It all starts with an explicit discussion on the best way to use the technology, i.e., online possibilities.

Virtual leadership requires accessibility, transparency, and dialogue concerning what is going on, who you are, and what you want as a leader. Otherwise, employees will draw their own conclusions as to what's going on, who you are, and what you want, regardless of the reality.

FUTURE RESEARCH DIRECTIONS

Research has been done over the years on trust in organizations, businesses, and communities, but research seems lacking in terms of effective working practices in virtual collaborations, including the importance and value of more informal communications.

In some cultures, informal communications are more important than in other cultures. Typically, a person from a Western culture in a business setting will be more direct and focus on task performance very quickly. In Asia, however, people spend more time establishing relationships in informal modes before focusing on the tasks at hand (Hofstede, 2010). There are indications that it is generally worth investing time and resources on virtual coffee breaks during a busy and hectic day, but the research is sparse.

Given that virtual work and virtual leadership are often defined by technological developments, it is important to envision scenarios and explore future developments in information and communication technology. Currently, a wide range of collaborative software systems are being developed. New software combines more formal and informal project systems to create a better synergy that focuses on building relationships and trust. Research could uncover how virtual reality and the possibilities inherent in that fledgling technology will affect the possibilities for online meetings and collaborations.

Younger generations, i.e., digital natives (Manafy & Gautschi 2011), have a completely different mindset and lifestyle in their approach to technology than previous generations, and the technological possibilities are more open-ended today. Future research could investigate the implications of this open-ended technology for future virtual work.

CONCLUSION

With technological and social developments constantly evolving, we likely won't be talking about virtual leadership in a few years as a separate discipline, but rather as an integrated part of workplace leadership in general, as well as the virtual skills that these leaders must master. Research on the future of leadership indicates that more technology is on the way to workplaces (Working Virtual, 2014). Employees at any hierarchical level must train themselves to use technology to create trust and good relationships, but they also need to be aware of the possibilities and limitations of working online. They should improve metacommunication about the usage of technology and the various forms of communication. Information technology allows organizations to create a global workplace that can break down walls and dispel cultural prejudices. Hence, virtual leaders and virtual teams should consciously and explicitly cultivate trust and never assume trust will come naturally in an online environment. Virtual leaders need to ask, "What is *trust* to me, and how do we create trust together?" This is especially important when work takes place across vast distances.

REFERENCES

Avolio, B. J., Walumbwa, F. O., & Weber, T. J. (2009). Leadership: Current theories, research and future directions. *Annual Review of Psychology*, *60*(1), 421–499. doi:10.1146/annurev.psych.60.110707.163621 PMID:18651820

Bateson, G. (1972). *Steps to an ecology of mind: collected essays in anthropology, psychiatry, evolution, and epistemology*. San Francisco: Chandler Publishing Co.

Bishop, A., Riopelle, K., Gluesing, J., Danowski, J., & Eaton, T. (2010). *Managing Global Compliance through Collaborative Innovation Networks* (Vol. 2). Procedia- Social and Behaivoural Sciences.

Bradley, E. W., & Vozikis, G. S. (2004). Trust in Virtual Teams. In H. S. Godar & P. S. Ferris (Eds.), *Virtual and Collaborative Teams: Process, Technologies and Practice*. Idea Group Publishing. doi:10.4018/978-1-59140-204-6.ch006

Brake, T. (2006). Leading Virtual Teams. *Industrial and Commercial Training*, *38*(3), 116–121. doi:10.1108/00197850610659364

Butler, J. K. Jr. (1991). Toward Understanding and Measuring Conditions of Trust: Evolution of a Conditions of Trust Inventory. *Journal of Management*, *17*(3), 643–663. doi:10.1177/014920639101700307

Butler, J. K. Jr. (1999). Trust Expectations, InformationSharing, Climate of Trust, and Negation Effectiveness and Efficiency. *Group & Organization Management*, *24*(2), 217–238. doi:10.1177/1059601199242005

Byron, K. (2008). Carrying Too Heavy A Load: The Communication and Miscommunication of Emotion by E-mail. *Academy of Management Review*, *33*(2), 309–327. doi:10.5465/AMR.2008.31193163

Chaulat, G. (2012). *Virtual leadership – learning to lead differently*. Libri Publishing.

Covey, S. M. R. (2006). *The Speed of Trust. The One Thing that Changes Everything*. New York: Simon & Schuster.

Ferran, C., & Watts, S. (2008). Videoconferencing in the field: A heuristic processing model. *Management Science*, *54*(9), 1565–1578. doi:10.1287/mnsc.1080.0879

Ferrazzi, K. (2015, March). How to run a great virtual meeting. *Harvard Business Review*. Retrieved May 28, 2017, from https://hbr.org/2015/03/how-to-run-a-great-virtual-meeting

Gibb, J. R. (1991). *Trust: A new Vision Of Human Relationships for Business, Education, Family, and Personal Living*. Newcastle Pub CO Inc.

Greenberg, P. S., Greenberg, R. H., & Antonucci, Y. L. (2007). Creating and sustaining trust in virtual teams. *Business Horizons*, *50*(4), 325–333. doi:10.1016/j.bushor.2007.02.005

Haraway, D. (1991). *A Cyborg Manifesto. In Simians, Cyborgs and Woman. The reinvention og Nature* (pp. 149–181). New York: Rotledge.

Hoch, J. (2014). Leading Virtual Teams: Hierachical Leadership, Structural Supports and shared Team Leadership. *The Journal of Applied Psychology*, *99*, 390–403. doi:10.1037/a0030264 PMID:23205494

Hofstede, G., Hofstede, G. J., & Minkov, M. (2010). Cultures and Organizations: Software of the Mind (3rd ed.). McGraw-Hill USA.

Hunsaker, P. L., & Hunsaker, J. S. (2008). Virtual Teams: A Leaders Guide. *Team Performance Management*, *14*(1/2), 86–101.

Jarvenpaa, S. L., & Leidner, D. E. (1999). Communication and trust in global virtual teams. *Organization Science*, *10*, 791–815. doi:10.1287/orsc.10.6.791

Knight, R. (2016). How to Make Your One-on-Ones with Employees More Productive. *Harvard Business Review.*

Lindholm, A. B. (2012). *Lederskab på distancen*. L&R Business.

Manafy, M., & Gautschi, H. (2011). *Dancing with digital natives*. Medford, NJ: CyberAge Books.

Mitchell, A. & Zigurs, I. (2009). Trust in virtual teams: solved or still a mystery? *ACM SIGMIS Database*, *40*(3).

Nyström, A. C., & Asproth, V. (2013). Virtual Teams – Support for Technical Communication? *Journal of Organisational Transformation & Social Change*, *10*(1), 64–80. doi:10.1179/1477963312Z.0000000006

Olson, J., & Olson, G. (2001). Distance matters. *Human-Computer Interaction*, *15*(2-3), 139–179. doi:10.1207/S15327051HCI1523_4

Pedersen, O. C., & Pedersen, K. L. (2006). *Mail på arbejde*. Børsens Forlag.

Rosen, B., Furst, S., & Blackburn, R. (2006). Training for virtual teams: An investigation of current practices and future needs. *Human Resource Management, 45*(2), 229–247. doi:10.1002/hrm.20106

Scannell, M., Abrams, M., & Mulvihill, M. (2012). *The big book of virtuel team building games*. McGraw-Hill Publishing.

Schramm, M., & Diederichsen, S. (2011). *Virtuel ledelse – skab synlighed, resultater og nærvær på afstand*. Dansk Psykologisk Forlag.

Schramm, M., & Lønsted, W. (2001). *www.nærkontakt.net: Netpsykologi, kommunikation og rådgivning*. Dansk Psykologisk Forlag.

Shriberg, A. (2009). Effectively leading and managing a virtual team. *Business Review (Federal Reserve Bank of Philadelphia), 12*(2), I–II.

Weisband, S. (2008). *Leadership at a distance*. Lawrence Erlbaum Associates.

Working Virtual. (2014). *VS14 (Report on Working Virtual)*. Retrieved September 21, 2017, from http://workingvirtual.dk/uploads/workingvirtual_dk/files/VS14.pdf

ADDITIONAL READING

Joinson, A. (1998). Causes and implications of Disinhibited Behavior on the internet. I: J. Gackenbach (Ed.) Psychology of the Internet. Academic Press.

Palner Stick, H. (2011). *Virtuelle team*. Gyldendal Business.

Parks, M. R., & Floyd, K. (1996). Making friends in cyberspace. *Journal of Communication, 46*(1), 80–97. doi:10.1111/j.1460-2466.1996.tb01462.x

Sproull, L. & Kiesler, S. (1991). *Computers, networks and work*. Scientific American, September.

Stone, S. (1991). Will the real body please stand up? Boundary stories about virtual cultures. I: M. Benedikt (Ed)., Cyberspace: First steps. Cambridge, MA: MIT Press.

Søndergaard, H. (2010). *Distanceledelse*. Forlaget Bios.

Wallace, P. (1999). *The Psychology of the Internet*. Cambridge University Press. doi:10.1017/CBO9780511581670

Wellman, B. (1997). *An electronic group is virtually a social network. I S. Kiesler (Red.), Culture of the Internet* (pp. 179–205). Mahwah, NJ: Lawrence Erlbaum.

KEY TERMS AND DEFINITIONS

Collaborative Software or Groupware: Is an application software designed to help people involved in a common task achieve their goals.

Emoticon: Is a pictorial representation of a facial expression using punctuation marks, numbers, and letters, usually written to express a person's feelings or mood.

Online Meeting: When people around the world meet, regardless of their location, use video, audio, and text to link up online. Online meetings allow people to share information and data in real time without being physically together. Online meetings use video-teleconference (VTC) software, such as Microsoft's Skype, Adobe's Connect, and Google's Hangouts, to name a few.

Virtual Team: Also known as a geographically dispersed team, distributed team, or remote team, usually refers to a group of individuals who work together from different geographic locations and rely on communication technology such as e-mail, video, or voice-conferencing services to collaborate.

Chapter 15
Professional Collaboration in a World Without Offices:
The Case of a Co-Working Space on Bali

Anders Sandborg Nielsen
Independent Researcher, Denmark

ABSTRACT

This chapter contributes to the understanding of how online collaboration and cooperation can take place between geographically dispersed people by providing insights into how people work from the Hubud co-working space on Bali in Indonesia. The chapter revolves around the presentation of an interview with a person working from this co-working space. The case represents an example of how people, who, in a radical way, embrace the new possibilities of information and communication technology (ICT) work and live their professional lives. The aims are to provide insight into how people in the co-working space collaborate and work in a primarily online world and to present material that other researchers, students, and people who are interested in online collaboration can use in their work. The chapter introduces a framework that is intended as a tool to explore the material by introducing some of the key concepts, a possible theoretical approach, and a context for the case.

ORGANIZATIONAL BACKGROUND

The co-working space Hubud is situated on the island of Bali in Indonesia. Hubud is a place where hundreds of people from all over the world meet and work for either a short or long period. The people who visit Hubud are in different lines of business and are not part of the same organization. What unites them is that they can all do their jobs online and are not bound to a specific physical workspace. The term *digital nomads* can be used to describe people who work in this way. Hubud consists of a series of modern bamboo buildings in the town of Ubud. It was founded in 2013 with the aim of creating an "unprecedented space that embodied the freedom paradigm of work + lifestyle" (Hubud, 2017). This ambition has led to a busy co-working space in which people meet and work. Hubud also hosts almost daily events, such as presentations about different topics ranging from technological issues, to business

DOI: 10.4018/978-1-5225-4094-6.ch015

issues, to lighter topics and courses. The people behind Hubud focus on making it a community in which people can learn, collaborate, and be inspired by each other. Hubud is part of a larger global trend in which thousands of co-working spaces are emerging around the world, offering people a physical setting from which to work. This case revolves around an interview with one of the people working from Hubud. He is not employed there, and from time to time, he works from other places. Hubud is, therefore, to be viewed as a setting in which many different individuals meet and create their own businesses. It is what these people do that is the focus of this chapter.

SETTING THE STAGE

Examining the field of technology and how it affects our ways of working and collaborating is a challenging task. Technology evolves quickly, and every day sees the development of new advances that were previously not thought of or accessible to a wider audience. These new technologies are incorporated into our private and professional lives with the same incredible speed, making it difficult to gain a solid understanding of what is happening and how. This challenge has been addressed by many scholars. In her work on digital competence, Tømte (2013, p. 76) argues that this notion constitutes a "moving target" because the development of new technology constantly changes what digital competence is. Online collaboration and cooperation constitute a moving target in the same way because of their technological component. In the early nineties, the Internet barely existed, and at the turn of the millennium, it had grown to slightly fewer than half a billion users. Today, more than 3 billion people use the Internet (Internet Live Stats, 2017). Many new technologies for communicating and collaborating are Internet based; not only are the technologies new, but they are also unfolding in a world of communication and interaction that did not exist 25 years ago. If we want to understand how we, as professionals, can lead, collaborate, and cooperate in such a context, we must be aware of the premise that we are examining a moving target in an emerging world that is constituted of billions of people.

This chapter dives into this whirlpool by presenting the case of a co-working space on Bali in Indonesia. The main element in the case is an interview with one of the people working from Hubud. The interview focuses on three themes: online collaboration, professional communities, and digital nomadism. The aim is to provide insight into how people at the co-working space collaborate and work in a primarily online world. The chapter presents a case that may inspire and motivate further research of the topic by presenting a glimpse of a very specific context—an approach advocated by Flyvbjerg (2015). In order to create a better understanding of the case, the chapter introduces some of the key concepts and a theoretical approach.

Why Is a Co-Working Space in Indonesia Interesting?

One approach to exploring online collaboration and cooperation is to look more closely at one of the many established organizations that seek to exploit the possibilities that ICT present. Organizations have to be "organized for constant change" (Drucker, 1992) in order to survive, and a great many organizations are adopting and experimenting with ICT to strengthen their ways of collaborating in order to be more efficient or innovative (Raschulte, 2011). However, traditional organizations sometimes evolve at a slow pace. Hauge and Norenes (2015) note that adopting ICT to support learning often proves to be a complicated matter, as the structures and frameworks of organizations do not necessarily fit the possibilities presented by ICT. In other words, many attempts to adopt ICT are often impeded by structures and frameworks that have been designed to facilitate organizations in a world without ICT. This is not the same as saying that traditional organizations do not evolve and adapt to new realities; they do and must (Bauman 2000). However, Hauge and Norenes' observation suggests that if we are looking for insights into online collaboration and leadership, it might be rewarding to look at places other than traditional organizations, which is an approach that is also advocated by Faraj, Jarvenpaa, and Majchrzak (2011). The case presented in this chapter is an attempt to do this.

There are some similarities between a more traditional organization and what goes on in the Hubud co-working space in Bali. First of all, Hubud is a professional setting. It is a place from which people in all lines of business work, and most people who stay there are owners or employees of companies who get paid for what they do. In this way, the people at Hubud are exactly like the people working in traditional organizations or the entrepreneurs creating new companies: They sustain themselves through some kind of work. However, in one way in particular—in their way of working— they represent something radically different, and that is in their way of working. They rely heavily on technology to do their jobs. More important is that technology is often their primary medium for collaborating with colleagues and superiors and for contacting customers and business partners worldwide. In this way, the companies they create have totally different traits from most organizations: They do not necessarily have any physical offices, they do not necessarily meet the people they work with face to face, and they do not necessarily stay in the same place while doing their jobs.

This way of working has arisen from the possibilities created by technology, and exploring how these people work constitutes an insight into a radical way of collaborating online within a setting that emerged from and was created through the possibilities that technology generates without the restraints and logic imposed by the structure of more traditional organizations. The insight into this specific setting has general perspectives that can inspire, create dialogue about, and contribute to understanding how we can collaborate online in a broader range of organizations. The general themes of technology, professionalism, and collaboration combine and create relevance across the different types of organizations, making the experiences of a person working in Indonesia relevant to a manager in a traditional organization in some other part of the world.

Background and Key Concepts

Digital Nomadism

Digital nomadism and digital nomads are concepts that require some introduction. The advent of technology in our professional lives has resulted in great changes to the way in which we work, and many academics have tried to describe and analyze the emerging phenomena. Sullivan's work on the broad concept of remote working (2012), Haddon and Brynin's work on teleworkers (2005), and the book at hand are examples of the efforts that have been made to understand this growing field. A constant challenge is the short life of the terminology used to label the different tendencies in the field of people working through technology. The technology constantly develops, creating new ways of living and working in its wake, thus making it impossible to describe or label movements in the field for more than a short time.

This chapter uses the term digital nomads to describe the people working from the Hubud co-working space in Bali. The primary reason for using this term is that they use it themselves. When they are at Hubud, their description of their situation focuses on being part of a community, but when they describe their overall situation, they often use the term digital nomads. The aim in this section is not to define the term but rather to expand on aspects of the notion and show how they can help in understanding what these people do. The term consists of two elements—the word digital and the word nomad. Digital refers to the importance of technology in what these people do. Hubud is located on the outskirts of a small city overlooking scenic rice paddies. It does not lie in the middle of an Indonesian Silicon Valley, and the business partners and clients of the people working at Hubud are almost always thousands of miles away. What Hubud has is a very fast Internet connection. Therefore, in most cases, if somebody wants to work from there, they have to do the work with the help of technology. If they are able to do so, they can enjoy the scenic surroundings, the low cost of living, and the pleasant climate. This attracts many people, such as programmers, marketers, and founders of web-based business startups. The jobs that these people do are done *with* technology, and their products are created and exist in a virtual world. At the same time, the place attracts many people who work *through* technology. The product of their work might still be virtual, but their work includes communicating, leading, and collaborating online. Some of these people work as project leaders, managers, or bloggers, for example. In real life, categorizing who works in which way is challenging, but in an attempt to explore online collaboration, it is essential to focus on ways of working online that entail a high degree of interpersonal interaction. Many people use technology in their work, but it is only when people interact through technology that we can grasp what it means to collaborate online. The other element of digital nomads refers to the concept of nomadism. Kakihara and Sorensen (2001) explore the concept of mobility and how technology has affected it. They claim that technology creates not only a greater spatial and temporal mobility but also greater contextual mobility, whereby people are not limited by the structural and contextual constraints of the physical world. Rather, to an increasing extent, they are able to communicate across "cultural background, shared mood and high degree of mutual recognition" (Kakihara & Sorensen, 2001), thus further increasing the possibilities of collaborating. The concept of nomadism is to be seen in this light. Digital nomads make use of this mobility, operating from different locations, sometimes asynchronously; across different time zones; and across structural barriers that, in the past, impeded professional collaboration. The concept of nomadism might give rise to the image of digital nomads as constantly moving, staying only briefly at each location, but this is a somewhat faulty image. The nomadic element underlines only that these people have chosen to make use of the increased mobility described by Kakihara and Sorensen (2001) in their professional lives, but this does not necessarily imply that they drift around constantly. They may be based at the same location for long periods. In this way, the digital nomad is someone who, to a great extent, makes use of the increased mobility but not necessarily on all three parameters all the time. In relation to the case presented in this chapter, the focus has been on interviewing people working in a way in which they make use of this mobility on all parameters, showing high degrees of mobility across time, space, and context. As mentioned in the introduction to this section, digital nomad is a label that many of the people at Hubud give themselves. As the case will show, the people at Hubud see digital nomadism not only as a way of working but also as a way of living, which grants the individual a high degree of freedom. Understanding digital nomadism can, therefore, not be reduced solely to looking exclusively into isolated work processes. To understand it, we must take the whole picture into account. For example, when the interviewee talks about freedom, it is not only the aspects that are closely related to work that occupy him. It is both the positive and negative sides of freedom in regard to all aspects of life—the freedom to experience foreign cultures, to travel, to be lonely, to live with a high degree of insecurity, and so forth. In this way, looking into digital nomadism does not only present an opportunity for traditional organizations to be inspired by concrete ways of collaborating online; rather, it also raises broader questions about how technology affects our work–life balance and how we can use technology to shape a life that contains more than work.

Connectivism: A Framework for Understanding Online Collaboration

The rationale for introducing a framework for reading the case is to present some of the newer trends within learning theory. The aim is not to discard theories made in an analogue world but rather to include and present some of the thoughts on online collaboration that are engendered by and build upon the premises of a digital world. Such considerations are found in a connectivist approach. This chapter does not deliver an analysis of the case from a connectivist point of view but merely presents some thoughts in order to create a more nuanced framework for further work using the information presented. A key premise for connectivism is that "Problems are becoming so complex that they cannot be contained in the mind of one individual—problems are held in a distributed manner across networks, with each node holding a part of the entire puzzle" (Siemens, 2006, p. 44). This changes the understanding of what it means to know something.

To know today means to be connected. Knowledge moves too fast for learning to be only a product. We used to acquire knowledge by bringing it close to ourselves. We were said to possess it—to have it exist in our heads. We can no longer seek to possess all needed knowledge personally. We must store it in our friends or within technology. (Siemens, 2006, p. 51)

This has some interesting implications for the way online collaboration is understood. Siemens argues that because it does not make sense to possess knowledge that might be outdated tomorrow, the key to navigating a digital world is not to possess knowledge but rather the ability to access it and activate it when it is needed. "The capacity to know more is more critical than what is currently known"(Siemens, 2006, p. 31). The relevance of connectivism in relation to this case is that there is a strong connection between collaboration and learning. To understand how to collaborate, it is necessary to understand this collaboration as a process of learning. Siemens states that "Learning needs and theories that describe learning principles and processes should be reflective of underlying social environments" (Siemens, 2005, Introduction Section, para 1). This implies an even greater focus on how technology affects the ways of interacting and living in a digital world. In this way, connectivism offers "a way of seeing trends developing in the world today" (Siemens, 2006, p. vi). Thus, this chapter offers insight into the vanguard of online collaboration (through the presentation of the case) and a framework for understanding it through theoretical work departing from the case's own reality (a digital world). The aim of connectivism is not to discard older learning theory (Siemens, 2005, 2006; Downes, 2005) but, rather, to complement understanding of living in a digital world.

The Research Setting of the Case Study

The focus of the interview has been affected by these ideas. The interviewee was informed that the interview was going to be about professional communities and collaboration through technology—which is to be seen as the operationalization of online collaboration and as something that is closely related to a collective and networked understanding of knowledge and value creation—and the importance of the ability to access knowledge rather than the ability to possess it. As the case will show, networks play a significant role in the professional life of the interviewee. The interviewee has his own company and two business partners, but many other people are involved on an ad hoc basis; they contribute to specific assignments both in an informal way—through, for example, sparring—and, more formally, via shorter contracts, in order to solve specific tasks. In this way, the company that the interviewee runs is of a very dynamic size and cannot always be understood as the three people constituting its core. The community at Hubud represents a mix of personal and professional relationships in which social interaction and professional collaboration go hand in hand. One aim of the interview was to explore how the interviewee navigates situations in which a random person who is sitting next to him at lunch might be a vital business partner who can help with a job in the future. It is in these circumstances that connectivism might prove a valuable approach, emphasizing the importance of being able to activate and access knowledge rather than possessing it oneself. The company that the interviewee owns would not be profitable if it had all the people working for it for shorter periods employed on a permanent basis. In this way, the dynamic organization of the company becomes a response to the changing qualifications needed for the different assignments that the company undertakes. This raises an interesting question in relation to more traditional organizations. As Drucker stated in 1992, organizations "organize for change"; but what if the setup of a traditional organization is built on possessing knowledge rather than having the ability to access and activate it? Online collaboration makes use of increased mobility across time, space, and context, as shown in the section on digital nomadism, making it possible for organizations to work in new ways. But the case and the connectivist approach show that reflections on how organizations establish connections and collaboration with potentially valuable knowledge and knowhow outside the organization might also be constructive. The aim of this chapter is not to advocate for more dynamic organizations with fluid borders but, rather, to draw attention to the premise that not only do the demands that organizations meet change rapidly but also that the knowledge needed to meet these demands is evolving constantly, thereby creating a situation in which it is impossible to possess the relevant knowledge at all times. The case presents insight into how this challenge can be addressed, but it does not deliver a universal answer to how it should be done.

The background for the case presented in this chapter was a visit to the Hubud co-working space in December 2016. The author was fortunate to visit the co-working space and to briefly work from there. The aim was to see the place and make connections with people who use technology as their primary tool in their professional work. To ensure that these people would be suitable, they had to meet certain criteria. First, they had to work mainly online, meaning that most of their professional contact with colleagues and customers had to be online. Second, it was important that this way of working was their primary method and did not solely constitute a break from their regular way of working. Third, they had to have been working this way for some time. Many people visit Hubud for periods of less than a month, and it was important to find and interview people who had been doing this for a longer period. These criteria were necessary to ensure that the interviewees would represent people with experience working online. One of these interviews forms the backbone of this chapter and contains a dialogue with one of the people who met these criteria. Two more people were interviewed, and these interviews were transcribed, but in order to present coherent material for further research, only the one interview is presented here. The content of the other interviews and the author's experience talking with other people from the co-working space support the general image of how digital nomads work that is presented in this chapter. To explore how the digital nomads work in an online world, it was decided that the interview would be a semi-structured life-world interview (Kvale, 2007). An exploratory approach was used to conduct the interview. These choices were made to create an interview process in which the interviewee had the best possible opportunity to talk about his reality and experience without being limited by a predefined, irrelevant structure. To maintain the focus on the topic, the interview centered on three topics—namely, the concept of digital nomads, professional communities in an online world, and collaboration through technology. Below each topic was a list of questions designed to get the interviewee to talk about that particular issue. The aim was not to ask all the questions but rather to use them in case the interview stalled. Before the interview was conducted, the interviewee was informed of the topics but was not told the questions. This was done so that the interviewee had the opportunity to direct attention to what was important to him within the three main topics.

Interview

The interview was completed using an online conference tool (Skype) and was recorded. The interviewee was calling from Bali and the interviewer from Argentina. Conducting interviews using teleconference tools has some limitations, but as the interviewee was used to communicating this way, it was decided that the tool would serve the interview purposes well. The interview was scheduled to take about one hour and was completed in approximately that time. After the interview was completed, it was transcribed. The next step in the process involved several considerations. The transcription contains the full interview, which includes an extensive amount of material, some of which is not specifically related to the three themes that were introduced. The aim of this chapter is to provide insight into the life world of the interviewee, and in this regard, it could prove valuable to present the interview in its entirety, showing the development of the process from beginning to end. However, the transformation of an oral dialogue into written text already constitutes an interpretation of the dialogue, thus creating a new product (Kvale, 2007). To increase the relevance and to facilitate communication to shed light on the three themes, the decision was made to do an initial coding, so only the passages related to the topics and their context are included in the chapter. Therefore, the case presented below does not constitute the whole interview but still contains an extensive part of it. This coding process was done in a very general way in order not to disconnect the elements from their contexts. The interviewee is not a native English speaker but works in an environment in which English is the common means of communication, and his English skills are good. Sometimes, there were small grammatical errors, which have been corrected to facilitate the reading. For the same reason, pauses, expressions of sentiment (laughing), and so forth have been omitted. The aim of these corrections is to create a more lyrical text that is clearer and more on topic. This chapter aims to present a case that other people can study and analyze without having any prior knowledge of the specific setting; therefore, the focus on communication is of great importance. Kvale (2007) discusses some of the dilemmas in the transcription of interviews, stating that "There is no true, objective transformation from the oral to the written mode. A more constructive question is: What is a useful transcription for my research purposes?" (Kvale, 2007, p. 98). It is this approach that has constituted the transformation of the interview into text in this case. Finally, the interview has been anonymized in order to protect the interviewee. The topics discussed in the interview are not highly controversial, but this decision was made to create a free and open atmosphere for the interview.

Interview With a Digital Nomad

The interviewee (A) is an Italian man in his mid-thirties, who has been working as a digital nomad for about three years. Before that, he worked for traditional organizations in his home country. Online marketing became a big part of his work life, and he, therefore, decided to come to the Hubud co-working space to learn more about this topic. Now, he runs his own company with two colleagues. We enter the dialogue when he is asked whether he has been on Bali ever since he started working remotely.

Interviewer: *Have you been at Hubud for the last three years or . . . ?*

A: *Yes. I came here in November 2014, and I have travelled also. I went to Chiang Mai, which is another very famous place for digital nomads and digital entrepreneurs, I went to Malaysia, and I went to some other places, but I still prefer it here, spending like 60% of my time during the year here. So, this is something like my base because I like it here, I like the vibe, everything here. I am very productive here. So, it is the place I stay when I have to work on something. I still have plans to move and explore other places; I don't know, maybe in the future I will find another place like this where I can work this way, but for now it is like my base.*

Interviewer: *In order to do your work, is there something special that draws you to Hubud, to Bali?*

A: *I think Hubud is different from... I do not know... There are a lot of people coming and going, spending one or two months here. I went to other co-working spaces around Asia and also in Europe, but they are still more like offices for rent. There are a lot of small companies based there and they stay there all the time and there are not so many people coming and going. For me, Hubud is the perfect mix of this; there are still some small companies here, but it is not like they have offices here. There are no offices here, like private offices. There are people coming and going, staying here one month or two months. It is enough to meet them, to talk to them and maybe to start collaborating with them. If this exchange of people is less than one or two weeks, you do not have the chance to meet people, but if it is like one or two months you can start to talk with them and maybe collaborate with them. It is this balance that this place has, and until now I have not seen any other places with this kind of network and community. Also, because this place is small, you start to talk with everybody. You know everybody here. It is not that difficult to talk to people here.*

Interviewer: *OK. You say that it takes some time before you can start collaborating with people; you need to know them, you need to know what they can contribute. You said that you had worked in a traditional organization before. How is it different to choose colleagues and jobs when you are not in a traditional organization?*

A: *I think that most of our countries have this concept of stability that sometimes... Sometimes stability is misunderstood as something that cannot change. I hear from my friends and people that I know in Italy; they always talk of the job of their life, they are going to like this job for their whole life, so their colleagues will be the same for their whole life. And also their goal in their job is fixed because they have a contract and after two years they can have a bit more on their salary and after three years a bit more. So, it is something like, everything is a package, where you already know the whole story. So, this thing makes you very lazy. You stop thinking about creating new solutions; your creativity is almost killed because you are always facing the same people with the same ideas. This is something that makes people feel safer because if I am talking with you every day I almost know your ideas. But if I always talk with different people I always get different feedback and different ideas and my creativity is influenced by this. So changing colleagues, changing people is for me very stimulating. It makes you active and after that you can choose, if you like, to work with some people because they are on the same page as you; the collaboration can be longer and you can collaborate with them. I chose my two business partners because we had the same way of thinking about work, common ideas about our future, our company so we decided to organize* *like that. But after that it is very nice to have different people, not every day, but every week, every month to talk with different people for sure. And this is the difference, the main difference between the old style job, the corporate job. I was not employed in a corporation, but it was the same. I was always working with the same people, always talking with the same people, having the same arguments. My creativity was killed by this. From the repetition.*

Interviewer: *So, if a big corporate firm offered you a job would you choose to stay working the way you are now?*

A: *Maybe three years ago, when my job was not working well. If I had that possibility, maybe I would have accepted the opportunity because I was not really aware about working in this way. Now it will be very, very difficult for me. But when you work for yourself it is also like… You are not sure to get paid every month. It can be like the tenth of the month and the salary... So there are still some points that are difficult to face, but it is much better from the freedom point of view. Freedom, well it is not only like work where you want and work when you want because I am working much more now, than when I was employed... But something like you do something for yourself and you can decide by yourself. And this is, you know… And you can decide if you want to take some jobs and not take other jobs. When you feel this freedom, it is very hard to step back. But I don't know maybe if a really good company like Google... It would be difficult for anybody to say no. For sure not a traditional corporation. If a company like Google or some other, like maybe... that would be an experience. But we still also prefer to be consultants; we have some companies that ask us "Maybe you can join us?" We still prefer to be independent so we can work together with some companies, sometimes also very strictly with them, but being at the same time independent. I mean a company is nice for this because you can say I have my own company so let us do a contract, like a retainer contract so we can work together for a year, two years or whatever, but we still keep our independence.*

Interviewer: *Why do you think people choose to work the way you do now? Why would people like to live with some uncertainty? You have mentioned the freedom, that it boosts your creativity. Are there any other reasons for working like you work?*

A: *I always say why do you do something you hate or you do not like for eight hours a day only to go to the gym or something like that? You can do something that can become your passion or you can transform your passion into a job, into a work. So, I think that the entrepreneur or the freelancer is making this transition. So, transforming... merging them. Merging the passion with the job, and the life with the job. There really is no boundary between life and job because you need to work in order to live, so your life is your job and your job is your life. So, it is something naturally integrated in your days. It is not like: Now I have the holidays so I do not want to think about the job. Many people try to make this switch; the job is there, holiday is another thing. But many people want to do something that is more natural and I think that is better from many points of view because they are following what they like, not what people tell them to do, or you know, what their boss tells them to do.*

Interviewer: *In a professional context, when you have to solve a problem... If I have to configure something, I will go to the IT department and ask them. How do you manage this? Not being in an organization?*

A: *When you are in a company, sometimes, no, many times... When you have a problem, you think that there is always somebody who can solve the problem for you or that can help you. And this is another killer of creativity, in my opinion because when you have a problem and you are alone or* *in a small group or are independent, the first thing that you have to do is try to solve the problem by yourself. And then… In this case you can analyze the problem and say OK I can learn to how to solve this problem. There is Google and YouTube and you can think about the problem first. Otherwise you arrive at a certain point where you say OK I need somebody because I cannot do this. So, for example, I am not a real developer. I can develop, but for example, I am not into Java, Javascript and whatever, so if I have a problem with Javascript, I have to hire somebody. So if you have a good network I can ask some programmers; OK, I have this problem, I cannot solve this problem. For me, it takes more time to learn than to hire somebody. So if I know that I cannot solve a problem I hire somebody to solve it. So, usually that is my process.*

Interviewer: *You said that when you are in a traditional organization you can become lazy because you don't need to find somebody who can solve something special. You know they are there and you can use them when you need them. For example, Hubud is... you used the word community. What do you do to keep these professional relations? How do you maintain a professional network in a place like Hubud?*

A: *It is so easy to maintain contact with people now, with Facebook or whatever. You do not need business cards. We still have business cards in my company, but I think I have distributed like three business cards in the last two years. Because it does not make sense. You have so many ways to reach a person and to communicate with somebody. And you can maintain your network. I am pretty organized so when I find somebody that can be interesting for our work I have a sort of database. So, I keep them in a database so I can say OK this is an App developer, so if we are going to develop an app for a client I can call him or her. I also made a system for remembering the professionals I think can be useful for my work and my company. After that you can maintain contact with everybody through the social network. LinkedIn is also very useful. When you meet somebody the first time it is Facebook and then if it is something interesting about a job you connect on LinkedIn. It is like a live database that is always updated.*

Interviewer: *When you meet someone at Hubud, do you see them as: Oh, there is someone nice to go out and have a beer with or like there is a possible business partner?*

A: *Both. When you like a person, you can say OK let us go for a beer and something pretty superficial. When you start to talk about work, and talking about work is pretty common here when you go out for a beer, and after that you start to go a bit deeper into the person and what they are doing and what their philosophy of work is.*

Interviewer: *Another question in a relation to a traditional organization. If you have a job, I know it is not always like that, but if you are in a traditional organization, if you are hired as a consultant, people would see it as a context in which they can do a career saying OK in three years I want to be a chief consultant in this firm. How do you see your future in terms of your professional life?*

A: *I do not know. I like what I am doing now, as we are starting to scale and to hire people. We are starting to dedicate ourselves to what we like and because when you start, we were three who started this company, and we were doing everything. Now we have started to hire people, so you can concentrate more on the job that you like. So, in my company, for example… As I told you I am the most organized and I like to study the project and to propose solutions to the project and follow the project, so I like the project management part more than the designer, branding part; that is more up to my business partners. So I see myself like managing different projects not executing them, but maybe collaborating in executing some part of the project, but more like managing things. This is the thing that I like.*

Interviewer: *So, you do not feel limited in terms of your perspectives on a career? You do not think that you have fewer possibilities because you are not in a traditional organization?*

A: *No, no, no. Well I think that my possibilities are all up to me, are all up to us, to become bigger or to stay where we are now. It is always up to you and your ambition. So, I do not feel limited; actually, I feel I have many more possibilities than in a regular job.*

Interviewer: *You say that you have colleagues in different parts of the world and that you work very different locations. If you do consultant work, is it also from a distance?*

A: *Yes*

Interviewer: *Could you tell me how you collaborate with people when they are not there? Very concretely, do you call them on Skype or what do you do?*

A: *In the beginning, it was a bit of challenge, but, I mean there are so many tools now, it is not so difficult. For sure, Skype calls are important, but they also take time, so it is not the best way. It is the best way to start a project and have constant updates about the project; so generally with a client we plan something like a Skype meeting once a month to have updates and share new ideas. The communication is always difficult when you are abroad because if it is messy, all the project will be a mess, so structuring the communication is the first thing you have to do. And then you have this kind of official communication, in my opinion... they are the emails so when you sign a proposal or do something in the project, it can be a long email with all these kinds of things. And then you need to, in my opinion, you need to keep the communication as simple as possible and as targeted as possible. The ongoing communication is the chat part. Usually, we use Slack to do chat. We are in different time zones, so sometimes my colleagues are working and I am sleeping. So not to have to wake up and see and read all the messages you also need a structure to focus on some topics and we use Trello for this, so we keep these three kinds of communication. Sometimes, I also involve the client in the tools. It can be useful because they can see the process of the project; if there are some problems or the project is stuck at a certain point, they can ask questions. I feel that when you collaborate in this way transparency is very important. To be transparent, you know, if I am doing something, tell them that you are doing something like this. We have share Google drive, so usually we work inside a Google drive folder, so my colleagues also can see that I am doing this proposal for a client, for example. So they can see the things that I did. Transparency and sharing the information with your colleagues... do not keep the information only for yourself. The hard part is not the tools, because the tools are pretty simple to learn, to understand, but it is the mindset that you have to build around working and sharing your work with your colleagues. If you are structuring a document with a presentation, they need to understand what you are saying, so be as clear as possible and more focused on certain tasks. When you are abroad and when you are in different time zones, this is really important. If you are in time zones with ten hours' difference, the work can be stuck because of this difference between you and your colleagues.*

Interviewer: *What are the differences between talking face to face and working online, in a professional context?*

A: *When you are working online and your colleagues are in another part of the world. First, and that is really stimulating for me, you try to solve the problem by yourself because if your colleagues are always there, it is normal to say: I cannot do this, you know how to do this. And then you do not think about the problem but you have a colleague to come and solve the problem for you. But if you are abroad, you have to get more responsible for your work. Also, you are more independent; when you have a problem, you start to think about the problem yourself and then maybe you will ask for* *help from your colleagues. And this is very important because it is stimulating and I always believe in this as a learning process. Learning to solve a problem is the most effective way to learn how to do something. And then… when you communicate and concentrate on a project while working remotely with your colleagues I think you need to have a better focus and make everything more simple than when you work face to face with a person...*

Interviewer: *How about the…*

A: *It gives you more responsibility. I mean it makes you more... If you are in an office with other people they have their own role, and if something is going wrong it is their fault and it is not your fault. But working remotely gives you more responsibility; you know that nobody is going do this job for you now, so it is up to you.*

Interviewer: *What about the social aspect? If you share an office with your colleagues then you can… on a Friday afternoon you can go out and have beer together. What do you do with your colleagues? Is it strictly a professional relationship or do share a beer on Skype?*

A: *No, of course we are not drinking a beer together, but our goal is to meet during the year. Not many times, but when we can. We try to meet as much as possible. And it is not only a work relationship. I already told you that work is not only work, but work is also life, so you do not choose your colleagues and your business partners only because they are good at their job but also because you share a mentality, a philosophy with them. I chose my business partners for these reasons. The social part is different of course; you are not going out with them for a beer, but you can still share a lot of things with them. The building of a project is pretty stimulating stuff. I mean it is different, but you go out with different kinds of people all the time here and that is also something that I like. It keeps your mind open. But with my colleagues.. when we have the chance to have a beer together we do it, but it can be something like three to four times in a year.*

Interviewer: *So, the relationship you have with your colleagues is this blend of… as you say you do not choose your colleagues only because they have some certain professional skills… it is also because you share a mindset and things like that?*

A: *Yes.*

Interviewer: *One last thing. Technology gives us these new possibilities for working together. How do you think that technology affects the way we work together? The way we collaborate?*

A: *I always think that technology needs to be simple and user-friendly in order not to complicate the project and the process more, but your mindset is always important. So, technology cannot substitute your mindset and your will to share and collaborate. Technology can help you in different ways with this and that, but for now technology cannot substitute your brain, your attitude and you. So, the main focus when you start to work remotely or with people remotely is to train yourself to be clear and simple and to communicate a lot and to ask questions twice or three times if you do not understand… Be humble… Technology does not solve every kind of problem, just some kinds of problems.*

Interviewer: *But what does it change in comparison to your work before?*

A: *One thing that is really important, it is pretty obvious. Technology made everything faster. Now you can do a job in one hour and before it was eight or nine hours. Other things take more time because you are collaborating with people in different countries. Time is less important, results are more important. You need to achieve a result and then if you do it in one hour or twelve hours it is more up to you. So, I do not think that it makes sense, for certain kinds of jobs of course, to work like eight hours a day or stay in the office for this amount of time. It is more like OK we need to do* *this, and after that you are free to go and do yoga, massage, or to focus and find another client or something. So everything becomes more flexible and with fewer boundaries between life and work.*

Interviewer: *OK.*

A: *But the tasks… We do not think about time. I mean I have to stay in this office eight hours a day. No. Sometimes I stay in this place ten hours a day, sometimes I stay four hours a day.*

Interviewer: *Is it easy for you to navigate in? If you are the one who decides if you work ten hours or two hours, how do you manage that?*

A: *You can always estimate how much time it takes you to do a job. But something can happen that takes much more time, so you need to be flexible in general. So, your life and your time schedule need be very flexible. Because if I have a problem with a client on the other side of the world, maybe I have to stay here during the night doing a Skype or trying to solve the problem. So you need to be very flexible. But I mean this flexibility can be from the work-side; maybe sometimes you need to work more and sometimes there is more flexibility for doing something else… going to the beach or whatever. So, both ways.*

Interviewer: *Great. Thank you. Is there any more you would like to say?*

A: *No I think, I... Nothing. Thank you and have a good night.*

Interviewer: *Thank you. And have a good morning*

SOLUTIONS AND RECOMMENDATIONS

The aim of this case presentation was to provide insight into how people collaborate online in a setting that was built entirely on the possibilities of a digital world. Such insight can inspire us while developing more traditional organizations and can give us a perspective on what might be ahead. Organizations have to adapt to a changing world; this does not mean that all knowledge-based organizations will structure themselves in a way that is similar to the digital nomads. However, their existence is an example of the increasing fluidity of what it means to be an organization and how people in an organization can work together and create value. There is no reason to believe that this fluidity and complexity will decrease in the years to come, and the case shows us that creating a product or a value does not necessarily have to be the result of traditional ways of working in a traditional organization. It can be done in a radically different way.

Presenting an extensive part of the interview makes it possible for students and researchers to engage with the material as a preliminary exercise for further studies. Below is a series of considerations and focus points for engaging with the material at hand. One of the topics that the interviewee focuses on is the importance of freedom. Technology gives the interviewee and other digital nomads great freedom when it comes to deciding how, where, and when they work. However, this freedom also demands taking great responsibility for your professional work, not being able to lean on the support structure of a traditional organization, and potentially feeling lonely. The freedom described in the interview thus contains both positive and negative aspects. Many might see having the freedom to decide their own work hours as attractive, but the required discipline and responsibility might not suit everyone. This dilemma does not exist solely because of technology, but technology creates possibilities that did not exist earlier, breaking down barriers of time, space, and culture, and it leaves us with greater freedom for better and for worse. One example of this multifaceted freedom is presented in the relationship between the interviewee's work and private life and how he balances the two. He mentions that the two merge, and he views this positively because he is in charge of defining how this coexistence should unfold. However, this coexistence might not unfold as harmoniously in other cases, and when organizations introduce technology to collaborate and lead, they should be aware of effects that extend beyond the specific task at hand. When technology helps us to break down barriers, we do not only create new possibilities for working more efficiently or better but also let the different domains of our lives flow together, raising the question of how we, as organizations and individuals, handle the intertwining and far-reaching effects of what might initially seem to be a technology that affects only the professional context.

Technology gives the interviewee an opportunity to get in contact and work with numerous different people, scaling his company in order to suit the needs of the specific job. This reflects the connectivist claim that "we can no longer seek to possess all needed knowledge personally. We must store it in our friends or within technology" (Siemens, 2006, p. 51). The interviewee describes a professional life in which the specific job defines which people and what knowledge are needed. The many people passing through Hubud constitute a flow of potential business partners with different competencies and skills. When the interviewee makes a contact, it is often kept alive through social media and lies ready to be activated when needed. The technology is a premise for this way of keeping a professional network. It makes it possible for the interviewee to access updated profiles and contact relevant people in an easy and instant way. However, it is interesting to note how the interviewee mentions that contact is made not only based on the skills and knowledge of the people he meets but also depends just as much on whether he likes them and shares their mind-set. The interview thereby draws attention to the importance of the relational aspect when it comes to activating knowledge and engaging in online collaboration. It is not enough to know where potential beneficial knowledge is located; it is the desire and ability to activate it that matters. In relation to more traditional organizations, this raises the question of how this desire and ability can be nourished. Organizations facing new demands in an ever-changing market might be attracted to the possibilities that technology presents for collaborating across time, space, and culture, thus activating the right resources and their relevant combinations. However, the interview illustrates that even in environments in which technology is the primary medium for professional interaction, such as the co-working space, the relational element has a high significance for the will to engage in such contact and collaboration.

From an academic perspective, the field is difficult to fixate and create lasting knowledge about, but this should not keep us from trying. However, the task of hitting the moving target that technology constitutes calls for an approach whereby critical reflections on method, academic genres, and basic understandings of learning and collaborating are essential. Regarding the latter, the connectivist approach in this chapter seems to hold some potential as an adjunct to more traditional theoretical approaches because it includes perspectives on how a digitally connected world affects our ways of learning. The use of connectivism in this chapter is merely introductory, and further research using a connectivist approach could prove valuable. The rapidly evolving pace of technology challenges the traditional genres of presenting academic work. It can often be a long process from conducting research to presenting results. The choice of presenting a case the way in which this chapter does reflects these considerations by leaving the material open for others to work with. That said, the academic explorations of the world of online collaboration might benefit from experimenting with new ways of presenting and working with knowledge.

REFERENCES

Bauman, Z. (2000). *Liquid modernity*. Cambridge, UK: Polity Press.

Downes, S. (2005). *An introduction to connective knowledge*. Retrieved May 27, 2017, from http://www.downes.ca/post/33034

Drucker, P. F. (1992). The new society of organizations. *Harvard Business Review*. Retrieved June 1, 2017, from https://hbr.org/1992/09/the-new-society-of-organizations

Faraj, S., Jarvenpaa, S. L., & Majchrzak, A. (2011). Knowledge collaboration in online communities. *Organization Science*, *22*(5), 1224–1239. doi:10.1287/orsc.1100.0614

Flyvbjerg, B. (2015). Fem misforståelser om casearbejdet. In Kvalitative metoder (pp. 497-520). Hans Reitzels Forlag.

Haddon, L., & Brynin, M. (2005). The character of telework and the characteristics of teleworkers. *New Technology, Work and Employment*, *20*(3), 34–46. doi:10.1111/j.1468-005X.2005.00142.x

Hauge, T. E., & Norenes, S. O. (2015). Collaborative leadership development with ICT experiences from three exemplary schools. *International Journal of Leadership in Education*, *18*(3), 340–364. doi:10.1080/13603124.2014.963689

Hubud homepage. (2017). Retrieved May 30, 2017, from https://www.hubud.org/

Internet Live Stats homepage. (2017). *Data gathered by International Telecommunication Union (ITU), World Bank, and United Nations Population Division*. Retrieved May 21, 2017, from http://www.internetlivestats.com/internet-users/

Kakihara, M., & Sørensen, C. (2001). *Organisational knowledge in mobile workplaces: A theoretical investigation*. Paper presented at Mobilize! Workshop on interventions in the social, cultural and interactional analysis of mobility, ubiquity and information and communication technology, Windsor, UK.

Kvale, S. (2007). *Doing interviews*. SAGE Publications Ltd. doi:10.4135/9781849208963

Raschulte, T. (2011). The evolution of collaborative work. In Distributed team collaboration in organizations (pp. 16-27). IGI Global. doi:10.4018/978-1-60960-533-9.ch002

Siemens, G. (2005). *Connectivism: A learning theory for the digital age*. Retrieved May 22, 2017, from http://www.itdl.org/journal/jan_05/article01.htm

Siemens, G. (2006). *Knowing knowledge*. Creative commons licensed edition. Retrieved May 25, 2017, from www.knowingknowledge.com

Sullivan, C. (2012) Remote working and work-life balance. In Work and Quality of Life. International Handbooks of Quality-of-Life (pp 275-290). Springer. doi:10.1007/978-94-007-4059-4_15

Tømte, C.E. (2013). Educating teachers for the new millennium? Teacher training, ICT and digital competence. *Nordic Journal of Digital Literacy*, *8*(1-2), 74-89.

KEY TERMS AND DEFINITIONS

Connectivism: A learning theory that emphasizes the role of networks and builds on the premises of a changing and digital world.

Co-Working Space: A shared working environment where people work and collaborate without necessarily being part of the same organization.

Digital Nomads: People working through technology and from changing locations, without the need to be physically present at the place where they deliver their product or render their services.

Hubud: A co-working space in Indonesia.

Mobility: In this chapter, the concept refers to the aspect of moving across spatial, temporal, and cultural distances.

Conclusion

INTRODUCTION

The chapters in this book share a common intent: to give the reader insight into the online work environment in order to apply best practices or improve the foundation for further research. The reader has been exposed to a variety of research settings: from a digital nomad in Bali, to a game provided by the World Bank, to teachers in a physiotherapy school in Denmark, as well as to medium-sized and large global companies. The authors have presented theories of and challenges for online collaboration and communication, and they have suggested best practices for working online. The authors agree that collaborating, communicating, and leading online requires knowledge of available IT and awareness of the work environment, which differs greatly from work environments with geographic proximity. The authors show pros and cons to collaborating and leading online, and the cases uncover emergent trends of independence and flexibility. This concluding chapter reviews the findings of the contributors to summarize theories for understanding the online work environment, best practices for online collaboration, the online work environment as a space for learning, and the role of being an online leader.

BACKGROUND

The literature review revealed a gap about online collaboration and communication in contemporary organizations, and this book is an attempt to fill that gap. According to the research, online leadership means leading geographical dispersed teams that rarely or never meet face-to-face, but cooperate and collaborate online to accomplish common objectives. Distance matters, as the communication is mediated by available IT stemming from video-conferences, intranet, phone calls etc. (Kirkman & Mathieu, 2005). Hence, online leadership includes a description of online collaboration and communication.

Online collaboration and communication are obviously dependent on IT, which becomes the backbone of the online work environment. Yet, this book does not explore or recommend specific tools or applications, because of the rapid development of information and communication technology. Instead, this book provides new theories for understanding the online work environment, explores best practices for online collaboration, communication, and leadership. The book describes the online work environment as a space of learning by presenting insightful cases that include various aspects of leading and collaborating online.

Theories for Understanding the Online Work Environment

The literature review uncovered several challenges of collaborating online across multiple boundaries (Brannen & Salk, 2000), including the need for a common language and shared technology, besides the necessity of cross-cultural competencies and strong communication skills (Wildman & Griffith, 2015). Jensen (2017) suggests an interdisciplinary approach to enhance *cultural awareness*, namely a postcultural prism that offers tools for reflection of online interactions in everyday situations. Online collaboration and leadership means lack of the same physical space and synchronous chronological time, which generates the ability to be present in more than one place at the same time. Nortvig (2017) presents the new term "multi-presence" to describe this particular phenomenon. The literature review revealed that lack of body language clues, along with high diversity, influence the ability to build *strong relations* between colleagues who never meet (Boh, Ren, Kiesler, & Bussjaeger, 2007; Delbridge & Sallaz, 2015). Sklaveniti (2017) builds upon Gergen's (2009) work when she describes the importance of "doing" relational practices to make technology collaborative.

Best Practices in Online Collaboration

According to the literature review, online collaboration and leadership require *flexibility* with regard to working hours and styles of communication and collaboration. Nielsen (2017) presents an interview with a digital nomad, who was interviewed during his stay at the Hubud co-working space in Bali. This digital nomad extends flexibility in both time and space, which supports Sklaveniti's (2017) conclusion about future generations: born and raised with IT, they may consider online presence differently than current generations do. Nearly all texts in the literature review point out that the ability to build *trust* is a challenge in the online work environment (Pinjani & Palvia, 2013; Richardson et al., 2016; Wildman & Griffith, 2015). Nielsen's interview (2017) also sheds light on the challenge of building trust, and Dalkir (2017) explains the necessity of a common goal, trust, and common working practices for online collaboration between groups that have never before worked together online. This is followed up by Cline (2017), who suggests using the Aristotelian concept of ethos to understand and build trust when working online. As online cooperation and collaboration have a tendency to neglect *informal communication*, practices for informal interactions are important to incorporate into online teamwork (Beugelsdijk & Mudambi, 2013; Korzynski, 2013; Richardson et al., 2016; Wildman & Griffith, 2015). Schramm (2017) suggests virtual coffee breaks to foster social relations and trust among online team members, and Pries-Heje and Pries-Heje (2017) incorporate six challenges in the framework for building social capital among online teams. This framework includes recommendations for supporting social ties: strong relations, feedback practices, building trust, establishing a mindset for working online, cultural awareness, and common language and concepts.

The Online Collaboration as a Space for Learning

The literature review found that the lack of physical clues often made it hard for members of online teams *to share and create knowledge* with colleagues far away (Pinjani & Palvia, 2013; Richardson et al., 2016; Wildman & Griffith, 2015). This challenge calls for methods to investigate learning in the context of work, and Kolbaek (2017a) suggests a revised format of design-based research as a suitable methodology for this purpose.

People cooperate and collaborate online to accomplish common objectives. This book shows that objectives may be defined by the workplace or as part of a computer game. In both cases, people involve themselves to accomplish the objectives, and they depend upon input and support from others for success. Francisco, Klein, Engeström and Sannino (2017) provide a six-step model for expansive learning based on a case study of mobile workers in an IT company established in southern Brazil. Wichmann (2017) describes how complete strangers share and create knowledge for the purpose of solving wicked problems. She emphasizes the importance of a caring atmosphere, and builds upon Nonaka's (1998) work on Ba to explain important behaviors for creating Ba. The literature review showed that online leaders require the same skills as traditional leaders who meet their teams face-to-face; additionally, online leaders must have the ability to lead online—whatever that takes (Wildman & Griffith, 2015). Kolbaek (2017c) presents a model that describes how online leaders may learn from work experience along with their teams. The model shows that learning may be initiated from either a top-down or a bottom-up approach, and that knowledge is created collaboratively among the online team members and their leader.

The Role of Online Leaders

Online leadership largely develops through practice, and online leaders must be flexible and adaptable to changes in technology and unforeseen events (Richardson, 2016); online leaders must adapt their *leadership styles* to fit the online work environment. Cheuk and McKenzie (2017) present the idea that leading face-to-face is based predominantly on command, control and competition, but that leading online depends on coordination, collaboration, and coaching. Schramm claims that the "space for interpretation" is bigger in an online work environment than it is when people work in a face-to-face setting, and he provides the "circle of trust," which online leaders can use to understand behaviors that support trust-building. But leadership style also includes power, and Fogsgaard, Elmholdt and Lindekilde (2017) discuss the phenomenon of power, then present a model for understanding and analyzing power in the online work environment. The role of online leaders includes considerations about their contributions to the organization, and Pries-Heje and Pries-Heje (2017) describe how online leaders are able to enhance social capital, whereas Kolbaek (2017b) explains that online leaders actually add value in three dimensions by enhancing cultural, social, and economic capital.

OVERCOMING CHALLENGES IN ONLINE COLLABORATION AND COMMUNICATION

This section describes understandings and recommendations to solve the main challenges in online collaboration and leadership a literature review revealed. The challenges mentioned in the introduction were: cultural awareness, lack of strong relations, flexibility, lack of trust, the need for informal communication, and different online leadership styles.

Cultural Awareness

The literature review and many of the authors address that working in geographically-dispersed teams requires cultural awareness, and Jensen (2017) problematizes postcolonial or restricted views on the term "culture." Instead she provides the suggestion of perceiving cultures as practices, i.e., how culture is "done," which enables readers of this book to see differences as well as similarities in practices that develop all the time. This approach to culture helps avoid perceiving colleagues as stereotypes. Dalkir (2017) raised awareness of organizational culture that influences how people interact, communicate and work in their organization—in other words, how they "do" culture. This approach is backed up by players in the game Urgent Evoke, who collaborate online with no chance of meeting face-to-face ever, and the individual players are challenged by ideas of others living very different lives with very different experiences (Wichmann 2017). Schramm (2017) and Pries-Heje and Pries-Heje (2017) present a very different view of culture based on Hofstede (2001), who claims that cultural differences are based on nationalities and cause different perceptions of risk, individuality, uncertainty, and power distance.

Strong Relations

Collaborating and communicating online seems to weaken the ability to build up strong relations, even though positive relationships among team members are important for success (Sklaveniti, 2017; Ahuja & Galvin, 2003. Current familiar ways of working will sometimes contradict requirements of online collaboration and communication (Cheuk & McKenzie, 2017). Thus, Schramm (2017) explains that the "space for interpretation" is bigger when people rarely see each other as part of virtual teams, and Wichmann (2017) emphasizes the importance of Ba, which enables collaborators to form relations by sharing knowledge, experiences, values, and ideas, and creating new knowledge together. As physical cues do not have the same significance or meaning in different cultures, asynchronous communication may actually enable, rather than hinder, dispersed collaboration, according to Sklaveniti, 2017. Nortvig (2017) discusses "place" and "space" and claims that active participation and social interactions in collaborative activities may transform online spaces into meaningful places. She stresses that it is difficult to design "placeness," because the precondition is taking part in collaborative interactions. These may include individual stories about processes that are shared and transformed into perceptions of successful project completion, according to Pries-Heje and Pries-Heje (2017). But if the collaborators hesitate to communicate, they are left to work alone lacking support, missing opportunities for building strong relationships, and probably feeling "invisible" in the organization, says Francicso et al. (2017).

Kolbaek (2017b) claims that ICT enables online leaders to build coherence and strong relationships with team members, given the communication includes emotional cues to create proximity, minimize awkward conversations, and share states of mind. Even though positional power-over is still common in the leader-employee relationship, it seems as if physical distance makes directive management techniques less effective, according to Fogsgaard et al. (2017). Instead, the leadership becomes more "equal," as equal relationships create good relations (Jensen, 2017; Nielsen, 2017). Self-employed leaders or consultants may decide "to work with some people because they are on the same page as you," as the digital nomad expressed himself in Nielsen (2017). In any case, it takes time to build up strong relations online.

Flexibility

Working online offers flexibility in regard to time and geographical space for work, and requires online tools for collaboration and communication. The digital nomad expressed enthusiasm for this type of work by saying it gives the freedom to "work where you want and work when you want" (Nielsen, 2017). This idea is supported by Francisco et al. (2017) in their report from Brazil that mobile workers perform their work practices anytime, anywhere. The downside may be a feeling of isolation and less ability to share knowledge or create solutions collaboratively. Nortvig (2017) points out the importance of professional reflection in action, which is only achievable online when collaborators have the ability to be multi-present, i.e., to be present in more than one space at the same time.

Trust

It is well-known that building and maintaining trust may be a challenge in online communication and collaboration; it emerged in the literature review, and several chapters in this book deal with the challenges of creating trust online. Trust can be described as "a willingness to risk being exposed to another person whose behavior is beyond control" (Butler, 1991, p. 650), which leads to the insight that trust is not only an individual attribution, but rather a relational characteristic (Gergen, 2009; Sklaveniti, 2017) that may be built up through social interactions such as communication (Pries-Heje & Pries-Heje, 2017). According to Cline (2017), this very communication must convince interacting people that the other person(s) knows what they are talking about (phronesis) and that they possess a "good moral character" (arete) (Cline, 2017). This opens up two forms of trust: role-based trust (e.g., credibility, reputation, and role) and personal trust (e.g., benevolence, altruism, and reciprocation) (Dalkir, 2017). "The circle of trust" describes nine leadership behaviors, which support the creation and maintenance of trust in online teams (Schramm, 2017). It is recommended that online leaders employ these leadership behaviors from the very beginning. According to Cheuk and McKenzie (2017), trust is fragile if controversy arises before a leader has established an online presence that manifests a pattern of open, authentic, and balanced behavior (Cheuk & McKenzie, 2017). Nielsen (2017) presents an alternative approach. He describes that the digital nomad often meets his collaborators in person before they start working together. This is not the case for the online leaders Kolbaek (2017b) presents. They rarely or never meet their team members face-to-face, and one of the top managers claimed that, by treating his team members as the intelligent human beings they are, his team was more autonomous and better able to make decisions on its own. Yet another top manager gave up micro-managing and also experienced very good results. These top managers built up trust with their team members, and this trust enabled the teams to efficiently achieve objectives.

Informal Communication

When working face-to-face, informal communication is important in building up trust. When people collaborate and communicate online, informal communication has a tendency to disappear, as online communication drifts naturally into more formal communication styles (Bradley & Vozikis, 2004; Chaulat, 2012). Francisco et al. (2017) highlight that lack of informal communication may negatively influence knowledge creation and sharing. Dalkir (2017) claims that formal roles seem less important in informal conversations than personal relationships. But what comes first—the relationship or the conversation?

Cheuk and McKenzie (2017) raise awareness of the importance of understanding the virtual environment in enabling informal communication, and Kolbaek (2017c) points to the lack of oral communication in online teams as an important reason for ineffective informal communication. When colleagues meet face-to-face, somebody may be silent, and this silence sometimes makes the colleague ignored. In online communication it is obvious when somebody is absent during a conversation, which may prompt colleagues to invite more silent collaborators into the conversations, formal or informal. If the collaborators "forget about" informal communication, a negative spiral may start, in which they do not build up personal relationships, trust, habits of support, or knowledge sharing and creation.

Share and Create Knowledge

When people collaborate and communicate online, they must develop applicable practices "where knowing how to go on and how to innovate the way of doing are produced, diffused, and kept" (Gherardi, 2015, p. 14). When studying knowledge sharing and creation in the context of work, Engeström (2001) advises that four questions be answered:

- What are employees learning?
- Why are they learning?
- Where are they learning?
- How are they learning?

What they are learning is dependent on context; however, online team members share knowledge from their own stances (Sklaveniti, 2017). Wichmann (2017) presents two different kinds of knowledge—tacit and explicit—which, according to Nonaka and Takeuchi (1998), may be described as follows:

- *Explicit knowledge* is knowledge people are aware they have, and it can be transmitted via words, numbers, pictures, etc.
- *Tacit knowledge* is internalized to an extent that makes it difficult to communicate, and it includes intuitions and hunches.

Cheuk and McKenzie (2017) explain that online leaders may learn from personalized support and coaching that represents both explicit and tacit knowledge, and Nielsen (2017) refers to explicit knowledge when he points to Facebook and LinkedIn as repositories for knowledge sharing.

Why they are learning is basically to participate in "culturally valued collaborative practices in which something useful is produced" (Engeström 2001, p. 141). Francisco et al. (2017) exemplifies the "why" with a group of mobile workers who faced challenges, such as multiple tasks, lack of standards, and difficulty meeting clients' needs. They realized why they needed to learn, defined what to learn, and created and implemented "the responsibility matrix," which allowed them to expand their view of the problematic situation and gain a holistic grasp of it.

When *where* is online, collaborators seem to have fewer opportunities for informal learning situations with peers (Francisco et al. 2017). Besides, the technology influences *how* people are learning. Thus, Fogsgaard et al. (2017) present three major barriers to knowledge sharing in online teams. First is language, second is differences in meaning, and third is differences in practice. Dalkir (2017) provides an example of differences in practices, as she describes how two different groups were supposed to share knowledge in a shared repository but instead set up separate repositories and restricted the other group's access. Additionally, she provides an example of differences in meaning, as one group was concerned about long-term assets such as peer-reviewed publication, and the other group was concerned about protecting new knowledge and innovations, typically through patents.

Kolbaek (2017c) suggests a learning spiral to describe the "what" and "how" behind the adevelopment of knowledge. She describes how online leaders may learn from direct reports of online proactive reviews, in which team members collaboratively reflect upon and answer seven open questions. Kolbaek (2017c) claims that proactive reviews include a bottom-up and a top-down approach to learning.

Online Leadership Style

Leadership may be described as "the ability to create direction, alignment and commitment across group boundaries in service of a higher vision and goal" (Ernst & Chrobot-Mason, 2011, p. 2). According to Fogsgaard et al. (2017), organizational leadership always includes power divided into three perspectives: formal power given by the formal position in the organization, personal power given by individual resources and properties, and discursive power that flourishes in informal processes. Fogsgaard et al. (2017) also found that online leaders in contemporary organizations should make everyone stronger in order to succeed. Cheuk and McKenzie (2017) support this point, as they state that online leaders move from a predominant command, control, and competition mindset to one in which they coordinate, collaborate, and involve direct reports in decision-making and strategy development. This change occurs because the online work environment undermines the hierarchical nature of followership and allows distributed individuals and groups to gain leadership influence. This point of view is backed up by Linda (Kolbaek, 2017b), who handed over some responsibility to members of her online team. Also, Francisco et al. (2017) found that restrictive control and supervision were weaker leadership styles compared with engagement, trust, commitment, and collaboration. These statements result in the conclusion that online leadership is both a paradox and an opportunity to open up the perception of leadership, as Sklaveniti (2017) puts it.

Practical approaches to online leadership styles include the advisable leadership behaviors in the "circle of trust" (Schramm, 2017), along with recommendations of creating a shared vision and shared habits of regular online meetings where the participants speak the same language and are able to see each other (Pries-Heje & Pries-Heje, 2017). Leading online may be perceived as an advantage, as the online team members must work more independently, according to Kolbaek (2017b), and the IT-mediated communication forces online leaders to set clear and understandable objectives and verbalize emotions to build relationships and trust with the team members.

FUTURE RESEARCH DIRECTIONS

Online collaboration, communication, and leadership will probably remain and develop over the next decades, which open up interesting research possibilities for the future. One well-known challenge that has been left out by the authors of this book, and which future research will hopefully explore, is feedback in the online work environment. This book has presented various methodologies and methods for studying online collaboration and communication, but this is still an area that needs attention.

Feedback

The literature review revealed the importance of feedback in online collaboration and leadership, but this book has a dearth of research about online feedback. Pries-Heje and Pries-Heje (2017) suggest celebration of successes, which is in itself a way of giving and receiving feedback, and Kolbaek (2017b) suggests that online teams give feedback and provide ideas to their online leaders. Wichmann describes how players in Urgent Evoke often provided feedback because they wanted to contribute to the community. Hence, future research could focus on giving and receiving feedback in the online work environment, where there are already challenges in creating trust and strong relationships. Maybe feedback could support the enhancement of both.

Methodologies for Studying Online Collaboration and Leadership

This book presents various methodologies for investigating online collaboration and leadership, including case studies (Flyvbjerg, 2006), action research (Baskerville & Myers, 2004), design-based research (Anderson & Shattock, 2012), design science research (Walls, Widmeyer & El Sawy, 1992), and the vignette method (Langer, 2016; Jenkins et al., 2010).

Many authors suggested case study as a methodology for their research because the aim of the case study is to produce a rich, context-dependent description of the situation that provides a first-hand understanding of how online collaboration and communication develop in an organizational context (Flyvbjerg, 2006). In this book, examples of case studies may be found in Wichmand's (2017) investigation of the game Urgent Evoke, Nortvig's (2017) role as participant observer in the e-learning sessions for student physiotherapists, Cline's (2017) exploration of ethos in the IT application Slack, Nielsen's (2017) study of a digital hub at Bali, Kolbaek's (2017b) research of Oracle as an example of online leaders' abilities to develop three types of capital, and the observations of Francisco et al. (2017) of mobile workers in an IT company in southern Brazil. Wichmand, Nortvig, Cline, Nielsen, and Kolbaek mainly gathered data online, whereas Francisco et al. observed by physical presence.

Baskerville and Myers (2004) describe action research this way: "Action research simultaneously assists in practical problem-solving and expands scientific knowledge, as well as enhances the competencies of the respective actors" (p. 247). This means that people under study actively participate in research decisions and apply research results as they become available (Kimiz, 2017).

Fogsgaard et al. (2017) conducted an action research from January 2012 to October 2013 in a large multinational Danish engineering manufacturer (DEM) with 17,000 employees. Mainly, the data was gathered in close physical proximity to the people under study, even though the aim was to develop skills and practices in online leadership.

Kimitz (2017) conducted action research on a project aimed at sharing expensive research equipment across different universities and companies. Kimitz (2017) explains that the participants held regular meetings face-to-face and created a shared repository accessible to all participants. Interviews were conducted face-to-face, and online data were scarce. These two examples of action research included limited amounts of online data.

Design-based research (DBR) was originally developed to study learning in the classroom (Anderson & Shattock, 2012). Kolbaek (2017a) describes and recommends DBR as a methodology for studying learning in the context of work, which is a new innovation. Kolbaek (2017a) suggests DBR whether or not the work environment is online. DBR at work is based on iterations of interventions in educational designs that include possibilities for exploring the context defined by physical surroundings, historical background, and sociological aspects. The chapter provides guidelines for DBR at work, including online work environments.

Design science research handles the design product as well as the design process (Walls et al. 1992), and the research may look into four elements: constructs, models, methods, and instantiations (March & Smith, 1995). Pries-Heje and Pries-Heje (2017) turned to design science research, which they utilized for studying concrete process improvement of online collaboration between Danske Bank teams in Denmark and India. The data collection consisted of literature review and on-site interviews; no online methods were utilized.

One chapter of this book (Jensen, 2017) presents two vignettes that work as text material for a post-cultural analysis (Langer, 2016; Jenkins et al., 2010). The first vignette is a single utterance condensed from an interview, whereas the second vignette is a broader description of Scandinavians in a cross-cultural relation (Jensen 2017). The vignettes mirror life experiences. Jensen (2017) gathered data for the first vignette from attending a course on "Intercultural Awareness", in which the participants met face-to- face. She gathered data for the second vignette from the article, "Globalization Denmark: Open-minded while ethnocentric" by Selmer and Lauring (2012). Hence, the first vignette is the result of a face-to-face interview, and the second vignette is the product of data gathered online.

The book presents a variety of research methodologies for studying online collaboration and communication in contemporary organizations, and it presents methods such as interviews and observations conducted in geographical proximity. Future research may explore and utilize online methods to a much greater extent.

CONCLUSION

There is still a lot to be done in the area of online collaboration and communication in contemporary organizations. We have seen theories for understanding the online work environment, and we have learned about best practices for online collaboration. We now better understand how online collaboration may be seen as a space for learning, and we have gained insight in the complex role of the online leader.

Recommended future research directions begin with continued exploration of challenges in online collaboration and leadership that embrace reflections of our cultural understanding and awareness. We should also deepen insights into practices for building strong relationships online, including how to start up and maintain informal communication online, because it seems to be the backbone of building trust between people that rarely or never meet. Another possible area of future research consists of envisioning more methods of sharing and creating knowledge when working online. Flexibility—what it really means to work independent of time and space—is another area to explore. Feedback is as important in online collaboration as when people meet face-to-face, and future research should include the topic of feedback in the online work environment.

The role of online leadership remains an important dimension, and research can shed light on various online leadership styles that could establish and continue beneficial online collaboration, while taking advantage of the flexibility the online work environment provides.

The chapters have revealed several methodologies for studying online collaboration, communication, and leadership, but only a few authors have utilized online data gathering. Future research directions could be the ongoing development of methods that may craft, review, and refine data available on the Internet or the intranet of researched organizations, or by employing traditional methods such as interview—although online.

ACKNOWLEDGMENT

The editor would like to acknowledge the authors and their collaborative organizations and sources. Without their contributions, this book would not have become a reality. This project started just before the beginning of 2017, and some authors have faced difficulties with tough deadlines. But through personal challenges, the authors delivered insightful and interesting studies and very useful results. Thank you!

REFERENCES

Adler, P. S., & Kwon, S. W. (2002). Social Capital: Prospects for a new concept. *Academy of Management Review*, *27*(1), 17–40.

Ahuja, M. K., & Galvin, J. E. (2003). Socialization in virtual groups. *Journal of Management*, *29*(2), 161–185. doi:10.1177/014920630302900203

Anderson, T., & Shattuck, J. (2012). Design-based research: A decade of progress in education research? *Educational Researcher*, *41*(1), 16–25. doi:10.3102/0013189X11428813

Baskerville, R., & Myers, M. D. (2004). Special issue on action research in information systems: Making IS research relevant to practice: Foreword. *Management Information Systems Quarterly*, *28*(3), 329–335. doi:10.2307/25148642

Beugelsdijk, S., & Mudambi, R. (2013). MNEs as border-crossing multi-location enterprises: The role of discontinuities in geographic space. *Journal of International Business Studies*, *44*(5), 413–426. doi:10.1057/jibs.2013.23

Boh, W. F., Ren, Y., Kiesler, S., & Bussjaeger, R. (2007). Expertise and collaboration in the geographically dispersed organization. *Organization Science*, *18*(4), 595–612. doi:10.1287/orsc.1070.0263

Bradley, E. W., & Vozikis, G. S. (2004). Trust in Virtual Teams. In Virtual and Collaborative Teams: Process, Technologies and Practice (pp. 99–113). Idea Group Publishing. doi:10.4018/978-1-59140-204-6.ch006

Brannen, M. Y., & Salk, J. E. (2000). Partnering across borders: Negotiating organizational culture in a German-Japanese joint venture. *Human Relations*, *53*(4), 451–487. doi:10.1177/0018726700534001

Buber, M. (1984). *Das Dialogische Prinzip* (5th ed.). Heidelberg, Germany: Lambert Schneider.

Butler, J. K. Jr. (1991). Toward understanding and measuring conditions of *trust*: Evolution of a conditions of *trust* inventory. *Journal of Management*, *17*(3), 643–663. doi:10.1177/014920639101700307

Chaulat, G. (2012). *Virtual leadership: Learning to lead differently*. Libri Publishing.

Cheuk, B., & McKenzie, J. (2017). Developing the practice of online leadership. In D. Kolbaek (Ed.), *Online Collaboration and Communication in Contemporary Organizations*. Hershey, PA: IGI Global.

Cline, B. J. (2017). Using the Classical Rhetorical Concept of "Ethos" to Solve Online Collaboration Problems of Trust and Presence. In D. Kolbaek (Ed.), *Online Collaboration and Communication in Contemporary Organizations*. Hershey, PA: IGI Global.

Cramton, C. D. (2001). The mutual knowledge problem and its consequences for dispersed collaboration. *Organization Science*, *12*(3), 346–371. doi:10.1287/orsc.12.3.346.10098

Cummings, J. N. (2004). Work groups, structural diversity, and knowledge sharing in a global organization. *Management Science*, *50*(3), 352–364. doi:10.1287/mnsc.1030.0134

Dalkir, K. (2017). Challenges in Online Collaboration. In D. Kolbaek (Ed.), *Online Collaboration and Communication in Contemporary Organizations*. Hershey, PA: IGI Global.

Delbridge, R., & Sallaz, J. J. (2015). Work: Four worlds and ways of seeing. *Organization Studies, 36*(11), 1449–1462. doi:10.1177/0170840615612021

Engeström, Y. (2001). Expansive learning at work: Toward an activity theoretical reconceptualization. *Journal of Education and Work, 14*(1), 133–156. doi:10.1080/13639080020028747

Ernst, C., & Chrobot-Mason, D. (2010). *Boundary Spanning Leadership: Six Practices for Solving Problems, Driving Innovation, and Transforming Organizations*. New York: McGraw-Hill.

Faraj, S., Jarvenpaa, S. L., & Majchrzak, A. (2011). Knowledge collaboration in online communities. *Organization Science, 22*(5), 1224–1239. doi:10.1287/orsc.1100.0614

Flyvbjerg, B. (2006). Five misunderstandings about case-study research. *Qualitative Inquiry, 12*(2), 219–245. doi:10.1177/1077800405284363

Fogsgaard, M., Elmholdt, C., & Lindekilde, R. (2017). Power in online leadership. In D. Kolbaek (Ed.), *Online Collaboration and Communication in Contemporary Organizations*. Hershey, PA: IGI Global.

Francisco, R., Klein, A. Z., Engeström, Y., & Sannino, A. (2017). Knowledge on the Move. In D. Kolbaek (Ed.), *Online Collaboration and Communication in Contemporary Organizations*. Hershey, PA: IGI Global.

Gergen, K. J. (2009). *Relational Being: Beyond Self and Community*. Oxford, UK: Oxford University Press.

Gherardi, S. (2015). How Turn to Practice may contribute to Working Life Studies. *Nordic Journal of Working Life Studies, 5*(3a), 13. doi:10.19154/njwls.v5i3a.4831

Haas, M. R. (2006). Acquiring and applying knowledge in transnational teams: The roles of cosmopolitans and locals. *Organization Science, 17*(3), 367–384. doi:10.1287/orsc.1060.0187

Hofstede, G. (2001). *Culture's consequences: Comparing Values, behaviors, Institutions, and Organizations Across Nations* (2nd ed.). Thousand Oaks, CA: Sage Publications.

Jenkins, H., Ford, S., & Green, J. (2013). *Spreadable media: Creating value and meaning in a networked culture*. New York: NYU Press.

Jensen, I. (2017). Online Leadership and Communication across Cultures. In D. Kolbaek (Ed.), *Online Collaboration and Communication in Contemporary Organizations*. Hershey, PA: IGI Global.

Kirkman, B. L., & Mathieu, J. E. (2005). The dimensions and antecedents of team virtuality. *Journal of Management, 31*(5), 700–718. doi:10.1177/0149206305279113

Kolbaek, D. (2017a). Design-based research as a methodology for studying learning in the context of work. In D. Kolbaek (Ed.), *Online Collaboration and Communication in Contemporary Organizations*. Hershey, PA: IGI Global.

Kolbaek, D. (2017b). Online Leaders Increase Three Types of Capital. In D. Kolbaek (Ed.), *Online Collaboration and Communication in Contemporary Organizations*. Hershey, PA: IGI Global.

Kolbaek, D. (2017c). Online Leadership and Learning. In D. Kolbaek (Ed.), *Online Collaboration and Communication in Contemporary Organizations*. Hershey, PA: IGI Global.

Korzynski, P. (2013). Online social networks and leadership: Implications of a new online working environment for leadership. *International Journal of Manpower*, *34*(8), 975–994. doi:10.1108/IJM-07-2013-0173

Langer, P. C. (2016). The research vignette: Reflexive writing as interpretative representation of qualitative inquiry—a methodological proposition. *Qualitative Inquiry*, *22*(9), 735–744. qix.sagepub.com. doi:10.1177/1077800416658066

Lyons, A., Reysen, S., & Pierce, L. (2012). Video lecture format, student technological efficacy, and social presence in online courses. *Computers in Human Behavior*, *28*(1), 181–186. doi:10.1016/j.chb.2011.08.025

March, S. T., & Smith, G. F. (1995). Design and natural science research on information technology. *Decision Support Systems*, *15*(4), 251–266. doi:10.1016/0167-9236(94)00041-2

Nonaka, I., & Konno, N. (1998). The concept of "ba": Building a foundation for knowledge creation. *California Management Review*, *40*(3), 40–54. doi:10.2307/41165942

Nonaka, I., & Takeuchi, H. (1995). *The Knowledge-Creating Company: How Japanese Companies Create the Dynamics of Innovation*. New York: Oxford University Press.

Nortvig, A.-M. (2017). Technology and Presence. In D. Kolbaek (Ed.), *Online Collaboration and Communication in Contemporary Organizations*. Hershey, PA: IGI Global.

Pinjani, P., & Palvia, P. (2013). Trust and knowledge sharing in diverse global virtual teams. *Information & Management*, *50*(4), 144–153. doi:10.1016/j.im.2012.10.002

Pries-Heje, J., & Pries-Heje, L. (2017). Improving online collaboration in contemporary IT development teams. In D. Kolbaek (Ed.), *Online Collaboration and Communication in Contemporary Organizations*. Hershey, PA: IGI Global.

Richardson, J. W., Beck, D., LaFrance, J., & McLeod, S. (2016). Job attainment and perceived role differences of cyberschool leaders. *Journal of Educational Technology & Society*, *19*(1), 211–222.

Schramm, M. (2017). The virtual coffee break. In D. Kolbaek (Ed.), *Online Collaboration and Communication in Contemporary Organizations*. Hershey, PA: IGI Global.

Selmer, J., & Lauring, J. (2013). Globalizing Denmark: Open-minded while ethnocentric. *International Studies of Management & Organization*, *43*(1), 9–25. doi:10.2753/IMO0020-8825430101

Sklaveniti, C. (2017). Theorizing virtual teams. In D. Kolbaek (Ed.), *Online Collaboration and Communication in Contemporary Organizations*. Hershey, PA: IGI Global.

Walls, J. G., Widmeyer, G. R., & El Sawy, O. A. (1992). Building an information system design theory for vigilant EIS. *Information Systems Research*, *3*(1), 36–59. doi:10.1287/isre.3.1.36

Wichmand, M. (2017). Urgent Evoke a digital ba? How a game can make space for knowledge creation. In D. Kolbaek (Ed.), *Online Collaboration and Communication in Contemporary Organizations*. Hershey, PA: IGI Global.

Wildman, J. L., & Griffith, R. L. (Eds.). (2015). *Leading Global Teams: Translating Multidisciplinary Science to Practice*. SpringerLink. Retrieved from http://dl.pgu.ac.ir/handle/Hannan/8470

KEY TERMS AND DEFINITIONS

Methodology: An approach to research that includes the theoretical approach or paradigm for the selection of relevant methods to be applied to the study.

Online Collaboration: A process in which participants are dependent on each other in order to create something together they would not be able to create one by one, and the participants communicate only online.

Online Leadership: A leadership that is mediated by IT because the leader rarely or never meets the online team members.

Online Methods: The specific way to gather data online (e.g., e-mail interview, observations of online interactions, or website analysis).

Online Work Environment: A virtual workplace where employees and leaders rarely or never meet face-to-face but utilize information technology (IT) for collaboration and communication.

Compilation of References

Abrams, L., Cross, R., Lesser, E., & Levin, D. (2003). Nurturing interpersonal trust in knowledge-sharing networks. *Academy of Management Review*, *17*(4), 64–78.

Adler, P. S., & Kwon, S. (2002). Social Capital: Prospects for a new concept. *Academy of Management Review*, *27*(1), 17–40.

Adler, R. B., & Proctor, R. F. (2011). *Looking out, Looking* (13th ed.). Boston: Wadsworth Cengage Learning.

Ahuja, M. K., & Galvin, J. E. (2003). Socialization in virtual groups. *Journal of Management*, *29*(2), 161–185. doi:10.1177/014920630302900203

Akkerman, S. F., & Bakker, A. (2011). Boundary crossing and boundary objects. *Review of Educational Research*, *81*(2), 132–169. doi:10.3102/0034654311404435

Al-Ani, B., Horspool, A., & Bligh, M. C. (2011). Collaborating with 'virtual strangers': Towards developing a framework for leadership in distributed teams. *Leadership*, *7*(3), 219–249. doi:10.1177/1742715011407382

Allan, J., & Lawless, N. (2005). Learning through online collaboration by SME staff: A scoping investigation into likely team-role stressors. *Education + Training*, *47*(8/9), 653–664. doi:10.1108/00400910510633189

Allee, V. (2000). Reconfiguring the value network. *The Journal of Business Strategy*, *21*(4), 36–40. doi:10.1108/eb040103

Allen, D. K., Brown, A., Karanasios, S., & Norman, A. (2013). How Should Technology-Mediated Organizational Change Be Explained? a Comparison of the Contributions of Critical Realism and Activity Theory. *Management Information Systems Quarterly*, *37*(3), 835–854. doi:10.25300/MISQ/2013/37.3.08

Altrichter, H., Kemmis, S., McTaggart, R., & Zuber-Skerritt, O. (2002). The concept of action research. *The Learning Organization*, *9*(3), 125–131. doi:10.1108/09696470210428840

Amadeo, R. (2017, March 9). *Google is Turning Hangouts Into a Slack Competitor.* Retrieved from Ars Technica: https://arstechnica.com/gadgets/2017/03/google-is-turning-hangouts-into-a-slack-competitor/

Andersen, C., & Nissley, C. (2007). *The long tail: How endless choice is creating unlimited demand.* New York, NY: Random House Business Books.

Andersen, J. F. (1979). Educator immediacy as a predictor of educator effectiveness. New Brunswick, NJ: Transaction Books. doi:10.1080/23808985.1979.11923782

Andersen, M. F., & Prætorius, N. (2016). Går skønne kræfter til spilde? In *Tæller vi det der tæller? Målstyring og standardisering i arbejdslivet.* Aarhus: Klim.

Anderson, T., & Shattuck, J. (2012). Design-based research: A decade of progress in education research? *Educational Researcher*, *41*(1), 16–25. doi:10.3102/0013189X11428813

Andriessen, D. (2007). Combining design-based research and action research to test management solutions. Paper presented at the 7th World Congress on Action Learning, Action Research and Process Management, Groningen.

Angrosino, M., & de Pérez, K. (2001). Rethinking Observation: From Method to Context. In The American Tradition in Qualitative Research (pp. 673–702). SAGE Publications Ltd. doi:<ALIGNMENT.qj></ALIGNMENT>10.4135/9781446263570

Arasaratnam, L. A. (2015). Research in intercultural communication: Reviewing the past decade. *Journal of International and Intercultural Communication*, *8*(4), 290–310. doi:10.1080/17513057.2015.1087096

Arvid, K. (2013). The Social Regulation of Emotion. Messy Layers. *Frontiers in Psychology*, *4*, 51. Retrieved from http://journal.frontiersin.org/article/10.3389/fpsyg.2013.00051 PMID:23424049

Augustine, A. (2017, May 10). *A review of the Slack Communication Service: Slack lets you do without email.* Retrieved from Lifewire: https://www.lifewire.com/slack-sets-standard-team-communication-online-771603

Avolio, B. J., & Kahai, S. S. (2003). Adding the "e" to e-leadership. *Organizational Dynamics*, *31*(4), 325–338. doi:10.1016/S0090-2616(02)00133-X

Avolio, B. J., Kahai, S., & Dodge, G. E. (2000). E-leadership: Implications for theory, research, and practice. *The Leadership Quarterly*, *11*(4), 615–668. doi:10.1016/S1048-9843(00)00062-X

Avolio, B. J., Sosik, J. J., Kahai, S. S., & Baker, B. (2014). E-leadership: Re-examining transformations in leadership source and transmission. *The Leadership Quarterly*, *25*(1), 105–131. doi:10.1016/j.leaqua.2013.11.003

Avolio, B. J., Walumbwa, F. O., & Weber, T. J. (2009). Leadership: Current theories, research and future directions. *Annual Review of Psychology*, *60*(1), 421–499. doi:10.1146/annurev.psych.60.110707.163621 PMID:18651820

Baaz, M. E. (2002). *The White Wo/Man's Burden in the Age of Partnership – A Postcolonial Reading of Identity in Development Aid.* Göteborg: Department of Peace and Development Research, Göteborg University.

Baba, M. L., Gluesing, J., Ratner, H., & Wagner, K. H. (2004). The contexts of knowing: Natural history of a globally distributed team. *Journal of Organizational Behavior*, *25*(5), 547–587. doi:10.1002/job.259

Baden-Fuller, C., & Haefliger, S. (2013). Business models and technological innovation. *Long Range Planning*, *46*(6), 419–426. doi:10.1016/j.lrp.2013.08.023

Bailey, C., & Madden, A. (2015). Time reclaimed: Temporality and the experience of meaningful work. *Work, Employment and Society*, *31*(1), 3–18. doi:10.1177/0950017015604100

Bailey, D. E., Leonardi, P. M., & Barley, S. R. (2012). The lure of the virtual. *Organization Science*, *23*(5), 1485–1504. doi:10.1287/orsc.1110.0703

Bailey, D., & Koney, K. (2000). *Strategic alliances among health and human services organizations: From affiliations to consolidations.* Thousand Oaks, CA: Sage. doi:10.4135/9781483328546

Baker, L. (2006). Observation: A Complex Research Method. *Library Trends*, *55*(1), 171–189. doi:10.1353/lib.2006.0045

Bakken, T., Holt, R., & Zundel, M. (2013). Time and play in management practice: An investigation through the philosophies of McTaggart and Heidegger. *Scandinavian Journal of Management*, *29*(1), 13–22. doi:10.1016/j.scaman.2012.09.003

Barab, S., & Squire, K. (2004). Design-based research: Putting a stake in the ground. *Journal of the Learning Sciences*, *13*(1), 1–14. doi:10.1207/s15327809jls1301_1

Baralou, E., & Tsoukas, H. (2015). How is new organizational knowledge created in a virtual context? An ethnographic study. *Organization Studies, 36*(5), 593–620. doi:10.1177/0170840614556918

Bardach, E. (1998). *Getting agencies to work together: The practice and theory of managerial craftsmanship*. Brookings Institution Press.

Barth, F. (1994). *Manifestasjon og prosess* [Manifestation and process]. Oslo: Universitetsforlaget.

Bartunek, J. M., & Louis, M. R. (1996). *Insider/outsider team research* (Vol. 40). Thousand Oaks, CA: Sage Publications.

Bartunek, J., & Louis, M. (1996). *Insider/outsider team research* (Vol. 40). Sage Publications.

Bartunek, J., & Rynes, S. (2014). Academics and practitioners are alike and unlike. The paradoxes of academic-practitioner relationships. *Journal of Management, 40*(5), 1181–1201. doi:10.1177/0149206314529160

Baskerville, R., & Myers, M. D. (2004). Special issue on action research in information systems: Making IS research relevant to practice: Foreword. *Management Information Systems Quarterly, 28*(3), 329–335. doi:10.2307/25148642

Baskerville, R., & Wood-Harper, A. T. (1998). Diversity in information systems action research methods. *European Journal of Information Systems, 7*(2), 90–107. doi:10.1057/palgrave.ejis.3000298

Bass, B. M., & Bass, R. (2008). *The Bass handbook of leadership theory: Research and managerial applications* (3rd ed.). New York: Free Press: Simon and Schuster.

Bateson, G. (1972). *Steps to an ecology of mind: collected essays in anthropology, psychiatry, evolution, and epistemology*. San Francisco: Chandler Publishing Co.

Bauman, Z. (2000). *Liquid modernity*. Cambridge, UK: Polity Press.

Beaudoin, M. F. (2002). Learning or lurking? Tracking the "invisible" online student. *The Internet and Higher Education, 5*(2), 147–155. doi:10.1016/S1096-7516(02)00086-6

Beck, J. (2017, May 9). *Slack Doesn't Want People Using it For Romance*. Retrieved from The Atlantic: https://www.theatlantic.com/technology/archive/2017/05/slack-doesnt-want-people-using-it-for-romance/525975/

Becker, H. S. (1958). Problems of Inference and Proof in Participant Observation. *Source American Sociological Review, 23*(6), 652–660. doi:10.2307/2089053

Beebe, S. A. (1974). Eye contact: A nonverbal determinant of speaker credibility. *The Speech Educator, 1*(23), 21–25. doi:10.1080/03634527409378052

Benbasat, I., & Zmud, R. (1999). Empirical Research in Information Systems: The Practice of Relevance. *Management Information Systems Quarterly, 23*(1), 3–16. doi:10.2307/249403

Benkler, Y. (2006). *The Wealth of Networks: How social production transforms markets and freedom*. Retrieved November 2, 2016, from http://www.benkler.org/Benkler_Wealth_Of_Networks.pdf

Berson, Y., Nemanich, L. A., Waldman, D. A., Galvin, B. M., & Keller, R. T. (2006). Leadership and organizational learning: A multiple levels perspective. *The Leadership Quarterly, 17*(6), 577–594. doi:10.1016/j.leaqua.2006.10.003

Beugelsdijk, S., & Mudambi, R. (2013). MNEs as border-crossing multi-location enterprises: The role of discontinuities in geographic space. *Journal of International Business Studies, 44*(5), 413–426. doi:10.1057/jibs.2013.23

Beyes, T., & Steyaert, C. (2012). Spacing organization: Non-representational theory and performing organizational space. *Organization, 19*(1), 45–61. doi:10.1177/1350508411401946

Bhat, J. M., Gupta, M., & Murthy, S. N. (2006). Overcoming Requirements Engineering Challenges: Lessons from Offshore Outsourcing. *IEEE Software*, *23*(5), 38–44. doi:10.1109/MS.2006.137

Biocca, F., Harms, C., & Burgoon, J. K. (2003). Toward a more robust theory and measure of social presence: Review and suggested criteria. *Presence (Cambridge, Mass.)*, *12*(5), 456–480. doi:10.1162/105474603322761270

Bishop, A., Riopelle, K., Gluesing, J., Danowski, J., & Eaton, T. (2010). *Managing Global Compliance through Collaborative Innovation Networks* (Vol. 2). Procedia- Social and Behaivoural Sciences.

Bjelland, O. M., & Wood. (2008). An inside view of IBM's 'innovation jam'. *MIT Sloan Management Review*, *50*(1), 32.

Bjorn, P., Scupola, A., & Fitzgerald, B. (2006). Expanding technological frames towards mediated collaboration. *Scandinavian Journal of Information Systems*, *18*(2), 1.

Blumer, H. (1969). *Symbolic interactionism: Perspective and method*. Los Angeles, CA: University of California Press.

Boehm, B. (1981). *Software engineering economics. Advances in computing science and technology*. Prentice-Hall.

Boh, W. F., Ren, Y., Kiesler, S., & Bussjaeger, R. (2007). Expertise and collaboration in the geographically dispersed organization. *Organization Science*, *18*(4), 595–612. doi:10.1287/orsc.1070.0263

Bolden, R. (2011). Distributed leadership in organizations: A review of theory and research. *International Journal of Management Reviews*, *13*(3), 251–269. doi:10.1111/j.1468-2370.2011.00306.x

Bonabeau, E. (2009). Decisions 2.0: The power of collective intelligence. *Sloan Management Review*, *50*(2), 45–52.

Borgatti, S., & Cross, R. (2003). A social network view of organizational learning: Relational and structural dimensions of "know wh0.". *Management Science*, *49*(4), 432–445. doi:10.1287/mnsc.49.4.432.14428

Børhaug, F. B. (2016). How to challenge a culturalization of human existence? Promoting interculturalism and ethical thinking in education. FLEKS. *Scandinavian Journal of Intercultural Theory and Practice*, *3*(1), 1–15.

Borup, J., West, R. E., & Graham, C. R. (2012). Improving online social presence through asynchronous video. *The Internet and Higher Education*, *15*(3), 195–203. doi:10.1016/j.iheduc.2011.11.001

Bosch-Sijtsema, P. M., Ruohomäki, V., & Vartiainen, M. (2010). Multi-locational knowledge workers in the office: Navigation, disturbances and effectiveness. *New Technology, Work and Employment*, *25*(3), 183–195. doi:10.1111/j.1468-005X.2010.00247.x

Bourdieu, P. (2011). The forms of capital. In Cultural theory: An anthology (pp. 81–93). Academic Press.

Bourdieu, P. (1986). The forms of capital. In J. Richardson (Ed.), *Handbook of theory and research for the sociology of education* (pp. 241–258). Westport, CT: Greenwood.

Bourdieu, P. (2003). Participant objectivation. *Journal of the Royal Anthropological Institute*, *9*(2), 281–294. doi:10.1111/1467-9655.00150

Bourdieu, P. (Ed.). (1983). *Ökonomisches Kapital, Kulturelles Kapital, Soziales Kapital*. Goettingen, Germany: Otto Schartz & Co.

Bowden, C., & Galindo-Gonzalez, S. (2015). Interviewing when you're not face-to-face: The use of email interviews in a phenomenological study. *International Journal of Doctoral Studies*, *10*, 79–92. doi:10.28945/2104

Bradbury Huang, H. (2010). What is good action research? Why the resurgent interest? *Action Research*, *8*(1), 93–109. doi:10.1177/1476750310362435

Bradbury, H., & Lichtenstein, B. M. B. (2000). Relationality in organizational research: Exploring the space between. *Organization Science*, *11*(5), 551–564. doi:10.1287/orsc.11.5.551.15203

Bradley, E. W., & Vozikis, G. S. (2004). Trust in Virtual Teams. In H. S. Godar & P. S. Ferris (Eds.), *Virtual and Collaborative Teams: Process, Technologies and Practice*. Idea Group Publishing. doi:10.4018/978-1-59140-204-6.ch006

Brake, T. (2006). Leading Virtual Teams. *Industrial and Commercial Training*, *38*(3), 116–121. doi:10.1108/00197850610659364

Brandi, U., & Elkjacr, B. (2011). Tidsskrift for arbejdsliv. *Organisatorisk læring i organisatorisk forandring*, *13*(2), 73.

Brandi, U., & Elkjaer, B. (2011). Organizational learning viewed from a social learning perspective. In *Handbook of Organizational Learning and Knowledge Management* (pp. 23–41). Chichester, UK: Wiley.

Brannen, M. Y., & Salk, J. E. (2000). Partnering across borders: Negotiating organizational culture in a German-Japanese joint venture. *Human Relations*, *53*(4), 451–487. doi:10.1177/0018726700534001

Brinkmann, S., & Tanggaard, L. (2010). *Kvalitative metoder: En grundbog. "Qualitative Methods: The foundations*. Copenhagen, Denmark: Hans Reitzels Forlag.

Brinkmann, S., & Tanggaard, L. (2010). *Kvalitative metoder: En grundbog*. København: Hans Reitzels Forlag.

Brown, A. L. (1992). Design experiments: Theoretical and methodological challenges in creating complex interventions in classroom settings. *Journal of the Learning Sciences*, *2*(2), 141–178. doi:10.1207/s15327809jls0202_2

Brown, M. E., & Gioia, D. A. (2002). Making things click: Distributive leadership in an online division of an offline organization. *The Leadership Quarterly*, *13*(4), 397–419. doi:10.1016/S1048-9843(02)00123-6

Bruns, A. (2008). *Blogs, Wikipedia, Second Life, and beyond: From production to produsage*. Bern, Switzerland: Peter Lang.

Bruns, H. C. (2013). Working alone together: Coordination in collaboration across domains of expertise. *Academy of Management Journal*, *56*(1), 62–83. doi:10.5465/amj.2010.0756

Buber, M. (1984). *Das Dialogische Prinzip* (5th ed.). Heidelberg, Germany: Lambert Schneider.

Buber, M. (1984). *Das dialogische Prinzip / Martin Buber* (5th ed.). Heidelberg, Germany: Lambert Schneider.

Buch-Hansen, H., & Nielsen, P. (2005). *Kritisk realisme*. Frederiksberg: Samfundslitteratur/Roskilde Universitetsforlag.

Burke, K. (1962). *A Grammer of Motives and A Rhetoric of Motives*. New York: The World Publishing Company.

Burke, K. (1966). *Language As Symbolic Action*. Berkeley, CA: Univeristy of California Press.

Burt, R. S. (1987). Social contagion and innovation: Cohesion versus structural equivalence. *American Journal of Sociology*, *92*(6), 1287–1335. doi:10.1086/228667

Butler, J. K. Jr. (1991). Toward Understanding and Measuring Conditions of Trust: Evolution of a Conditions of Trust Inventory. *Journal of Management*, *17*(3), 643–663. doi:10.1177/014920639101700307

Butler, J. K. Jr. (1999). Trust Expectations, InformationSharing, Climate of Trust, and Negation Effectiveness and Efficiency. *Group & Organization Management*, *24*(2), 217–238. doi:10.1177/1059601199242005

Byram, M. (1997). *Teaching and Assessing Intercultural Communicative Competence*. Clevedon, UK: Multilingual Matters LTD.

Byram, M. (2000). Assessing Intercultural Competence in Language Teaching. *Sprogforum, 6*(18), 8–13.

Byron, K. (2008). Carrying Too Heavy A Load: The Communication and Miscommunication of Emotion by E-mail. *Academy of Management Review, 33*(2), 309–327. doi:10.5465/AMR.2008.31193163

Carlile, P. R. (2002). A pragmatic view of knowledge and boundaries: Boundary objects in new product development. *Organization Science, 13*(4), 442–455. doi:10.1287/orsc.13.4.442.2953

Carlile, P. R. (2004). Transferring, translating and transforming knowledge: An integrative framework for managing knowledge across boundaries. *Organization Science, 15*(5), 555–568. doi:10.1287/orsc.1040.0094

Charlier, S. D., Stewart, G. L., Greco, L. M., & Reeves, C. J. (2016). Emergent leadership in virtual teams: A multilevel investigation of individual communication and team dispersion antecedents. *The Leadership Quarterly, 27*(5), 745–764. doi:10.1016/j.leaqua.2016.05.002

Chaulat, G. (2012). *Virtual leadership – learning to lead differently.* Libri Publishing.

Chaulat, G. (2012). *Virtual leadership: Learning to lead differently.* Libri Publishing.

Checkland, P. (1991). From framework through experience to learning: The essential nature of action research. *Information Systems Research: Contemporary Approaches and Emergent Traditions,* 397–403.

Chen, L. (2015). A Model of Mobile Work Continuance of Knowledge Workers: Evidences from China. In *The Third International Conference on E-Technologies and Business on the Web (EBW2015)* (pp. 43–52). Academic Press.

Chen, G., & Starosta, W. J. (2008). Intercultural communication competence. A Synthesis. In M. K. Asante, Y. Miike, & J. Yin (Eds.), *The global intercultural communication reader* (pp. 215–237). New York: Routledge.

Chen, L., & Nath, R. (2008). A socio-technical perspective of mobile work. *Information, Knowledge, Systems Management, 7,* 41–60. Retrieved from http://iospress.metapress.com/index/b41248867220l268.pdf

Cheuk, B., & Dervin, B. (2011). Leadership 2.0 in action: A journey from knowledge management to "knowledging". *Knowledge Management & E-Learning: An International Journal, 3*(2), 119–138.

Cheuk, B., & Li, H. (2013). Using social media to co-create new business strategies at environmental resources management. *Global Business and Organizational Excellence, 32*(4), 49–58. doi:10.1002/joe.21494

Cheuk, B., & McKenzie, J. (2017). Developing the practice of online leadership. In D. Kolbaek (Ed.), *Online Collaboration and Communication in Contemporary Organizations.* Hershey, PA: IGI Global.

Chiasson, M., Germonprez, M., & Mathiassen, L. (2009). Pluralist action research: A review of the information systems literature. *Information Systems Journal, 19*(1), 31–54. doi:10.1111/j.1365-2575.2008.00297.x

Chidambaram, L., & Jones, B. (1993). Impact of communication medium and computer support on group perceptions and performance: A comparison of face-to-face and dispersed meetings. *Management Information Systems Quarterly, 17*(4), 465–491. doi:10.2307/249588

Clark, H. H., & Brennan, S. E. (1991). Grounding in communication. *Perspectives on Socially Shared Cognition, 13,* 127–149. doi:10.1037/10096-006

Clegg, S., Courpasson, D., & Phillips, N. (2006). *Power and organizations.* Thousand Oaks, CA: Sage Publications.

Cline, B. (2012). The Future of Reading/Thinking: Epistemological Construction in the Age of the Kindle. *Communication +1, 1*(1), Article 2.

Cline, B. (2014). The Concept of Ethos as a Tool for Assessing Civility. In J. Bauer & C. Stevenson (Eds.), *Building Online Communities in Higher Education Institutions* (pp. 191–200). Hershey, PA: Information Science Reference (An Imprint of IGI Global). doi:10.4018/978-1-4666-5178-4.ch010

Cline, B. J. (2016). *E-thinking: How Thinking Changes in the Age of the Kindle*. Saarbrücken, Germany: Lambert Academic Publishing.

Cline, B. J. (2017). Using the Classical Rhetorical Concept of "Ethos" to Solve Online Collaboration Problems of Trust and Presence. In D. Kolbaek (Ed.), *Online Collaboration and Communication in Contemporary Organizations*. Hershey, PA: IGI Global.

Coenen, M., & Kok, R. W. (2014). Workplace flexibility and new product development performance: The role of telework and flexible work schedules. *European Management Journal*, *32*(4), 564–576. doi:10.1016/j.emj.2013.12.003

Coghlan, D., & Coughlan, P. (2015). Effecting change and learning in networks through network action learning. *The Journal of Applied Behavioral Science*, *51*(3), 375–400. doi:10.1177/0021886314540210

Coghlan, D., Shani, A. B., Roth, J. M., & Sloyan, R. (2014). Executive development through insider action research: Voices of insider action researchers. *Journal of Management Development*, *33*(10), 991–1003. doi:10.1108/JMD-06-2012-0072

Cohen, R. L. (2010). Rethinking "mobile work": Boundaries of space, time and social relation in the working lives of mobile hairstylists. *Work, Employment and Society*, *24*(1), 65–84. doi:10.1177/0950017009353658

Cole, J. I., Suman, M., Lunn, R., Aquino, J.-S., Firth, D., Fortier, D., . . . Lebo, H. (2013). *The UCLA Internet Report: Surveying the Digital Future-- Year Three.* Retrieved February 14, 2017, from Center for the Digital Future: http://www.digitalcenter.org/wp-content/uploads/2013/02/2003_digital_future_report-year3.pdf

Cole, M., & Engeström, Y. (1993). A cultural-historical approach to distributed cognition. In G. Slaomon (Ed.), *Distributed cognitions–psychological and educational considerations* (pp. 1–46). Melbourne, Australia: Cambridge University Press.

Coleman, J. S. (1988). Social capital in the creation of human capital. *American Journal of Sociology*, *94*, 95–120. doi:10.1086/228943

Collier, M. J. (1998). Researching cultural identity: Reconciling interpretive and postcolonial perspectives. In D. V. Tanno & A. Gonzales (Eds.), *Communication and Identity across Cultures* (pp. 122–147). Thousand Oaks, CA: Sage.

Collins, A. (2010). *Design experiments. In International Encyclopedia of Education.* Elsevier Ltd.

Collins, A., Joseph, D., & Bielaczyc, K. (2004). Design Research: Theoretical and methodological issues. *Journal of the Learning Sciences*, *13*(1), 15–42. doi:10.1207/s15327809jls1301_2

Commisso, T. H., & Pries-Heje, J. (2011). *Optimér dit projektteam. Håndbog i teamudvikling.* Samfundslitteratur.

Connolly, T. M., Boyle, E. A., MacArthur, E., Hainey, T., & Boyle, J. M. (2012). A systematic literature review of empirical evidence on computer games and serious games. *Computers & Education*, *59*(2), 661–686. doi:10.1016/j.compedu.2012.03.004

Corradi, G., Gherardi, S., & Verzelloni, L. (2010). Through the practice lens: Where is the bandwagon of practice-based studies heading? *Management Learning*, *41*(3), 265–283. doi:10.1177/1350507609356938

Covey, S. M. R. (2006). *The Speed of Trust. The One Thing that Changes Everything*. New York: Simon & Schuster.

Cramton, C. D. (2001). The mutual knowledge problem and its consequences for dispersed collaboration. *Organization Science*, *12*(3), 346–371. doi:10.1287/orsc.12.3.346.10098

Cramton, C. D. (2002). Finding common ground in dispersed collaboration. *Organizational Dynamics*, *30*(4), 356–367. doi:10.1016/S0090-2616(02)00063-3

Cramton, C. D., & Hinds, P. J. (2014). An embedded model of cultural adaptation in global teams. *Organization Science*, *25*(4), 1056–1081. doi:10.1287/orsc.2013.0885

Cranshaw, J., & Kittur, A. (2011, May). The polymath project: lessons from a successful online collaboration in mathematics. In *Proceedings of the SIGCHI Conference on Human Factors in Computing Systems* (pp. 1865-1874). ACM. doi:10.1145/1978942.1979213

CRESCENDO Consortium. (2009). *Enterprise collaborative state-of-art review report.* CRESCENDO FP7- 234344. Retrieved from http://cordis.europa.eu/docs/results/234344/final1-crescendo-d017-final report-20130430-r1.pdf

Crisp, B. C., & Jarvenpaa, S. L. (2013). Swift trust in global virtual teams. *Journal of Personnel Psychology*, *12*(1), 45–56. doi:10.1027/1866-5888/a000075

Crosby, A., & Notley, T. (2014). Using Video and Online Subtitling to Communicate across Languages from West Papua. *The Australian Journal of Anthropology*, *25*(2), 138–154. doi:10.1111/taja.12085

Cummings, J. N. (2004). Work groups, structural diversity, and knowledge sharing in a global organization. *Management Science*, *50*(3), 352–364. doi:10.1287/mnsc.1030.0134

Czarniawska, B. (2004). On time, space, and action nets. *Organization*, *11*(6), 773–791. doi:10.1177/1350508404047251

Czarniawska, B. (2007). *Shadowing: And other techniques for doing fieldwork in modern societies.* Copenhagen: Copenhagen Business School Press DK.

d'Iribarne, P. (2009). National cultures and organizations in search of a theory: An interpretative approach. *International Journal of Cross Cultural Management*, *9*(3), 309–321. doi:10.1177/1470595809346601

Daft, R., & Engel, R. (1984). Information richness: A new approach to managerial behaviour and organizational design. *Research in Organizational Behavior*, *6*, 191–233.

Dahl, R. A. (1957). The concept of power. In Behavioural science (vol. 2). Academic Press. doi:10.1002/bs.3830020303

Dahl, Ø. (2016). *Human encounters. Introduction to intercultural communication.* Oxford, UK: Peter Lang.

Daim, T., Ha, A., Reutiman, S., Hughes, B., Pathak, U., Bynum, W., & Bhatla, A. (2012). Exploring the communication breakdown in global virtual teams. *International Journal of Project Management*, *30*(2), 199–221. doi:10.1016/j.ijproman.2011.06.004

Dal Fiore, F., Mokhtarian, P. L., Salomon, I., & Singer, M. E. (2014). "Nomads at last"? A set of perspectives on how mobile technology may affect travel. *Journal of Transport Geography*, *41*, 97–106. doi:10.1016/j.jtrangeo.2014.08.014

Dalkir, K. (2011). *Knowledge management in theory and practice* (2nd ed.). Boston, MA: MIT Press.

Dalkir, K. (2017). Challenges in Online Collaboration. In D. Kolbaek (Ed.), *Online Collaboration and Communication in Contemporary Organizations*. Hershey, PA: IGI Global.

Dalzel-Job, O., Oberlander, J., & Smith, T. J. (2011). Don't Look Now: The relationship between mutual gaze, task performance and staring in Second Life. In *Proceedings of the 33rd Annual Conference of the Cognitive Science Society* (pp. 832–837). Academic Press.

Dasgupta, M., & Gupta, R. (2009). Innovation in organizations A review of the role of organizational learning and knowledge management. *Global Business Review*, *10*(2), 203–224. doi:10.1177/097215090901000205

DasGupta, P. (2011). Literature review: Online leadership. *Emerging Leadership Journeys*, *4*(1), 1–36.

Davidson, C. N. (2011). *Now you see it: How the brain science of attention will transform the way we live, work, and learn*. New York: Viking New York.

Davies, B., & Harré, R. (1990). Positioning: The discursive production of selves. *Journal for the Theory of Social Behaviour*, *20*(1), 43–63. doi:10.1111/j.1468-5914.1990.tb00174.x

Deardorff, D. K. (2006). Identification and assessment of intercultural competence as a student outcome of internationalization. *Journal of Studies in International Education*, *10*(3), 241–266. doi:10.1177/1028315306287002

Dede, C. (2004). If design-based research is the answer, what is the question? A commentary on Collins, Joseph, and Bielaczyc; diSessa and Cobb; and Fishman, Marx, Blumenthal, Krajcik, and Soloway in the JLS Special Issue on design-based research. *Journal of the Learning Sciences*, *13*(1), 105–114. doi:10.1207/s15327809jls1301_5

Delbridge, R., & Sallaz, J. J. (2015). Work: Four worlds and ways of seeing. *Organization Studies*, *36*(11), 1449–1462. doi:10.1177/0170840615612021

Dell & Intel. (2016). *Future workforce study*. Austin, TX: Authors.

DeMarco, T., & Lister, T. (1999). *Peopleware: Productive projects and teams* (2nd ed.). Dorset House Publishing Company.

Denyer, D., Parry, E., & Flowers, P. (2011). “Social”, “open” and “participative”? Exploring personal experiences and organisational effects of enterprise 2.0 use. *Long Range Planning*, *44*(5-6), 375–396. doi:10.1016/j.lrp.2011.09.007

Denzin, N. K. (2012). Triangulation 2.0. *Journal of Mixed Methods Research*, *6*(2), 80–88. doi:10.1177/1558689812437186

deRue, D. S. (2011). Adaptive leadership theory: Leading and following as a complex adaptive process. *Research in Organizational Behavior*, *31*, 125–150. doi:10.1016/j.riob.2011.09.007

Dervin, B. (2008). Interviewing as dialectical practice: Sense-Making methodology as exemplar. In *Proceedings of the International Association for Media and Communication Research (IAMCR) 2008 Annual Meeting* (pp. 20-25). Stockholm: Academic Press.

Dervin, B. (1998). Sensemaking theory and practice: An overview of user interests in knowledge seeking and use. *Journal of Knowledge Management*, *2*(2), 36–46. doi:10.1108/13673279810249369

Dervin, F. (2011). A plea for change in research on intercultural discourses: A “liquid” approach to the study of the acculturation of Chinese students. *Journal of Multicultural Discourses*, *6*(1), 37–52. doi:10.1080/17447143.2010.532218

Design-Based Research Collective. (2003). Design-based research: An emerging paradigm for educational inquiry. *Educational Researcher*, *32*(1), 5–8. doi:10.3102/0013189X032001005

Develotte, C., Guichon, N., & Vincent, C. (2010). The use of the webcam for teaching a foreign language in a desktop videoconferencing environment. *ReCALL*, *22*(3), 293–312. doi:10.1017/S0958344010000170

Dewett, T., & Jones, G. R. (2001). The role of information technology in the organization: A review, model, and assessment. *Journal of Management*, *27*(3), 313–346. doi:10.1177/014920630102700306

Dewey, J., & Boydston, J. A. (1976). *The Middle Works: 1899–1924* (Vol. 9). Carbondale, IL: Southern Illinois Univ. Press.

Di Domenico, M., Daniel, E., & Nunan, D. (2014). “Mental mobility” in the digital age: Entrepreneurs and the online home-based business. *New Technology, Work and Employment*, *29*(3), 266–281. doi:10.1111/ntwe.12034

Ding, W. W., Levin, S. G., Stephan, P. E., & Winkler, A. E. (2010). The impact of information technology on academic scientists’ productivity and collaboration patterns. *Management Science*, *56*(9), 1439–1461. doi:10.1287/mnsc.1100.1195

Dirks, K., & Ferrin, D. (2001). The role of trust in organizational settings. *Organization Science*, *12*(4), 450–467. doi:10.1287/orsc.12.4.450.10640

Dohn, N. B., Hansen, J. J., Jensen, J., Johnsen, L., Lauridsen, E., Møller, M. H., & Yepez-Reyes, V. (2016). *Digital educational design - process, product, and practice*. Paper presented at the Designs for Learning, Fifth International Conference, København. Retrieved from http://www.designsforlearning2016.aau.dk/digitalAssets/198/198492_dfl_short-papers-final-2.pdf

Dooley, K. J., Tsoukas, H., Lorino, P., Tricard, B., & Clot, Y. (2011). Research methods for non-representational approaches to organizational complexity: The dialogical mediated inquiry. *Organization Studies*, *32*(6), 769–801. doi:10.1177/0170840611410807

Dourish, P. (2006). Re-space-ing place: place and space ten years on. In *Proceedings of the 2006 20th anniversary conference on Computer supported cooperative work* (pp. 299–308). ACM. doi:10.1145/1180875.1180921

Downes, S. (2005). *An introduction to connective knowledge*. Retrieved May 27, 2017, from http://www.downes.ca/post/33034

Drath, W. H., McCauley, C. D., Palus, C. J., Van Velsor, E., O'Connor, P. M. G., & McGuire, J. B. (2008). Direction, alignment, commitment: Toward a more integrative ontology of leadership. *The Leadership Quarterly*, *19*(6), 635–653. doi:10.1016/j.leaqua.2008.09.003

Drexler, A. B., Sibbet, D., & Forrester, R. H. (1988). The team performance model. *Team building: Blueprints for productivity and satisfaction*, 45-61.

Drucker, P. F. (1992). The new society of organizations. *Harvard Business Review*. Retrieved June 1, 2017, from https://hbr.org/1992/09/the-new-society-of-organizations

Duarte, D. L., & Snyder, N. T. (2006). *Mastering virtual teams: Strategies, tools and techniques that succeed* (3rd ed.). San Francisco, CA: Jossey-Bass.

Durugbo, C., Hutabarat, W., Tiwari, A., & Alcock, J. R. (2011). Modelling collaboration using complex networks. *Information Sciences*, *181*(15), 3143–3161. doi:10.1016/j.ins.2011.03.020

Dziuban, C., & Moskal, P. (2001). Emerging research issues in distributed learning. In *The 7th Sloan-C International Conference on Asynchronous Learning Networks*. Academic Press.

Easterby-Smith, M., & Lyles, M. A. (2011). *Handbook of Organizational Learning and Knowledge Management* (2nd ed.). Chichester, UK: Wiley.

Easterby-Smith, M., & Prieto, I. M. (2008). Dynamic capabilities and knowledge management: An integrative role for learning? *British Journal of Management*, *19*(3), 235–249. doi:10.1111/j.1467-8551.2007.00543.x

Economist, I. U. (2009). *Organizational agility. How business can survive and thrive in turbulent times*. Retrieved from London: http://www.emc.com/collateral/leadership/organisational-agility-230309.pdf

Edmondson, A., & Harvey, J. F. (2017). Cross-boundary teaming for innovation: Integrating research on teams and knowledge in organizations. *Human Resource Management Review*. doi:10.1016/j.hrmr.2017.03.002

Edwards, G. (2011). Concepts of community: A framework for contextualizing distributed leadership. *International Journal of Management Reviews*, *13*(3), 301–312. doi:10.1111/j.1468-2370.2011.00309.x

Eisenhardt, K. M. (1989). Building theories from case study research. *Academy of Management Review*, *14*(4), 532–550.

Elkjaer, B. (2003). Organizational learning with a pragmatic slant. *International Journal of Lifelong Education, 22*(5), 481–494. doi:10.1080/0260137032000102841

Engeström, Y. (2010). Activity theory and learning at work. In *The Sage handbook of workplace learning* (pp. 74–89). SAGE.

Engeström, Y. (1996). Developmental work research as educational research. Nordisk Pedagogik. *Journal of Nordic Educational Research, 16*(5), 131–143.

Engeström, Y. (2001). Expansive learning at work: Toward an activity theoretical reconceptualization. *Journal of Education and Work, 14*(1), 133–156. doi:10.1080/13639080020028747

Engeström, Y. (2007). From stabilization knowledge to possibility knowledge in organizational learning. *Management Learning, 38*(3), 271–275. doi:10.1177/1350507607079026

Engeström, Y. (2011). From design experiments to formative interventions. *Theory & Psychology, 21*(5), 598–628. doi:10.1177/0959354311419252

Engeström, Y. (2015). *Learning by expanding: An activity-theoretical approach to developmental research* (2nd ed.). Cambridge, UK: Cambridge University Press.

Engeström, Y. (2016). *Studies in expansive learning: Learning what is not yet there.* Cambridge University Press. doi:10.1017/CBO9781316225363

Engeström, Y., Miettinen, R., & Punamäki, R.-L. (1999). *Perspectives on activity theory.* New York: Cambridge University Press. doi:10.1017/CBO9780511812774

Engeström, Y., Rantavuori, J., & Kerosuo, H. (2013). Expansive learning in a library: Actions, cycles, and deviations from instructional intentions. *Vocations and Learning, 6*(1), 81–106. doi:10.1007/s12186-012-9089-6

Engeström, Y., & Sannino, A. (2010). Studies of expansive learning: Foundations, findings and future challenges. *Educational Research Review, 5*(1), 1–24. doi:10.1016/j.edurev.2009.12.002

Engeström, Y., & Sannino, A. (2011). Discursive manifestations of contradictions in organizational change efforts: A methodological framework. *Journal of Organizational Change Management, 24*(3), 368–387. doi:10.1108/09534811111132758

Ernst, C., & Chrobot-Mason, D. (2010). *Boundary Spanning Leadership: Six Practices for Solving Problems, Driving Innovation, and Transforming Organizations.* New York: McGraw-Hill.

Ernst, C., & Chrobot-Mason, D. (2011). *Boundary Spanning Leadership.* New York: McGraw-Hill.

Evans, M. (2012). *Knowledge Sharing: An empirical study of the role of trust and other social-cognitive factors in an organizational setting* (Doctoral thesis). Faculty of Information Studies, University of Toronto, Toronto, Canada.

Evans, M. M., & Wensley, A. K. P. (2009). Predicting the influence of network structure on trust in knowledge communities: Addressing the interconnectedness of four network principles and trust. *Electronic Journal of Knowledge Management, 7*(1), 41–54.

Evans, M. M., Wensley, A. K. P., & Frissen, I. (2015). The mediating effects of trustworthiness on social-cognitive factors and knowledge sharing in a large professional service firm. *Electronic Journal of Knowledge Management, 13*(3), 240–254.

Evans, W. R., & Carson, C. M. (2005). A social capital explanation of the relationship between functional diversity and group performance. *Team Performance Management, 11*(7/8), 302–315. doi:10.1108/13527590510635170

Fanon, F. (1967). *Black Skin, White Masks.* New York: Grove Press.

Faraj, S., Jarvenpaa, S. L., & Majchrzak, A. (2011). Knowledge collaboration in online communities. *Organization Science, 22*(5), 1224–1239. doi:10.1287/orsc.1100.0614

Faraj, S., & Johnson, S. L. (2011). Network exchange patterns in online communities. *Organization Science*, *22*(6), 1464–1480. doi:10.1287/orsc.1100.0600

Faraj, S., Kudaravalli, S., & Wasko, M. (2015). Leading collaboration in online communities. *Management Information Systems Quarterly*, *39*(2), 393–412. doi:10.25300/MISQ/2015/39.2.06

Fayard, A.-L., & Metiu, A. (2014). The role of writing in distributed collaboration. *Organization Science*, *25*(5), 1391–1413. doi:10.1287/orsc.2013.0893

Ferran, C., & Watts, S. (2008). Videoconferencing in the field: A heuristic processing model. *Management Science*, *54*(9), 1565–1578. doi:10.1287/mnsc.1080.0879

Ferrazzi, K. (2015, March). How to run a great virtual meeting. *Harvard Business Review*. Retrieved May 28, 2017, from https://hbr.org/2015/03/how-to-run-a-great-virtual-meeting

Festinger, L. (1957). *A Theory of Cognitive Dissonance*. Palo Alto, CA: Stanford University Press.

Festinger, L. (1962). Cognitive dissonance. *Scientific American*, *207*(4), 93–107. doi:10.1038/scientificamerican1062-93 PMID:13892642

Fichter, D. (2005). The many forms of e-collaboration: Blogs, wikis, portals, groupware, discussion boards, and instant messaging. *Online (Bergheim)*, *29*(4), 48–50.

Fineman, S., Maitlis, S., & Panteli, N. (2007). Virtuality and emotion. *Human Relations*, *60*(4), 555–560. doi:10.1177/0018726707078346

Flyvbjerg, B. (2015). Fem misforståelser om casearbejdet. In Kvalitative metoder (pp. 497-520). Hans Reitzels Forlag.

Flyvbjerg, B. (2004). Five misunderstandings about case-study research. In C. Seale, G. Gobo, J. F. Gubrium, & D. Silverman (Eds.), *Qualitative research practice* (pp. 420–434). London: Sage. doi:10.4135/9781848608191.d33

Flyvbjerg, B. (2006). Five misunderstandings about case-study research. *Qualitative Inquiry*, *12*(2), 219–245. doi:10.1177/1077800405284363

Flyverbom, M. (2012). Globalization as it happens: On globalizing assemblages in tax planning, *Distinktion. Scandinavian Journal of Social Theory*, *13*(3), 295–309.

Flyverbom, M. (2012). Globalization as it happens: On globalizing assemblages in tax planning. *Distinktion: Scandinavian Journal of Social Theory*, *13*(3), 295–309. doi:10.1080/1600910X.2012.723013

Fogsgaard, M. K., & Elmholdt, C. W. (2016). *Power and leadership. In Global Encyclopedia of Public Administration, Public Policy, and Governance* (pp. 1–8). Switzerland: Springer Publishing Company. DOI.

Fogsgaard, M., Elmholdt, C., & Lindekilde, R. (2017). Power in online leadership. In D. Kolbaek (Ed.), *Online Collaboration and Communication in Contemporary Organizations*. Hershey, PA: IGI Global.

Ford, J. D., Ford, L. W., & D'Amelio, A. (2008). Resistance to change: The rest of the story. *Academy of Management Review*, *33*(2), 362–377. doi:10.5465/AMR.2008.31193235

Fortenbaugh, W. W. (1992). Aristotle on Persuasion through Character. *Rhetorica*, *10*(3), 220. doi:10.1525/rh.1992.10.3.207

Foucault, M. (2002). *Overvågning og straf – fængslets fødsel*. Frederiksberg: Det lille Forlag.

Foucault, M. (1979). *Discipline and punish: the birth of the prison*. New York: Vintage Books.

Foucault, M. (1980). Two lectures. In M. Foucault (Ed.), *Power/knowledge*. Harlow: Pearson Education. (Original work published 1977)

Foucault, M. (1982). The subject and power. In H. L. Dreyfus & P. Rabinow (Eds.), *Michel Foucault—Beyond structuralism and hermeneutics*. Chicago: The University of Chicago Press. doi:10.1086/448181

Francisco, R., Klein, A. Z., Engeström, Y., & Sannino, A. (2017). Knowledge on the Move. In D. Kolbaek (Ed.), *Online Collaboration and Communication in Contemporary Organizations*. Hershcy, PA: IGI Global.

Fraser, H., & Michell, D. (2014). Feminist memory work in action: Method and practicalities. *Qualitative Social Work: Research and Practice*, *0*(00), 1–17.

French, J. R. P., Jr., & Raven, B. (1959). The bases of social power. In D. Cartwright (Ed.), Studies in social power. Ann Arbor, MI: University of Michigan.

Frey, B. B., Lohmeier, J. H., Lee, S. W., & Tollefson, N. (2006). Measuring collaboration among grant partners. *The American Journal of Evaluation*, *27*(3), 383–392. doi:10.1177/1098214006290356

Fried, C. B. (2008). In-class laptop use and its effects on student learning. *Computers & Education*, *50*(3), 906–914. doi:10.1016/j.compedu.2006.09.006

Friedman, T. L. (2006). The World is Flat: The Globalized World in the Twenty-First century (Updated and expanded ed.). London: Penguin Books.

Friesen, N. (2011). Media: Digital, ecological and epistemological. *E-Learning and Digital Media*, *8*(3), 175–180. doi:10.2304/elea.2011.8.3.175

Fuchs, C. (2015). *Culture and economy in the age of social media*. New York, NY: Routledge.

Gaible, E., & Dabla, A. (2010). *EVOKE project evaluation*. Retrieved from www.siteresources.worldbank.org/EDUCATION/Resources/ProjectEvoke-evaluation-final-16oct11.pdf

Gajda, R. (2004). Utilizing collaboration theory to evaluate strategic alliances. *The American Journal of Evaluation*, *25*(1), 65–77. doi:10.1177/109821400402500105

Gal, U., Jensen, T. B., & Lyytinen, K. (2014). Identity orientation, social exchange, and information technology use in interorganizational collaborations. *Organization Science*, *25*(5), 1372–1390. doi:10.1287/orsc.2014.0924

Gambetta, D. (1988). *Trust: Making and breaking cooperative relations*. Oxford, UK: Basil Blackwell.

Garrison, D. R. (2011). E-learning in the 21st century : a framework for research and practice / D. Randy Garrison (2nd ed.). New York, NY: Routledge.

Gartner Group (2002). *Gartner survey on managing information*. Note number COM-15-0871. Author.

Gartner Group. (2007). *Enterprise 2.0: Collaboration and Knowledge Management Renaissance*. Report number G00203367. Author.

Gergen, K. J. (2009). *Relational being: Beyond self and community*. Oxford, UK: Oxford University Press.

Gergen, K. J. (2009). *Relational Being: Beyond Self and Community*. Oxford, UK: Oxford University Press.

Gergen, K. J. (2010). Co-constitution, causality, and confluence: Organizing in a world without entities. In T. Hernes & S. Maitlis (Eds.), *Process, Sensemaking and Organization* (pp. 55–69). Oxford, UK: Oxford University Press. doi:10.1093/acprof:oso/9780199594566.003.0004

Gergen, K. J., & Thatchenkery, T. J. (2006). Organizational science and the promise of postmodernism. In D. M. Hosking & S. McNamee (Eds.), *The Social Construction of Organization* (pp. 34–53). Malmo, Sweden: Liber & Copenhagen Business School Press.

Gertsen, M. C. (1990). Intercultural competence and expatries. *International Journal of Human Resource Management*, *1*(3), 341–362. doi:10.1080/09585199000000054

Gertsen, M. C., Søderberg, A., & Zølner, M. (2012). Introduction and overview. In M. C. Gertsen, A. Søderberg, & M. Zølner (Eds.), *Global collaboration: Intercultural experiences and learning* (pp. 1–11). London: Palgrave Macmillan. doi:10.1057/9781137026064_1

Gherardi, S. (2015). To start practice theorizing anew: The contribution of the concepts of agencement and formativeness. Organization, 1-19.

Gherardi, S. (2009). Community of pratice or practices of a community? In S. Armstrong & C. Fukami (Eds.), *The sage handbook of management learning, education, and development* (pp. 514–530). London: Sage. doi:10.4135/9780857021038.n27

Gherardi, S. (2012). *How to conduct a practice-based study: Problems and methods*. Northampton, MA: Edward Elgar. doi:10.4337/9780857933386

Gherardi, S. (2015). How the turn to practice may contribute to working life studies. *Nordic Journal of Working Life Studies*, *5*(3a), 13–25. doi:10.19154/njwls.v5i3a.4831

Gherardi, S., Nicolini, D., & Odella, F. (1998). Toward a social understanding of how people learn in organizations: The notion of situated curriculum. *Management Learning*, *29*(3), 273–297. doi:10.1177/1350507698293002

Gibb, J. R. (1991). *Trust: A new Vision Of Human Relationships for Business, Education, Family, and Personal Living*. Newcastle Pub CO Inc.

Gibbs, J. (2009). Dialectics in a global software team: Negotiating tensions across time, space, and culture. *Human Relations*, *62*(6), 905–935. doi:10.1177/0018726709104547

Gibson, C. B., & Cohen, S. G. (Eds.). (2003). *Virtual teams that work: Creating conditions for virtual team effectiveness*. San Francisco, CA: John Wiley & Sons.

Gibson, C. B., & Gibbs, J. L. (2006). Unpacking the concept of virtuality: The effects of geographic dispersion, electronic dependence, dynamic structure, and national diversity on team innovation. *Administrative Science Quarterly*, *51*(3), 451–495. doi:10.2189/asqu.51.3.451

Gibson, C. B., Gibbs, J. L., Stanko, T. L., Tesluk, P., & Cohen, S. G. (2011). Including the "i" in virtuality and modern job design: Extending the job characteristics model to include the moderating effect of individual experiences of electronic dependence and copresence. *Organization Science*, *22*(6), 1481–1499. doi:10.1287/orsc.1100.0586

Giesbers, B., Rienties, B., Tempelaar, D. T., & Gijselaers, W. (2014). Why Increased Social Presence through Web Videoconferencing Does Not Automatically Lead to Improved Learning. *E-Learning and Digital Media*, *11*(1), 31–45. doi:10.2304/elea.2014.11.1.31

Gilson, L. L., Maynard, M. T., Jones Young, N. C., Vartiainen, M., & Hakonen, M. (2015). Virtual teams research: 10 years, 10 themes, and 10 opportunities. *Journal of Management*, *41*(5), 1313–1337. doi:10.1177/0149206314559946

Gittell, R., & Vidal, A. (1998). *Community Organizing: Building Social Capital as a Development Strategy*. Newbury Park, CA: Sage Publications.

Goetz, J. P., & LeCompte, M. D. (1981). Ethnographic Research and the Problem of Data Reduction. *Anthropology & Education Quarterly*, *12*(1), 51–70. doi:10.1525/aeq.1981.12.1.05x1283i

Goffman, E. (2005). *Interaction ritual: Essays in face to face behavior*. New Brunswick, NJ: Aldine Transaction.

Golden, A. G., & Geisler, C. (2007). Work-life boundary management and the personal digital assistant. *Human Relations*, *60*(3), 519–551. doi:10.1177/0018726707076698

Golden, J. L., Berquist, G. F., Coleman, W. E., & Sproule, J. M. (2011). *The Rhetoric of Western Thought*. Dubuque, IA: Kendall Hunt Publishing Company.

Golden, T. D., & Fromen, A. (2011). Does it matter where your manager works? Comparing managerial work mode (traditional, telework, virtual) across subordinate work experiences and outcomes. *Human Relations*, *64*(11), 1451–1475. doi:10.1177/0018726711418387

Golub, E. (2005). On audience activities during presentations. *Journal of Computing Sciences in Colleges*, *20*(3), 38–46.

Goodier, H. (2012). The Participation choice. *BBC Marketing and Audience*. Retrieved from http://www.bbc.co.uk/blogs/bbcinternet/2012/05/bbc_online_ briefing_spring2012_participation.pdf

Govindarajan, V., & Gupta, A. (2001). Building an Effective Global Business Team. *MIT Sloan Management Review*, *42*(4), 63–71.

Greckhamer, T. (2011). Cross-cultural Differences in Compensation Level and Inequality across Occupations: A Set-theoretic Analysis. *Organization Studies*, *32*(1), 85–115. doi:10.1177/0170840610380806

Greenberg, P. S., Greenberg, R. H., & Antonucci, Y. L. (2007). Creating and sustaining trust in virtual teams. *Business Horizons*, *50*(4), 325–333. doi:10.1016/j.bushor.2007.02.005

Greene, J. C. (2012). Engaging critical issues in social inquiry by mixing methods. *The American Behavioral Scientist*, *56*(6), 755–773. doi:10.1177/0002764211433794

Greenleaf, R. K. (1977). *Servant leadership*. New York: Paulist Press.

Greeno, J. G., & Engeström, Y. (2014). Learning in activity. In R. K. Sawyer (Ed.), *The Cambridge handbook of the learning sciences* (2nd ed.; pp. 128–147). Cambridge: Cambridge University Press. doi:10.1017/CBO9781139519526.009

Greenwood, D. J., & Levin, M. (2006). *Introduction to action research: Social research for social change*. SAGE publications.

Griffith, T. L. (2012). Technology and teams: The next ten years. In M. A. Neale & E. A. Mannix (Eds.), *Looking back, moving forward: A review of group and team-based research* (pp. 245–278). Bingley, UK: Emerald Group Publishing Ltd. doi:10.1108/S1534-0856(2012)0000015013

Gullestad, M. (2007). *Misjonsbilder – Bidrag til norsk selvforståelse* [Mission pictures – Contributions to Norwegian self-understanding]. Oslo: Universitetsforlaget.

Gumperz, J., Jupp, T. C., & Roberts, C. (1979). *Crosstalk – A study of cross-cultural communication: Background material and notes to accompany the BBC film*. London: The National Centre for Industrial Language Training.

Haas, M. R. (2006). Acquiring and applying knowledge in transnational teams: The roles of cosmopolitans and locals. *Organization Science*, *17*(3), 367–384. doi:10.1287/orsc.1060.0187

Habermas, J. (1991). *The Structural Transformation of the Public Sphere: An inquiry into a category of Bourgeois Society* (T. Burger, Trans.). Cambridge, MA: MIT University Press.

Haddon, L., & Brynin, M. (2005). The character of telework and the characteristics of teleworkers. *New Technology, Work and Employment*, *20*(3), 34–46. doi:10.1111/j.1468-005X.2005.00142.x

Hafner, K., & Lyon, M. (1996). *Where wizards stay up late: The origins of the Internet*. New York: Simon and Schuster.

Hall, E. (1976). *Beyond Culture*. New York, NY: Doubleday.

Hall, S. (1996). When was the "postcolonial?" Thinking at the limit. In I. Chambers & L. Curti (Eds.), *The postcolonial question: Common skies, divided horizons* (pp. 242–260). London: Routledge.

Halualani, R. T., Mendoza, S. L., & Drzewiecka, J. A. (2009). Critical junctures in intercultural communication studies: A review. *The Review of Communication Journal*, *9*(1), 17–35. doi:10.1080/15358590802169504

Hambley, L. A., O'Neill, T. A., & Kline, T. J. (2007). Virtual team leadership: The effects of leadership style and communication medium on team interaction styles and outcomes. *Organizational Behavior and Human Decision Processes*, *103*(1), 1–20. doi:10.1016/j.obhdp.2006.09.004

Hambley, L. A., O'Neill, T. A., & Kline, T. J. B. (2007). Virtual team leadership: Perspectives from the field. *International Journal of e-Collaboration*, *3*(1), 40–45, 47–48, 50–64. doi:10.4018/jec.2007010103

Haraway, D. (1991). *A Cyborg Manifesto. In Simians, Cyborgs and Woman. The reinvention og Nature* (pp. 149–181). New York: Rotledge.

Hardin, G. (1968). The tragedy of the commons. *Science*, *162*(3859), 1243–1248. doi:10.1126/science.162.3859.1243 PMID:5699198

Harmer, B. M., & Pauleen, D. J. (2012). Attitude, aptitude, ability and autonomy: The emergence of "offroaders", a special class of nomadic worker. *Behaviour & Information Technology*, *31*(5), 439–451. doi:10.1080/0144929X.2010.489117

Harreld, J. B., O'Reilly, C. A. III, & Tushman, M. L. (2007). Dynamic capabilities at IBM: Driving strategy into action. *California Management Review*, *49*(4), 21–43. doi:10.2307/41166404

Harré, R., & Langenhove, L. (1998). *Positioning Theory*. Oxford, UK: Blackwell.

Hatch, M. J. (2002). Essai: Doing time in organization theory. *Organization Studies*, *23*(6), 869–875. doi:10.1177/0170840602236008

Hatch, M. J. (2011). *Organization theory – modern symbolic and postmodern perspectives*. Oxford, UK: Oxford University Press.

Hattie, J., & Yates, G. C. R. (2014). *Visible learning and the science of how we learn*. New York: Routledge.

Hauge, T. E., & Norenes, S. O. (2015). Collaborative leadership development with ICT experiences from three exemplary schools. *International Journal of Leadership in Education*, *18*(3), 340–364. doi:10.1080/13603124.2014.963689

Haug, F. (2008). Memory work. *Australian Feminist Studies*, *23*(58), 537–541. doi:10.1080/08164640802433498

Heede, D. (2002). *Det tomme menneske – Introduktion til Michel Foucault* (2nd ed.). København: Museum Tusculanums Forlag.

Helin, J., Hernes, T., Hjorth, D., & Holt, R. (2014). Process is how process does. In J. Helin, T. Hernes, D. Hjorth, & R. Holt (Eds.), *The Oxford Handbook of Process Philosophy and Organization Studies* (pp. 1–16). Oxford, UK: Oxford University Press.

Hembrooke, H., & Gay, G. (2003). The laptop and the lecture: The effects of multitasking in learning environments. *Journal of Computing in Higher Education*, *15*(1), 46–64. doi:10.1007/BF02940852

Henfridsson, O., & Bygstad, B. (2013). The generative mechanisms of digital infrastructure evolution. *Management Information Systems Quarterly*, *37*(3), 907–931. doi:10.25300/MISQ/2013/37.3.11

Herbsleb, J. D., & Mockus, A. (2003). An empirical study of speed and communication in globally distributed software development. *IEEE Transactions on Software Engineering*, *29*(6), 481–494. doi:10.1109/TSE.2003.1205177

Hernes, T. (2004). Studying composite boundaries: A framework of analysis. *Human Relations*, *57*(1), 9–29. doi:10.1177/0018726704042712

Hernes, T., Simpson, B., & Söderlund, J. (2013). Managing and temporality. *Scandinavian Journal of Management*, *29*(1), 1–6. doi:10.1016/j.scaman.2012.11.008

Hevner, A. R., March, S. T., Park, J., & Ram, S. (2004). Design Science In Information Systems Research. *Management Information Systems Quarterly*, *28*(1), 75–105. doi:10.2307/25148625

Hinds, P., & Mortensen, M. (2002). Understanding antecedents to conflict in geographically distributed research and development teams. ICIS 2002 Proceedings, 38, 415-424.

Hinds, P. J., & Bailey, D. E. (2003). Out of sight, out of sync: Understanding conflict in distributed teams. *Organization Science*, *14*(6), 615–632. doi:10.1287/orsc.14.6.615.24872

Hinds, P. J., & Kiesler, S. (2002). *Distributed work*. Cambridge, MA: MIT Press.

Hinds, P. J., & Mortensen, M. (2005). Understanding conflict in geographically distributed teams: The moderating effects of shared identity, shared context, and spontaneous communication. *Organization Science*, *16*(3), 290–307. doi:10.1287/orsc.1050.0122

Hinds, P., Liu, L., & Lyon, J. (2011). Putting the global in global work: An intercultural lens on the practice of cross-national collaboration. *The Academy of Management Annals*, *5*(1), 135–188. doi:10.1080/19416520.2011.586108

Hoch, J. (2014). Leading Virtual Teams: Hierachical Leadership, Structural Supports and shared Team Leadership. *The Journal of Applied Psychology*, *99*, 390–403. doi:10.1037/a0030264 PMID:23205494

Hofstede, G., Hofstede, G. J., & Minkov, M. (2010). Cultures and Organizations: Software of the Mind (3rd ed.). McGraw-Hill USA.

Hofstede, G. (2001). *Culture's Consequences: Comparing Values, bahaviors, Institutions, and Organizations Across nations* (2nd ed.). Thousand Oaks, CA: Sage Publications.

Hofstede, G. (2001). *Culture's consequences: Comparing Values, behaviors, Institutions, and Organizations Across Nations* (2nd ed.). Thousand Oaks, CA: Sage Publications.

Hofstede, G. H. (1980). *Culture's consequences: International differences in workrelated values*. London: Sage.

Hogue, T. (1993). *Community-based collaboration: Community wellness multiplied*. Bend, OR: Chandler Center for Community Leadership. Oregon Center for Community Leadership, Oregon State University.

Holmgren, A. (2006). *Ledelse med et narrativt og poststrukturalistisk perspektiv. I: Helth, P. Lederskabelse – det personligevlederskab*. Roskilde: Samfundslitteratur.

Holsapple, C. W., & Wu, J. (2011). An elusive antecedent of superior firm performance: The knowledge management factor. *Decision Support Systems*, *52*(1), 271–283. doi:10.1016/j.dss.2011.08.003

Honan, M. (2014, August 8). *The Most Fascinating Profile You'll ever Read About a Guy and His Boring Startup.* Retrieved from Wired.com: https://www.wired.com/2014/08/the-most-fascinating-profile-youll-ever-read-about-a-guy-and-his-boring-startup/

Hosking, D. M. (2006). Discourses of relations and relational processes. In O. Kyriakidou & M. Ozbilgin (Eds.), *Relational Perspectives in Organizational Studies: A Research Companion* (pp. 265–277). Cheltenham, UK: Edward Elgar Publishing Ltd. doi:10.4337/9781781950548.00020

Hosking, D. M. (2011). Telling tales of relations: Appreciating relational constructionism. *Organization Studies*, *32*(1), 47–65. doi:10.1177/0170840610394296

Hosking, D. M., Dachler, H. P., & Gergen, K. J. (1995). *Management and Organization: Relational alternatives to individualisation.* Aldershot, UK: Avenbury.

How to play. (2010, May 12). Retrieved from http://www.urgentevoke.com/page/how-to-play

Hoyer, P., & Steyaert, C. (2015). Narrative identity construction in times of career change: Taking note of unconscious desires. *Human Relations*, *68*(12), 1837–1863. doi:10.1177/0018726715570383

Hsiao, R.-L., Tsai, D.-H., & Lee, C.-F. (2012). Collaborative knowing: The adaptive nature of cross-boundary spanning. *Journal of Management Studies*, *49*(3), 463–491. doi:10.1111/j.1467-6486.2011.01024.x

Hubud homepage. (2017). Retrieved May 30, 2017, from https://www.hubud.org/

Huffaker, D. (2010). Dimensions of leadership and social influence in online communities. *Human Communication Research*, *36*(4), 593–617. doi:10.1111/j.1468-2958.2010.01390.x

Hughes, J. (Ed.). (2012). *SAGE internet research methods*. London, UK: Sage. doi:10.4135/9781446263327

Hult, M., & Lennung, S.-Å. (1980). Towards a definition of action research: A note and bibliography. *Journal of Management Studies*, *17*(2), 241–250. doi:10.1111/j.1467-6486.1980.tb00087.x

Humphrey, W. (2006). *TSP: Coaching Development Teams*. Addison-Wesley Professional.

Hunsaker, P. L., & Hunsaker, J. S. (2008). Virtual teams: A leader's guide. *Team Performance Management*, *14*(1/2), 86–101. doi:10.1108/13527590810860221

Hunsaker, P. L., & Hunsaker, J. S. (2008). Virtual Teams: A Leaders Guide. *Team Performance Management*, *14*(1/2), 86–101.

Huston, T. L. (2009, September 30). What's love got to do with it? Why some marriages succeed and others fail. *Personal Relationships*, *16*(3), 301–327. doi:10.1111/j.1475-6811.2009.01225.x

Hwang, E. H., Singh, P. V., & Argote, L. (2015). Knowledge sharing in online communities: Learning to cross geographic and hierarchical boundaries. *Organization Science*, *26*(6), 1593–1611. doi:10.1287/orsc.2015.1009

Internet Live Stats homepage. (2017). *Data gathered by International Telecommunication Union (ITU), World Bank, and United Nations Population Division.* Retrieved May 21, 2017, from http://www.internetlivestats.com/internet-users/

Isaacs, W. (1999). *Dialogue and the art of thinking together.* New York: Currency Doubleday.

Iversen, J. H., Mathiassen, L., & Nielsen, P. A. (2004). Managing risk in software process improvement: An action research approach. *Management Information Systems Quarterly*, *28*(3), 395–433. doi:10.2307/25148645

Jarrahi, M. H., & Thomson, L. (2017). The interplay between information practices and information context: The case of mobile knowledge workers. *Journal of the Association for Information Science and Technology*, *68*(5), 1073–1089. doi:10.1002/asi.23773

Jarvenpaa, S. L., & Leidner, D. E. (1999). Communication and trust in global virtual teams. *Organization Science*, *10*(6), 791–815. doi:10.1287/orsc.10.6.791

Jenkins, H. (1992). *Textual poachers: Television fans and participatory culture*. New York, NY: Routledge.

Jenkins, H., Ford, S., & Green, J. (2013). *Spreadable media: Creating value and meaning in a networked culture*. New York: New York University Press.

Jensen, A. F. (2005). Mellem ting – Foucaults filosofi. Det lille Forlag.

Jensen, I. (2017). *Unpublished fieldnotes from Course in Intercultural awareness in Kolding Municipality*. Academic Press.

Jensen, I. (2006a). The practice of intercultural communication: Reflections for professionals in cultural meetings. In L. A. Samovar, R. E. Porter, & E. R. McDaniel (Eds.), *Intercultural communication – a reader* (pp. 39–48). Wadsworth.

Jensen, I. (2006b). The Aspect of Power in Intercultural Communication Practice. In Ø. Dahl, I. Jensen, & P. Nynäs (Eds.), *Bridges of Understanding* (pp. 85–100). Oslo: Unipub.

Jensen, I. (2011a). If culture is practice …? A practice theoretical perspective on intercultural communication and mediation. In J. Finch & P. Nynäs (Eds.), *Transforming Otherness* (pp. 9–33). Transaction.

Jensen, I. (2011b). *Jobsamtaler med etniske minoriteter* [Job interviews with ethnic minorities]. Frederiksberg: Roskilde University Press.

Jensen, I. (2017). Online Leadership and Communication across Cultures. In D. Kolbaek (Ed.), *Online Collaboration and Communication in Contemporary Organizations*. Hershey, PA: IGI Global.

Jones, Q., Ravid, G., & Rafaeli, S. (2004). Information overload and the message dynamics of online interaction spaces: A theoretical model and empirical exploration. *Information Systems Research*, *15*(2), 194–210. doi:10.1287/isre.1040.0023

Jonsen, K., Butler, C. L., Makela, K., Piekkari, R., Drogendijk, R., Lauring, J., ... Zander, L. (2013). Processes of international collaboration in management research: A reflexive, autoethnographic approach. *Journal of Management Inquiry*, *22*(4), 394–413. doi:10.1177/1056492612464429

Josefy, M., Kuban, S., Ireland, R. D., & Hitt, M. A. (2015). All things great and small: Organizational size, boundaries of the firm, and a changing environment. *The Academy of Management Annals*, *9*(1), 715–802. doi:10.1080/19416520.2015.1027086

Joshi, A., Lazarova, M. B., & Liao, H. (2009). Getting everyone on board: The role of inspirational leadership in geographically dispersed teams. *Organization Science*, *20*(1), 240–252. doi:10.1287/orsc.1080.0383

Justesen, L., & Mik-Meyer, N. (2012). *Qualitative research methods in organisation studies*. Copenhagen, Denmark: Hans Reitzels Forlag.

Kahai, S. S., & Cooper, R. B. (2003). Exploring the core concepts of media richness theory: The impact of cue multiplicity and feedback immediacy on decision quality. *Journal of Management Information Systems*, *20*(1), 263–299. doi:10.1080/07421222.2003.11045754

Kahneman, D. (2011). *Thinking fast and slow*. London: Allen Lane an imprint of Penguin Books.

Kakihara, M., & Sørensen, C. (2001). *Organisational knowledge in mobile workplaces: A theoretical investigation*. Paper presented at Mobilize! Workshop on interventions in the social, cultural and interactional analysis of mobility, ubiquity and information and communication technology, Windsor, UK.

Kanawattanachai, P., & Yoo, Y. (2002). Dynamic nature of trust in virtual teams. *The Journal of Strategic Information Systems*, *11*(3), 187–213. doi:10.1016/S0963-8687(02)00019-7

Karanasios, S., & Allen, D. (2014). Mobile technology in mobile work: Contradictions and congruencies in activity systems. *European Journal of Information Systems*, *23*(5), 529–542. doi:10.1057/ejis.2014.20

Kasper-Fuehrer, E. C., & Ashkanasy, N. M. (2001). Communicating trustworthiness and building trust in interorganizational virtual organizations. *Journal of Management*, *27*(3), 235–254. doi:10.1016/S0149-2063(01)00090-3

Kayes, D. C. (2002). Experiential learning and its critics: Preserving the role of experience in management learning and education. *Academy of Management Learning & Education*, *1*(2), 137–149. doi:10.5465/AMLE.2002.8509336

Kelley, E. (2001). Keys to effective virtual global teams. *The Academy of Management Executive*, *15*(2), 132–133. doi:10.5465/AME.2001.4615036

Keltner, D. (2016). *The Power Paradox: How We Gain and Lose Influence*. New York: Penguin Press.

Kemmis, S., McTaggart, R., & Nixon, R. (2013). *The action research planner: Doing critical participatory action research*. Springer Science & Business Media.

Kemmis, S., & Mutton, R. (2012). Education for sustainability (EfS): Practice and practice architectures. *Environmental Education Research*, *18*(2), 187–207. doi:10.1080/13504622.2011.596929

Kenney, M. (1987). The Ethical Dilemmas of University – Industry Collaborations. *Journal of Business Ethics*, *6*(2), 127–135. doi:10.1007/BF00382026

Kerth, N. L. (2001). *Project Retrospectives: A Handbook for Team Reviews*. Dorset House Publishing Company, Incorporated.

Kiesler, S., Siegel, J., & McGuire, W. T. (1984). Social psychological aspects of computer-mediated communication. *The American Psychologist*, *39*(10), 1123–1134. doi:10.1037/0003-066X.39.10.1123

Kietzmann, J., Plangger, K., Eaton, B., Heilgenberg, K., Pitt, L., & Berthon, P. (2013). Mobility at work: A typology of mobile communities of practice and contextual ambidexterity. *The Journal of Strategic Information Systems*, *22*(4), 282–297. doi:10.1016/j.jsis.2013.03.003

Kim, E. (2016, September 28). *Slack, the red hot $3.8 billion startup, has a hidden meaning behind its name*. Retrieved from Business Insider UK: http://uk.businessinsider.com/where-did-slack-get-its-name-2016-9?_ga=1.138509299.1175501442.1475074973?r=US&IR=T

Kim, A. J. (2000). *Community building on the web: Secret strategies for successful online communities*. Berkley, CA: Addison-Wesley Longman Publishing.

Kirkman, B. L., Cordery, J. L., Mathieu, J., Rosen, B., & Kukenberger, M. (2013). Global organizational communities of practice: The effects of nationality diversity, psychological safety, and media richness on community performance. *Human Relations*, *66*(3), 333–362. doi:10.1177/0018726712464076

Kirkman, B. L., & Mathieu, J. E. (2005). The dimensions and antecedents of team virtuality. *Journal of Management*, *31*(5), 700–718. doi:10.1177/0149206305279113

Kirkman, B. L., Rosen, B., Gibson, C. B., Tesluk, P. E., & McPherson, S. O. (2002). Five challenges to virtual team success: Lessons from Sabre, Inc. *The Academy of Management Executive*, *16*(3), 67–79. doi:10.5465/AME.2002.8540322

Kleinbaum, A. M., & Stuart, T. E. (2014). Network responsiveness: The social structural microfoundations of dynamic capabilities. *The Academy of Management Perspectives*, *28*(4), 353–367. doi:10.5465/amp.2013.0096

Klitmøller, A., & Lauring, J. (2013). When global virtual teams share knowledge: Media richness, cultural distance and language commonality. *Journal of World Business*, *48*(3), 398–406. doi:10.1016/j.jwb.2012.07.023

Knight, R. (2016). IIow to Makc Your Onc-on-Ones with Employees More Productive. *Harvard Business Review*.

Kolb, D. G., Prussia, G., & Francoeur, J. (2009). Connectivity and leadership: The influence of online activity on closeness and effectiveness. *Journal of Leadership & Organizational Studies, 15*(4), 342-352.

Kolbaek, D. (2012). Proactive review – How to make your organisation learn from experience. BoD.

Kolbaek, D. (2017). Design-based research as a "smart" methodology for studying learning in the context of work. *Proceedings of IRIS40, Information Systems Research Seminar in Scandinavia.* Retrieved from https://pdfs.semanticscholar.org/26b5/411610955317c57761f7902dd6ed704b0671.pdf

Kolbæk, D. (2011). *Proactive review*. Copenhagen, Denmark: DJØF Forlag.

Kolbaek, D. (2012). *Proactive reviews: How to make your organization learn from experience*. Copenhagen: Books on Demand GmbH.

Kolbaek, D. (2014). *Learning from experience in the context of work*. Aalborg, Denmark: Aalborg University Press.

Kolbaek, D. (2014). *Learning From Experience in the Context of Work*. Aalborg: Aalborg University Press.

Kolbaek, D. (2015). Proactive Review. In S. McIntyre, K. Dalkir, P. Paul, & I. C. Kitimbo (Eds.), *Utilizing Evidence-Based Lessons Learned for Enhanced Organizational Innovation and Change* (pp. 82–100). Hershey, PA: IGI Global. doi:10.4018/978-1-4666-6453-1.ch005

Kolbaek, D. (2017). Proactive Review som laeringsrum for online-ledere. In H. D. Keller & S. Frimann (Eds.), *Lederes laeringsrum* (pp. 149–179). Aalborg, Denmark: Aalborg Universitetsforlag.

Kolbaek, D. (2017a). Design-based research as a methodology for studying learning in the context of work. In D. Kolbaek (Ed.), *Online Collaboration and Communication in Contemporary Organizations*. Hershey, PA: IGI Global.

Kolbaek, D. (2017b). Online Leaders Increase Three Types of Capital. In D. Kolbaek (Ed.), *Online Collaboration and Communication in Contemporary Organizations*. Hershey, PA: IGI Global.

Kolbaek, D. (2017c). Online Leadership and Learning. In D. Kolbaek (Ed.), *Online Collaboration and Communication in Contemporary Organizations*. Hershey, PA: IGI Global.

Kolb, D. G., Caza, A., & Collins, P. D. (2012). States of connectivity: New questions and new directions. *Organization Studies*, *33*(2), 267–273. doi:10.1177/0170840611431653

Kollock, P., & Smith, M. (1996). Managing the Virtual Commons: Cooperation and conflict in computer communities. In S. Herring (Ed.), *Computer-Mediated Communication: Linguistic, Social, and Cross-Cultural Perspectives* (pp. 109–128). Amsterdam: John Benjamins. doi:10.1075/pbns.39.10kol

Koroma, J., Hyrkkänen, U., & Vartiainen, M. (2014). Looking for people, places and connections: Hindrances when working in multiple locations: A review. *New Technology, Work and Employment*, *29*(2), 139–159. doi:10.1111/ntwe.12030

Korthagen, F. A. J., Noordewier, S. A., & Zwart, R. C. (2014). Educator–student contact: Exploring a basic but complicated concept. *Teaching and Educator Education*, *40*, 22–32. doi:10.1016/j.tate.2014.01.006

Korzynski, P. (2013). Online social networks and leadership: Implications of a new online working environment for leadership. *International Journal of Manpower*, *34*(8), 975–994. doi:10.1108/IJM-07-2013-0173

Kozinets, R. (2015). *Netnography: Redefined*. London: Sage. doi:10.1002/9781118767771.wbiedcs067

Kramer, R. M., & Tyler, T. R. (1996). *Trust in Organizations: Frontiers of Theory and Research*. Thousand Oaks, CA: SAGE Publications.

Kramer, R., & Tyler, T. (Eds.). (1996). *Trust in organizations: Frontiers of theory and research*. Thousand Oaks, CA: Sage.

Kraushaar, J. M., & Novak, D. C. (2010). Examining the Affects of Student Multitasking with Laptops during the Lecture. *Journal of Information Systems Education*, *21*(2), 241–251.

Kristula, D. (2009). *History of the Internet*. Retrieved 1 2014, 10, from History of the Internet: http://www.davesite.com/webstation/net-history1.shtml

Kumparak, G. (2015, February 5). *Slack's Co-Founders Take Home The Crunchie For Founder Of The Year.* Retrieved from Tech Crunch: https://techcrunch.com/2015/02/05/slacks-co-founders-take-home-the-crunchie-for-founder-of-the-year/

Kvale, S. (2002). *Interview: En introduction til det kvalitative forskningsinterview*. København: Hans Reitzels Forlag.

Kvale, S. (2007). *Doing interviews*. SAGE Publications Ltd. doi:10.4135/9781849208963

Lacity, M. C., Willcocks, L. P., & Rottman, J. W. (2008). Global Outsourcing of Back Office Services: Lessons, Trends and Enduring Challenges. *Strategic Outsourcing*, *1*(1), 13–34. doi:10.1108/17538290810857457

LaFrance, A. (2016, June 21). *Slack, The Facebook Slayer.* Retrieved from The Atlantic: https://www.theatlantic.com/technology/archive/2016/06/slack-eats-internet/488033/

Landow, G. P. (1992). *Hypertext: The Convergence of Contemporary Critical Theory and Technology*. Baltimore, MD: Johns Hopkins University Press.

Langer, P. (2016). The Research Vignette: Reflexive Writing as Interpretative Representation of Qualitative Inquiry—A Methodological Proposition. *Qualitative Inquiry, 22*(9), 735–744. Retrieved from qix.sagepub.com

Langer, P. C. (2016). The research vignette: Reflexive writing as interpretative representation of qualitative inquiry—a methodological proposition. *Qualitative Inquiry*, *22*(9), 735–744. qix.sagepub.com. doi:10.1177/1077800416658066

Larsson, A. (2005). *Engineering know-how: Why social connectedness matters to global design teams* (Doctoral thesis). Department of Applied Physics and Mechanical Engineering, Lulea University of Technology, Lulea, Sweden.

Latour, B. (2005). *Reassembling the social: An introduction to actor-network-theory*. Oxford, UK: Oxford University Press.

Lauring, J. (2008). Rethinking social identity theory in international encounters: Languages as a negotiated object for identity making. *International Journal of Cross Cultural Management*, *8*(3), 343–361. doi:10.1177/1470595808096673

Lave, J. (1996). Situated learning in communities of practice. In L. B. Resnick, J. M. Levine, & S. D. Teasley (Eds.), *Perspectives on Socially Shared Cognition* (pp. 63–81). Washington, DC: American Psychological Association. Retrieved from http://www.ecologyofdesigninhumansystems.com/wp-content/uploads/2012/12/Lave-Situating-learning-in-communities-of-practice.pdf

Lave, J., & Wenger, E. (1991). *Situated learning: Legitimate peripheral participation*. Cambridge, UK: Cambridge University Press. doi:10.1017/CBO9780511815355

Lawrence, T., Hardy, C., & Phillips, N. (2002). Institutional effects of interorganisational collaboration: The emergence of proto-institutions. *Academy of Management Journal*, *45*(1), 281–290. doi:10.2307/3069297

Leadbeater, C. (2008). *We-think*. London, UK: Profile Books.

Ledell, G. H., & Scott, R. (1925). *Arete*. Retrieved May 15, 2017, from A Greek-English Lexicon: http://www.perseus.tufts.edu/hopper/text?doc=LSJ%20a)reth/&lang=original

Lee, H., & Liebenau, J. (1999). Time in organizational studies: Towards a new research direction. *Organization Studies*, *20*(6), 1035–1058. doi:10.1177/0170840699206006

Lee, M. R. (2009). E-ethical leadership for virtual project teams. *International Journal of Project Management*, *27*(5), 456–463. doi:10.1016/j.ijproman.2008.05.012

Leonardi, P. M. (2011). Innovation blindness: Culture, frames, and cross-boundary problem construction in the development of new technology concepts. *Organization Science*, *22*(2), 347–369. doi:10.1287/orsc.1100.0529

Leonardi, P. M., & Bailey, D. E. (2008). Transformational technologies and the creation of new work practices: Making implicit knowledge explicit in task-based offshoring. *Management Information Systems Quarterly*, *32*(2), 411–436. doi:10.2307/25148846

Leung, K., & Wang, J. (2015). Social processes and team creativity in multicultural teams: A socio-technical framework. *Journal of Organizational Behavior*, *36*(7), 1008–1025. doi:10.1002/job.2021

Levesque, V. R., Calhoun, A. J., Bell, K. P., & Johnson, T. R. (2017). Turning contention into collaboration: Engaging power, trust, and learning in collaborative networks. *Society & Natural Resources*, *30*(2), 245–260. doi:10.1080/08941920.2016.1180726

Levi, D. (2015). *Group dynamics for teams* (5th ed.). Thousand Oaks, CA: Sage Publications Inc.

Levinsen, K., Ørngreen, R., & Buhl, M. (2013). Telepresence as Educational Practice in the Third Teaching-Room–a Study in Advanced Music Education. In *Proceedings of the 12th European Conference of E-learning. Sonning Common: Academic Conferences and Publishing International Limited* (pp. 250–257). Academic Press.

Lévy, P., & Bonomo, R. (1999). *Collective intelligence: Mankind's emerging world in cyberspace*. Cambridge, MA: Perseus Publishing.

Lin, C. A., & Creswell, K. W. (1989). Effects of televised lecture presentation styles on student learning. *Learning, Media and Technology*, *15*(1), 37–52.

Lindholm, A. B. (2012). *Lederskab på distancen*. L&R Business.

Lipnack, J., & Stamps, J. (1997). *Virtual teams: reaching across space, time and organization with technology*. London: John Wiley & Sons.

Lipnack, J., & Stamps, J. (1999). Virtual teams: The new way to work. *Strategy and Leadership*, *27*(1), 14–19. doi:10.1108/eb054625

Lock, J. V., & Redmond, P. (2006). International online collaboration: Modeling online learning and teaching. *Journal of Online Learning and Teaching / MERLOT*, *2*(4), 233–247.

Lund Dean, K., & Jolly, J. P. (2012). Student identity, disengagement, and learning. *Academy of Management Learning & Education*, *11*(2), 228–243. doi:10.5465/amle.2009.0081

Lundin, J., & Magnusson, M. (2003). Collaborative learning in mobile work. *Journal of Computer Assisted Learning*, *19*(May), 273–283. doi:10.1046/j.0266-4909.2003.00029.x

Lüscher, L. S., & Lewis, M. W. (2008). Organizational change and managerial sensemaking: Working through paradox. *Academy of Management Journal*, *51*(2), 221–240. doi:10.5465/AMJ.2008.31767217

Lyons, A., Reysen, S., & Pierce, L. (2012). Video lecture format, student technological efficacy, and social presence in online courses. *Computers in Human Behavior*, *28*(1), 181–186. doi:10.1016/j.chb.2011.08.025

Lyotard, J.-F. (1997). *The Postmodern Condition: A report on Knowledge* (G. Bennington & B. Massumi, Trans.). Minneapolis, MN: University of Minnesota Press.

Mailhot, C., Gagnon, S., Langley, A., & Binette, L.-F. (2016). Distributing leadership across people and objects in a collaborative research project. *Leadership*, *12*(1), 53–85. doi:10.1177/1742715014543578

Maitlis, S., & Christianson, M. (2014). Sensemaking in organizations: Taking stock and moving forward. *The Academy of Management Annals*, *8*(1), 57–125. doi:10.1080/19416520.2014.873177

Majchrzak, A., Rice, R. E., Malhotra, A., King, N., & Ba, S. L. (2000). Technology adaptation: The case of a computer-supported inter-organizational virtual team. *Management Information Systems Quarterly*, *24*(4), 569–600. doi:10.2307/3250948

Mäkinen, S. (2012). Mobile work and its challenges to personal and collective information management. *Information Research, 17*(3). Retrieved from http://www.informationr.net/ir/17-3/paper522.html#.VpgtuRHCoUF

Malhotra, A., & Majchrzak, A. (2014). Enhancing performance of geographically distributed teams through targeted use of information and communication technologies. *Human Relations*, *67*(4), 389–411. doi:10.1177/0018726713495284

Malhotra, A., Majchrzak, A., Carman, R., & Lott, V. (2001). Radical innovation without collocation: A case study at Boeing-Rocketdyne. *Management Information Systems Quarterly*, *25*(2), 229–249. doi:10.2307/3250930

Maloney, S., Storr, M., Morgan, P., & Ilic, D. (2013). The effect of student self-video of performance on clinical skill competency: A randomised controlled trial. *Advances in Health Sciences Education: Theory and Practice*, *18*(1), 81–89. doi:10.1007/s10459-012-9356-1 PMID:22354337

Manafy, M., & Gautschi, H. (2011). *Dancing with digital natives*. Medford, NJ: CyberAge Books.

Mäntymäki, M., & Riemer, K. (2014). Digital natives in social virtual worlds: A multi-method study of gratifications and social influences in Habbo Hotel. *International Journal of Information Management*, *34*(2), 210–220. doi:10.1016/j.ijinfomgt.2013.12.010

March, S. T., & Smith, G. F. (1995). Design and natural science research on information technology. *Decision Support Systems*, *15*(4), 251–266. doi:10.1016/0167-9236(94)00041-2

Mark, G. (2002). Conventions and commitments in distributed CSCW groups. *Computer Supported Cooperative Work*, *11*(3), 349–387. doi:10.1023/A:1021289427473

Martin, J. N. (2015). Revisiting intercultural communication competence: Where to go from he- re. *International Journal of Intercultural Relations*, *48*, 6–8. doi:10.1016/j.ijintrel.2015.03.008

Martins, L. L., Gilson, L. L., & Maynard, M. T. (2004). Virtual teams: What do we know and where do we go from here? *Journal of Management*, *30*(6), 805–835. doi:10.1016/j.jm.2004.05.002

Mathiassen, L. (2002). Collaborative practice research. *Scandinavian Journal of Information Systems*, *14*, 57–76.

Mattessich, P. W., Murry-Close, M., & Monsey, B. R. (2001). *Collaboration: What makes it work* (2nd ed.). St. Paul, MN: Wilder Publishing Center.

Mayer, R., Davis, J., & Schoorman, S. (1995). An integration model of organizational trust. *Academy of Management Review, 20*, 709–734.

Maynard, M. T., Mathieu, J. E., Rapp, T. L., & Gilson, L. L. (2012). Something(s) old and something(s) new: Modeling drivers of global virtual team effectiveness. *Journal of Organizational Behavior, 33*(3), 342–365. doi:10.1002/job.1772

Mazmanian, M., Orlikowski, W. J., & Yates, J. (2013). The Autonomy Paradox: The Implications of Mobile Email Devices for Knowledge Professionals. *Organization Science, 24*(5), 1337–1357. doi:10.1287/orsc.1120.0806

Maznevski, M. L., & Chudoba, K. M. (2000). Bridging space over time: Global virtual team dynamics and effectiveness. *Organization Science, 11*(5), 473–492. doi:10.1287/orsc.11.5.473.15200

McAdam, R., O'Hare, T., & Moffett, S. (2008). Collaborative knowledge sharing in composite new product development: An aerospace study. *Technovation, 28*(5), 245–256. doi:10.1016/j.technovation.2007.07.003

McAfee, A. (2006). Enterprise 2.0: The dawn of emergent collaboration. *Sloan Management Review, 47*(3), 21–28.

McCall, M. W. Jr. (2004). Leadership development through experience. *The Academy of Management Executive, 18*(3), 127–130. doi:10.5465/AME.2004.14776183

McDaniel, T., Reitsma, R., & Duan, X. (2016). *Forrester's Global Business Technographics Telecommunications And Mobility Workforce Survey, 2016: Overview*. Retrieved January 20, 2017, from https://www.forrester.com/report/Forresters+Global+Business+Technographics+Telecommunications+And+Mobility+Workforce+Survey+2016+Overview/-/E-RES136155

McGonigal, J. (2008). Engagement economy: The future of massively scaled collaboration and participation. *Institute for the Future*. Retrieved from www.iftf.org/uploads/media/Engagement_Economy_sm_0.pdf

McGonigal, J. (2011). *Reality is broken: Why games make us better and how they can change the world*. London, UK: Jonathan Cape.

McKenzie, J., & Aitken, P. (2012). Learning to lead the knowledgeable organization. *Strategic HR Review, 11*(6), 329–334. doi:10.1108/14754391211264794

McKenzie, J., & Van Winkelen, C. (2004). *Understanding the knowledgeable organization: Nurturing knowledge competence*. London: Cengage Learning EMEA.

McKenzie, J., Woolf, N., van Winkelen, C., & Morgan, C. (2009). Cognition in strategic decision-making: A model of nonconventional thinking capacities for complex situations. *Management Decision, 47*(2), 209–232. doi:10.1108/00251740910938885

McKnight, D. H., Cummings, L. L., & Chervany, N. L. (1998). Initial trust formation in new organizational relationships. *Academy of Management Review, 23*, 473–490.

McLaughlin, K. (2004). Towards a "modernized" voluntary and community sector? *Public Management Review, 6*(4), 555–562. doi:10.1080/1471903042000303337

McLuhan, M. (1964). *Understanding Media: The Extensions of Man*. New York: McGraw Hill.

McNamee, S. (1998). Re-inscribing organizational wisdom and courage: The relationally engaged organization. In S. Srivastva & D. L. Cooperrider (Eds.), *Organizational wisdom and executive courage* (pp. 101–117). San Francisco: New Lexington Press.

McNiff, J. (2013). *Action research: Principles and practice*. Routledge.

Mead, G. H. (1913). The social self. *The Journal of Philosophy, Psychology, and Scientific Methods*, *10*(14), 374–380. doi:10.2307/2012910

Mead, G. H. (1934). *Mind, self, and society: From the standpoint of a social behaviorist*. Chicago: University of Chicago press.

Mehrabian, A. (1971). *Silent Messages: Implicit Communication of Emotions and Attitude*. Belmont, CA: Wadsworth Publishing Compan.

Michailova, S., & Hutchings, K. (2006). National cultural influences on knowledge sharing: A comparison of China and Russia. *Journal of Management Studies*, *43*(3), 383–405. doi:10.1111/j.1467-6486.2006.00595.x

Miettinen, R., Samra-Fredericks, D., & Yanow, D. (2009). Re-turn to practice: An introductory essay. *Organization Studies*, *30*(12), 1309–1327. doi:10.1177/0170840609349860

Mik-Meyer, N., & Villadsen, K. (2007). *Magtens former: Sociologiske perspektiver på statens møde med borgeren*. København: Hans Reitzel.

Miles, M. B., & Huberman, A. M. (1994). *Qualitative data analysis: An expanded sourcebook*. London: Sage.

Miles, M. B., & Huberman, A. M. (1994). *Qualitative Data Analysis: An Expanded Sourcebook*. Thousand Oaks, CA: Sage Publications Inc.

Mingers, J. (2001). Combining IS research methods: Towards a pluralist methodology. *Information Systems Research*, *12*(3), 240–259. doi:10.1287/isre.12.3.240.9709

Mitchell, A. & Zigurs, I. (2009). Trust in virtual teams: solved or still a mystery? *ACM SIGMIS Database*, *40*(3).

Mohrman, S. A. (1999). The contexts for geographically dispersed teams and networks. *Journal of Organizational Behavior*, *6*, 63–80.

Mondada, L. (2006). Participants' online analysis and multimodal practices: Projecting the end of the turn and the closing of the sequence. *Discourse Studies*, *8*(1), 117–129. doi:10.1177/1461445606059561

Montoya-Weiss, M. M., Massey, A. P., & Song, M. (2001). Getting it together: Temporal coordination and conflict management in global virtual teams. *Academy of Management Journal*, *44*(6), 1251–1262. doi:10.2307/3069399

Moyer, M. W. (2016). Eye Contact: How Long Is Too Long? *Scientific American Mind*, *27*(1), 8. doi:10.1038/scientificamericanmind0116-8

Mulder, M. (1977). *The daily power came*. Leiden, The Netherlands: Maritinus Nijhoff Social Science. doi:10.1007/978-1-4684-6951-6

Muukkonen, H., Hakkarinen, K., Li, S., & Vartiainen, M. (2014). Tracking mobile workers ' daily activities with the Contextual Activity Sampling System. *Design, User Experience, and Usability. User Experience Design for Diverse Interaction Platforms and Environments*, 289–300.

Myers, M. D. (2009). *Qualitative Research in Business & Management*. London: Sage Publications.

Myers, M. D., & Avison, D. (2002). An Introduction to Qualitative Research in Information Systems. In M. D. Myers & D. Avison (Eds.), *Qualitative Research in Information Systems - A Reader* (pp. 3–12). London: Sage Publications. doi:10.4135/9781849209687.n1

Nahapiet, J., & Ghoshal, S. (1998). Social capital, intellectual capital, and the organizational advantage. *Academy of Management Review*, *23*(2), 242–266.

Nahapiet, J., & Ghoshal, S. (1998). Social Capital, Intellectual Capital, and the Organizational Advantage. *Academy of Management Review, 23*(2), 242–266.

Nakayama, T. K., & Halualani, R. T. (Eds.). (2010). *The handbook of critical intercultural communication.* West Sussex, UK: Wiley-Blackwell. doi:10.1002/9781444390681

Neeley, T. B., Hinds, P. J., & Cramton, C. D. (2012). The (un)hidden turmoil of language in global collaboration. *Organizational Dynamics, 41*(3), 236–244. doi:10.1016/j.orgdyn.2012.03.008

Nelson, S. B., Jarrahi, M. H., & Thomson, L. (2017). Mobility of knowledge work and affordances of digital technologies. *International Journal of Information Management, 37*(2), 54–62. doi:10.1016/j.ijinfomgt.2016.11.008

New York Stock Exchange. (n.d.). *Oracle stock.* Retrieved 20 February 2017 from https://www.nyse.com/quote/XNYS:ORCL

Newitz, A. (2016, March 9). *What Slack is doing to our offices—and our minds.* Retrieved from Ars Technica: https://arstechnica.com/information-technology/2016/03/what-slack-is-doing-to-our-offices-and-our-minds/

Nicolini, D. (2012). *Practice theory, work, and organization: An introduction.* Oxford university press.

Nonaka, I., Hirose, A., & Takeda, Y. (2016). 'Meso'-foundations of dynamic capabilities: Team-Level synthesis and distributed leadership as the source of dynamic creativity. *Global Strategy Journal, 6*(3), 168–182. doi:10.1002/gsj.1125

Nonaka, I., & Konno, N. (1998). The concept of "ba": Building a foundation for knowledge creation. *California Management Review, 40*(3), 40–54. doi:10.2307/41165942

Nonaka, I., & Takeuchi, H. (1995). *The knowledge creation company: How Japanese companies create the dynamics of innovation.* New York: Oxford University Press.

Nonaka, I., & Takeuchi, H. (1995). *The knowledge-creating company.* New York, NY: Oxford University Press.

Nonaka, I., & Takeuchi, H. (1995). *The Knowledge-Creating Company.* New York: Oxford University Press, Inc.

Nonaka, I., & Takeuchi, H. (1995). *The Knowledge-Creating Company: How Japanese Companies Create the Dynamics of Innovation.* New York: Oxford University Press.

Nonaka, I., Toyama, R., & Byosière, P. (2001). A theory of organizational knowledge creation: Understanding the dynamic process of creating knowledge. In M. Dierkes, A. B. Antal, J. Child, & I. Nonaka (Eds.), *Handbook of organizational learning and knowledge* (pp. 491–517). Oxford, UK: Oxford University Press.

Nonaka, I., Toyama, R., & Konno, N. (2000). ECI, Ba and leadership: A unified model of dynamic knowledge creation. *Long Range Planning, 33*(1), 5–34. doi:10.1016/S0024-6301(99)00115-6

Nørskov, S. V., & Rask, M. (2011). Observation of Online Communities: A Discussion of Online and Offline Observer Roles in Studying Development, Cooperation and Coordination in an Open Source Software Environment. *Forum Qualitative Sozialforschung / Forum: Qualitative Social Research, 12*(3).

Northouse, P. G. (2016). *Leadership: Theory and practice.* Thousand Oaks, CA: Sage Publications.

Nortvig, A.-M. (2015b). "Sometimes, I feel a bit decoupled": Strategies in Videoconference Teaching. In *Proceedings of the 1th D4|Learning international Conference Innovations in Digital Learning for Inclusion* (pp. 81–89). Academic Press.

Nortvig, A.-M. (2014). E-learning in poly-topic settings. *Electronic Journal of E-Learning, 12*(2), 206–214.

Nortvig, A.-M. (2015a). *"At sidde på skolebænken i egen sofa": En undersøgelse af hvordan e-læring ses at have indflydelse på muligheder for deltagelse og tilstedevær i undervisning og dermed udvikling af professionel identitet i relation til professionsuddannelse.* Aalborg: Aalborg Universitetsforlag.

Nortvig, A.-M. (2017). Technology and Presence. In D. Kolbaek (Ed.), *Online Collaboration and Communication in Contemporary Organizations*. Hershey, PA: IGI Global.

Nyberg, D. (2009). Computers, customer service operatives and cyborgs: Intra-actions in call centres. *Organization Studies*, *30*(11), 1181–1199. doi:10.1177/0170840609337955

Nyström, A. C., & Asproth, V. (2013). Virtual Teams – Support for Technical Communication? *Journal of Organisational Transformation & Social Change*, *10*(1), 64–80. doi:10.1179/1477963312Z.0000000006

O'Leary, M. B., & Cummings, J. N. (2007). The spatial, temporal, and configurational characteristics of geographic dispersion in teams. *Management Information Systems Quarterly*, *31*(3), 433–452. doi:10.2307/25148802

O'Leary, M. B., & Mortensen, M. (2010). Go (con)figure: Subgroups, imbalance, and isolates in geographically dispersed teams. *Organization Science*, *21*(1), 115–131. doi:10.1287/orsc.1090.0434

Ogden, C. K., & Richads, I. A. (1989). *The Meaning of Meaning: A Study of the Influence of Language Upon Thought and Of the Science of Symbolism*. New York: Harcourt Brace Javonovich Publishers.

Olson, G. M., & Olson, J. S. (2000). Distance matters. *Human-Computer Interaction*, *15*(2), 139–178. doi:10.1207/S15327051HCI1523_4

Ong, W. J. (1977). *Interfaces of the Word: Studies in the evolution of consiousness and culture*. Cornell University Press.

Onyx, J., & Small, J. (2004). Memory-work: The method. In J. Small (Ed.), Memory-work. Qualitative research in tourism: Ontologies, epistemologies and methodologies. Routledge.

Orlikowski, W. J. (1992). The duality of technology: Rethinking the concept of technology in organizations. *Organization Science*, *3*(3), 398–427. doi:10.1287/orsc.3.3.398

Orlikowski, W. J. (2000). Using technology and constituting structures: A practice lens for studying technology in organizations. *Organization Science*, *11*(4), 404–428. doi:10.1287/orsc.11.4.404.14600

Orlikowski, W. J., & Scott, S. V. (2008). Sociomateriality: Challenging the separation of technology, work and organization. *The Academy of Management Annals*, *2*(1), 433–474. doi:10.1080/19416520802211644

Orman, E. K., & Whitaker, J. A. (2010). Time usage during face-to-face and synchronous distance music lessons. *American Journal of Distance Education*, *24*(2), 92–103. doi:10.1080/08923641003666854

Oswick, C., & Robertson, M. (2009). Boundary Objects Reconsidered: From Bridges and Anchors to Barricades and Mazes. *Journal of Change Management*, *9*(2), 179–193. doi:10.1080/14697010902879137

Panteli, N., & Chiasson, M. (2008). Rethinking virtuality. In Exploring Virtuality within and beyond Organizations (pp. 1-17). Palgrave Macmillan UK. doi:10.1057/9780230593978_1

Parkinson, C. (2006). Building successful collaborations: A guide to collaboration among non-profit agencies and between nonprofit agencies and businesses. Cambridge, Canada: Cambridge & North Dumfries Community Foundation.

Parry, K. W. (1998). Grounded theory and social process: A new direction for leadership research. *The Leadership Quarterly*, *9*(1), 85–105. doi:10.1016/S1048-9843(98)90043-1

Pasin, F., Jobin, M. H., & Cordeau, J. F. (2002). An application of simulation to analyse resource sharing among healthcare organisations. *International Journal of Operations & Production Management*, *22*(4), 381–393. doi:10.1108/01443570210420395

Pasmore, B., & O'Shea, T. (2010). Leadership agility: A business imperative for a VUCA world. *People and Strategy, 33*(4), 32.

Pauleen, D., Campbell, J., Harmer, B., & Intezari, A. (2015). Making Sense of Mobile Technology: The Integration of Work and Private Life. *SAGE Open, 5*(2), 1–10. doi:10.1177/2158244015583859

Payne, M. (2000). *Teamwork in multiprofessional care*. Basingstoke, UK: Macmillan.

Pearce, L. D. (2012). Mixed methods inquiry in sociology. *The American Behavioral Scientist, 56*(6), 829–848. doi:10.1177/0002764211433798

Pedersen, M., Klitmøller, J., & Nielsen, K. (2012). *Deltagerobservation*. København: Hans Reitzels Forlag.

Pedersen, O. C., & Pedersen, K. L. (2006). *Mail på arbejde*. Børsens Forlag.

Peterson, N. (1991). Interagency collaboration under part H: The key to comprehensive, multidisciplinary, coordinated infant/toddler intervention services. *Journal of Early Intervention, 15*(1), 89–105. doi:10.1177/105381519101500111

Pettigrew, A. M. (1990). Longitudinal Field Research on Change: Theory and Practice. *Organization Science, 1*(3), 267–292. doi:10.1287/orsc.1.3.267

Pines, A. M., & Zaidman, N. (2014). Stress and burnout in bicultural teams in hi-tech industry. *British Journal of Management, 25*(4), 819–832. doi:10.1111/1467-8551.12056

Pinjani, P., & Palvia, P. (2013). Trust and knowledge sharing in diverse global virtual teams. *Information & Management, 50*(4), 144–153. doi:10.1016/j.im.2012.10.002

Poellhuber, B., Racette, N., & Chirchi, M. (2012). De la présence dans la distance par la visioconférence Web. *Revue Internationale Des Technologies En Pédagogie universitaire/International Journal of Technologies in Higher Education, 9*(1–2), 63–77.

Preece, J., Nonnecke, B., & Andrews, D. (2004). The top five reasons for lurking: Improving community experiences for everyone. *Computers in Human Behavior, 20*(2), 201–223. doi:10.1016/j.chb.2003.10.015

Pries-Heje, J., & Commisso, T. (2010). *Improving Team Performance*. Paper presented at the Proceedings of the 33rd IRIS Seminar, Rebild Bakker.

Pries-Heje, J., & Pries-Heje, L. (2017). Improving online collaboration in contemporary IT development teams. In D. Kolbaek (Ed.), *Online Collaboration and Communication in Contemporary Organizations*. Hershey, PA: IGI Global.

Pugsley, M. (2010). *An Exploratory Study of Educator Perception of Social Presence: Design and Instructional Practices for New Online Educators*. Academic Press.

Punch, K. (2009). *Introduction to research methods in education*. Thousand Oaks, CA: Sage.

Puranam, P., Alexy, O., & Reitzig, M. (2014). What's "new" about new forms of organizing? *Academy of Management Review, 39*(2), 162–180. doi:10.5465/amr.2011.0436

Putnam, R. (1995). Bowling Alone: America's Declining Social Capital. *Journal of Democracy, 6*(1), 63–78. doi:10.1353/jod.1995.0002

Putnam, R. (2000). *Bowling Alone: The Collapse and Revival of American Community*. Simon and Schuster. doi:10.1145/358916.361990

Putnam, R. D. (2001). *Bowling alone: The collapse and revival of American community*. New York, NY: Simon and Schuster.

Pytash, K. E., & O'Byrne, W. I. (2014). Research on Literacy Instructions and Learning in Virtual, Blended, and Hybrid Environments. ETC Press.

Quan-Haas, A., Martin, K., & McCay-Peet, L. (2015). Networks of Digital Humanities Scholars: The informational and social uses and gratifications of Twitter. *Big Data and Society*, *2*(1), 1–12.

Quintillian. (1922). *Quintilian, Institutio Oratoria.* Retrieved from http://www.perseus.tufts.edu/hopper/text?doc=Perseus:text:2007.01.0069:book=12

Qvortrup, L. (2000). *Det hyperkomplekse samfund: 14 fortællinger om informationssamfundet* (2nd ed.). København: Gyldendal.

Rabin, R. (2013). *Blended learning for leadership: The ccl approach.* Retrieved from https://www.ccl.org/wp-content/uploads/2015/04/BlendedLearningLeadership.pdf

Raelin, J. A. (2016). Imagine there are no leaders: Reframing leadership as collaborative agency. *Leadership*, *12*(2), 131–158. doi:10.1177/1742715014558076

Raffnsøe, S., Gudmand-Høyer, M., & Thaning, M. (2008). *Foucault.* Frederiksberg: Forlaget Samfundslitteratur.

Raghuram, S., Garud, R., Wiesenfeld, B., & Gupta, V. (2001). Factors contributing to virtual work adjustment. *Journal of Management*, *27*(3), 383–405. doi:10.1177/014920630102700309

Ranney, J. D., & Troop-Gordon, W. (2012). Computer-Mediated Communication With Distant Friends: Relations With Adjustment During Students' First Semester in College. *Journal of Educational Psychology*, *104*(3), 848–861. doi:10.1037/a0027698

Rao, R., Dalkir, K., & Wiseman, E. (2004). Public access to information and the creation of an 'information commons.' In *ASIST 2004: Proceedings of the 67th ASIST annual meeting.* Chicago, IL: Information Today, Inc.

Rao, M. (2003). *Leading with knowledge.* Delhi: TATA McGraw-Hill.

Raschulte, T. (2011). The evolution of collaborative work. In Distributed team collaboration in organizations (pp. 16-27). IGI Global. doi:10.4018/978-1-60960-533-9.ch002

Rau, C., Möslein, K. M., & Neyer, A.-K. (2016). Playing possum, hide-and-seek, and other behavioral patterns: Knowledge boundaries at newly emerging interfaces. *R & D Management*, *46*(2), 341–353. doi:10.1111/radm.12185

Ravlin, E. C., Ward, A.-K., & Thomas, D. C. (2014). Exchanging social information across cultural boundaries. *Journal of Management*, *40*(5), 1437–1465. doi:10.1177/0149206314531623

Reckwitz, A. (2002). Toward a theory of social practices. A development in culturalist theorizing. *European Journal of Social Theory*, *5*(2), 243–263. doi:10.1177/13684310222225432

Reynolds, N. (2015). Making sense of new technology during organisational change. *New Technology, Work and Employment*, *30*(2), 145–157. doi:10.1111/ntwe.12045

Richardson, J. W., Beck, D., LaFrance, J., & McLeod, S. (2016). Job attainment and perceived role differences of cyberschool leaders. *Journal of Educational Technology & Society*, *19*(1), 211–222.

Riegelsbereger, J., & Sasse, J. M. (2003). The Researcher's Dillema: Evaluating Trust in Computer Mediated Communication. *International Journal of Human-Computer Studies*, *58*(6), 759–781. doi:10.1016/S1071-5819(03)00042-9

Roberts, L. G. (1968/2001). *Resource Sharing Computer Networks.* Retrieved October 1, 2015, from Packet.cc: http://www.packet.cc/files/res-share-comp-net.html

Rorty, A. O. (1980). Structuring Rhetoric. In A. O. Rorty (Ed.), *Essays on Aristotle's Rhetoric* (p. 9). University Of California Press.

Rosen, B., Furst, S., & Blackburn, R. (2006). Training for virtual teams: An investigation of current practices and future needs. *Human Resource Management*, *45*(2), 229–247. doi:10.1002/hrm.20106

Rouse, J. (2007). Social practices and normativity. *Philosophy of the Social Sciences*, *37*(1), 1–11. doi:10.1177/0048393106296542

Sackmann, S., & Phillips, M. (2004). Contextual influence on culture research; Shifting assumptions for workplace realities. *International Journal of Cross Cultural Management*, *4*(3), 370–390. doi:10.1177/1470595804047820

Said, E. W. (1978). *Orientalism*. London: Routledge & Kegan Paul.

Salvucci, D. D., & Bogunovich, P. (2010). Multitasking and monotasking: the effects of mental workload on deferred task interruptions. In *Proceedings of the SIGCHI conference on human factors in computing systems* (pp. 85–88). ACM. doi:10.1145/1753326.1753340

Sannino, A., Engeström, Y., & Lemos, M. (2016). Formative interventions for expansive learning and transformative agency. *Journal of the Learning Sciences*, *25*(4), 599–633. doi:10.1080/10508406.2016.1204547

Satar, M. H. (2013). Multimodal Language Learner Interactions via Desktop Videoconferencing within a Framework of Social Presence: Gaze. *ReCALL: The Journal of EUROCALL*, *25*(1), 122–122. doi:10.1017/S0958344012000286

Saunders, C., Van Slyke, C., & Vogel, D. R. (2004). My time or yours? Managing time visions in global virtual teams. *The Academy of Management Executive*, *18*(1), 19–31. doi:10.5465/AME.2004.12691177

Sayah, S. (2013). Managing work-life boundaries with information and communication technologies: The case of independent contractors. *New Technology, Work and Employment*, *28*(3), 179–196. doi:10.1111/ntwe.12016

Scannell, M., Abrams, M., & Mulvihill, M. (2012). *The big book of virtuel team building games*. McGraw-Hill Publishing.

Schatzki, T. (1996). *Social Practices. A Wittgensteinian Approach to Human Activity and the Social*. Cambridge, UK: Cambridge University Press. doi:10.1017/CBO9780511527470

Schatzki, T. (2002). *The Site of the Social: a philosophical account of the constitution of social life and change*. Pennsylvania State University Press.

Scheff, T. J. (2005). Looking-Glass Self: Goffman as Symbolic Interactionist. *Symbolic Interaction*, *28*(2), 147–166. doi:10.1525/si.2005.28.2.147

Schön, D. A. (1991). *The reflective practitioner: How professionals think in action*. Avebury.

Schrage, M. (1990). *Shared minds*. New York, NY: Random House.

Schramm, M., & Diederichsen, S. (2011). Virtuel ledelse: Skab synlighed, resultater og nærvær på afstand. Viborg, Denmark: Dansk Psykologisk Forlag.

Schramm, M., & Lønsted, W. (2001). *www.nærkontakt.net: Netpsykologi, kommunikation og rådgivning*. Dansk Psykologisk Forlag.

Schramm, M. (2017). The virtual coffee break. In D. Kolbaek (Ed.), *Online Collaboration and Communication in Contemporary Organizations*. Hershey, PA: IGI Global.

Schramm, M., & Diederichsen, S. (2011). *Virtuel ledelse – skab synlighed, resultater og nærvær på afstand*. Dansk Psykologisk Forlag.

Schratz-Hadwich, B. (2005). Collective memory-work: The self as a re/source for re/search. In M. Schratz & R. Walker (Eds.), Research as social change: New opportunities for qualitative research. New York, NY: Routledge.

Sedley, D. (2003). *The Cambridge Companion to Greek and Roman Philosophy*. Cambridge, UK: Cabridge University Press. doi:10.1017/CCOL0521772850

Selmer, J., & Lauring, J. (2013). Globalization Denmark: Open-minded while ethnocentric. *International Studies of Management & Organization*, *43*(1), 9–25. doi:10.2753/IMO0020-8825430101

Semple, E. (2012). *Organizations don't tweet, people do: A manager's guide to the social web*. Chichester, UK: John Wiley and Sons.

Sewell, G., & Taskin, L. (2015). Out of sight, out of mind in a new world of work? Autonomy, control, and spatiotemporal scaling in telework. *Organization Studies*, *36*(11), 1507–1529. doi:10.1177/0170840615593587

Shah, C. (2009). Toward collaborative information seeking (CIS). *Proceedings of JCDL 2008 Workshop on Collaborative Exploratory Search.*

Shah, C. (2012). *Collaborative information seeking: The art and science of making the whole greater than the sum of all* (Vol. 32). Springer Information Retrieval Series. doi:10.1007/978-3-642-28813-5

Shapira, Z., Lewin, A. W., Schoonhoven, C. B., Argote, L., & Levinthal, D. (2014). Editorial: Organization Science at 25. *Organization Science*, *25*(6), 1573–1578. doi:10.1287/orsc.2014.0943

Shirkey, C. (2008). *Here comes everybody: The power of organizing without organizations*. London: The Penguin Press.

Shirky, C. (2010). *Cognitive surplus. Creativity and generosity in a connected age*. London: Allen Lane - The Penguin Group.

Short, J. A., Williams, E., & Christie, B. (1976). *The social psychology of telecommunications*. London, UK: Wiley.

Shove, E., & Pantzar, M. (2005). Consumers, Producers and Practices – Understanding the invention and reinvention of Nordic walking. *Journal of Consumer Culture*, *5*(1), 43–64. doi:10.1177/1469540505049846

Shove, E., Pantzar, M., & Watson, M. (2012). *The dynamics of social practice – Everyday life and how it changes*. Los Angeles, CA: Sage. doi:10.4135/9781446250655

Shriberg, A. (2009). Effectively leading and managing a virtual team. *Business Review (Federal Reserve Bank of Philadelphia)*, *12*(2), I–II.

Sibbet, D. (2002). Best Practices for Facilitation. San Francisco, CA: The Grove Consultants International.

Sieben, B. (2007). Doing research on emotion and virtual work: A compass to assist orientation. *Human Relations*, *60*(4), 561–580. doi:10.1177/0018726707078347

Siemens, G. (2005). *Connectivism: A learning theory for the digital age*. Retrieved May 22, 2017, from http://www.itdl.org/journal/jan_05/article01.htm

Siemens, G. (2006). *Knowing knowledge*. Creative commons licensed edition. Retrieved May 25, 2017, from www.knowingknowledge.com

Sklaveniti, C. (2017). Theorizing virtual teams. In D. Kolbaek (Ed.), *Online Collaboration and Communication in Contemporary Organizations*. Hershey, PA: IGI Global.

Slack Help Center. (2017). *Customize your team's profiles*. Retrieved from Slack.com: https://get.slack.help/hc/en-us/articles/212281478-Customize-your-team-s-profiles

Slater, M., Usoh, M., & Steed, A. (1994). Depth of presence in virtual environments. *Presence (Cambridge, Mass.)*, *3*(2), 130–144. doi:10.1162/pres.1994.3.2.130

Smith, T. L. (1999). *Decolonizing Methodologies – Research and Indigenous Peoples*. Dunedin: Zed Books, University of Otago Press.

Snowden, D. J., & Boone, M. E. (2007). A leader's framework for decision making. *Harvard Business Review*, *85*(11), 69–76. PMID:18159787

Sole, D., & Edmondson, A. (2002). Situated knowledge and learning in dispersed teams. *British Journal of Management*, *13*(S2), 17–34. doi:10.1111/1467-8551.13.s2.3

Sørensen, C. (2011). Enterprise Mobility: Tiny Technology with Global Impact on Work. In *The Computing Handbook Set – Information Systems and Information Technology* (Vol. 2). Retrieved from http://digitalinfrastructures.org/publications/Sorensen2012-CS-ChapterPreprint.pdf

Spradley, J. P. (2016). Participant Observation (Reissue ed.). Long Grove: Waveland Press.

Spradley, J. P. (1980). *Participant observation*. New York: Holt Rinehart and Winston.

Star, S. L., & Griesemer, J. R. (1989). Institutional ecology, 'translations' and boundary objects: Amateurs and professionals in Berkeley's Museum of Vertebrate Zoology. *Social Studies of Science*, *19*(3), 387–420. doi:10.1177/030631289019003001

Stieglitz, S., Lattemann, C., & Brockmann, T. (2015). Mobile Applications for Knowledge Workers and Field Workers. *Mobile Information Systems*, *2015*, 1–8. doi:10.1155/2015/372315

Sullivan, C. (2012) Remote working and work-life balance. In Work and Quality of Life. International Handbooks of Quality-of-Life (pp 275-290). Springer. doi:10.1007/978-94-007-4059-4_15

Sun, P. C., & Cheng, H. K. (2007). The design of instructional multimedia in e-Learning: A Media Richness Theory-based approach. *Computers & Education*, *49*(3), 662–676. doi:10.1016/j.compedu.2005.11.016

Surowiecki, J. (2005). *The wisdom of crowds*. Garden City, NY: Anchor Books.

Sutherland, J., & Schwaber, K. (2010). *The Scrum Papers: Nut, Bolts, and Origins of an Agile Framework* (Vol. Draft - 14 May 2010). Retrieved from http://jeffsutherland.com/ScrumPapers.pdf

Swan, K., & Shih, L. F. (2005). On the nature and development of social presence in online course discussions. *Journal of Asynchronous Learning Networks*, *9*(3), 115–136.

Swidler, A. (2001). What Anchors Cultural Practices. In T. Schatzki, K. Knorr-Cetina, & E. von Savigny (Eds.), *The Practice Turn in Contemporary Theory*. London: Routledge.

Szreter, S., & Woolcock, M. M. (2004). Health by association? Social capital, social theory, and the political economy of public health. *International Journal of Epidemiology*, *33*(4), 650–667. doi:10.1093/ije/dyh013 PMID:15282219

Tabrizi, B. (2015). 75% of cross-functional teams are dysfunctional. *Harvard Business Review*, (June): 2–4.

Tan, E. (2008). *Leadership 2.0*. Retrieved from http://greenchameleon.com/gc/blog_detail/leadership_20/

Tange, H., & Jensen, I. (2012). Good teachers and deviant learners? The meeting of practices in university level international education. *Journal of Research in International Education*, *11*(2), 181–193. doi:10.1177/1475240912447849

Tapscott, D., & Williams, A. D. (2006). *Wikinomics. How mass collaboration changes everything*. New York: Portfolio.

Teece, D. J. (2007). Explicating dynamic capabilities: The nature and microfoundations of (sustainable) enterprise performance. *Strategic Management Journal*, *28*(13), 1319–1350. doi:10.1002/smj.640

Teece, D. J. (2014). The foundations of enterprise performance: Dynamic and ordinary capabilities in an (economic) theory of firms. *The Academy of Management Perspectives*, *28*(4), 328–352. doi:10.5465/amp.2013.0116

Teece, D. J., Pisano, G., & Shuen, A. (1997). Dynamic capabilities and strategic management. *Strategic Management Journal*, *18*(7), 509–533. doi:10.1002/(SICI)1097-0266(199708)18:7<509::AID-SMJ882>3.0.CO;2-Z

Thorpe, R., Gold, J., & Lawler, J. (2011). Locating distributed leadership. *International Journal of Management Reviews*, *13*(3), 239–250. doi:10.1111/j.1468-2370.2011.00303.x

Tømte, C.E. (2013). Educating teachers for the new millennium? Teacher training, ICT and digital competence. *Nordic Journal of Digital Literacy*, *8*(1-2), 74-89.

Tseng, H., Wang, C. H., Ku, H. Y., & Sun, L. (2009). Key factors in online collaboration and their relationship to teamwork satisfaction. *Quarterly Review of Distance Education*, *10*(2), 195.

Tyre, M., & Orlikowski, W. (1994). Windows of opportunities: Temporal patterns of technological adaptation in organizations. *Organization Science*, *5*(1), 98–118. doi:10.1287/orsc.5.1.98

Uhl-Bien, M., & Marion, R. (2009). Complexity leadership in bureaucratic forms of organizing: A meso model. *The Leadership Quarterly*, *20*(4), 631–650. doi:10.1016/j.leaqua.2009.04.007

Uhl-Bien, M., Marion, R., & McKelvey, B. (2007). Complexity leadership theory: Shifting leadership from the industrial age to the knowledge era. *The Leadership Quarterly*, *18*(4), 298–318. doi:10.1016/j.leaqua.2007.04.002

US Army, . (1993). *A leader's guide to after-action reviews*. Washington, DC: Department of the Army. Retrieved from http://www.au.af.mil/au/awc/awcgate/army/tc_25-20/tc25-20.pdf

Valsiner, J., & Rudolph, L. (2012). Who shall survive? Psychology that replaces quantification with qualitative mathematics. In E. Abbey & S. Surgan (Eds.), *Emerging methods in psychology* (pp. 121–140). New Brunswick, NJ: Transaction Publishers.

Van Dierendonck, D. (2011). Servant leadership: A review and synthesis. *Journal of Management*, *37*(4), 1228–1261. doi:10.1177/0149206310380462

Van Dijck, J. (2013). *The culture of connectivity: A critical history of social media*. New York: Oxford University Press. doi:10.1093/acprof:oso/9780199970773.001.0001

Van Maanen, J. (1979). The fact of fiction in organizational ethnography. *Administrative Science Quarterly*, *24*(4), 539–550. doi:10.2307/2392360

Vartiainen, M. (2008). *Facilitating Mobile and Virtual Work. In 21st Century Management, A Reference Handbook* (Vol. 2, pp. 348–360). Sage. Retrieved from https://www.researchgate.net/profile/Matti_Vartiainen/publication/256498038_Facilitating_Mobile_and_Virtual_Work/links/00b7d52323adc5eef9000000.pdf

Vartiainen, M., Kakonen, M., Koivisto, S., Mannonen, P., Nieminen, M. P., Ruohomäki, V., & Vartola, A. (2007). *Distributed and Mobile Work: Places, People and Technology*. Springer.

Vesa, M., & Franck, H. (2013). Bringing strategy to time, studying strategy as experiential vectors. *Scandinavian Journal of Management*, *29*(1), 23–34. doi:10.1016/j.scaman.2012.11.005

Vetere, F., Green, J., Nisselle, A., Dang, X. T., Zazryn, T., & Deng, P. P. (2012). Inclusion during school absence: Using ambient technology to create a classroom presence for hospitalised children. *Telecommunications Journal of Australia, 62*(5), 3–27. doi:10.7790/tja.v62i5.353

Vignovic, J. A., & Thompson, L. F. (2010). Computer-Mediated Cross-Cultural Collaboration: Attributing Communication Errors to the Person Versus the Situation. *The Journal of Applied Psychology, 95*(2), 265–276. doi:10.1037/a0018628 PMID:20230068

Von Krogh, G., & Geilinger, N. (2014). Knowledge creation in the eco-system: Research imperatives. *European Management Journal, 32*(1), 155–163. doi:10.1016/j.emj.2013.04.002

Wageman, R., Gardner, H., & Mortensen, M. (2012). The changing ecology of teams: New directions for teams research. *Journal of Organizational Behavior, 33*(3), 301–315. doi:10.1002/job.1775

Wahlstedt, A., Pekkola, S., & Niemelä, M. (2008). From e-learning space to e-learning place. *British Journal of Educational Technology, 39*(6), 1020–1030. doi:10.1111/j.1467-8535.2008.00821_1.x

Walls, J. G., Widmeyer, G. R., & El Sawy, O. A. (1992). Building an information system design theory for vigilant EIS. *Information Systems Research, 3*(1), 36–59. doi:10.1287/isre.3.1.36

Walsham, G. (2006). Doing Interpretive Research. *European Journal of Information Systems, 15*(3), 320–330. doi:10.1057/palgrave.ejis.3000589

Wang, F., & Hannafin, M. J. (2005). Design-based research and technology-enhanced learning environments. *Educational Technology Research and Development, 53*(4), 5–23. doi:10.1007/BF02504682

Ward, L. (2001). Nobility and Necessity: The problem of Courage in Aristotle's Nicomachean Ethics. *Political Science Review, 95*(1), 71. Retrieved from http://go.galegroup.com/ps/i.do?p=AONE&sw=w&u=nm_a_wnmu&v=2.1&it=r&id=GALE%7CA73021417&asid=b295edb474117734b0f7643af4e2ac8a

Watson-Manheim, M. B., & Chudoba, K. (2002). Discontinuities and continuities: A new way to understand virtual work. *Information Technology & People, 15*(3), 191–209. doi:10.1108/09593840210444746

Weick, K. E. (2002). Leadership when events don't play by the rules. *Reflections: The SoL Journal, 4*(1), 30–32. doi:10.1162/152417302320467535

Weick, K. E., & Roberts, K. H. (1993). Collective mind in organizations - heedful interrelating on flight decks. *Administrative Science Quarterly, 38*(3), 357–381. doi:10.2307/2393372

Weick, K. E., Sutcliffe, K. M., & Obstfeld, D. (2005). Organizing and the process of sensemaking. *Organization Science, 16*(4), 409–421. doi:10.1287/orsc.1050.0133

Weisband, S. (2008). *Leadership at a distance*. Lawrence Erlbaum Associates.

Wenger, E., White, N., & Smith, J. D. (2009). *Digital habitats: Stewarding technology for communities.* Portland, OR: CPsquare.

Wenger, E. (1998). *Communities of Practice: Learning, Meaning, and Identity*. New York: Cambridge University Press. doi:10.1017/CBO9780511803932

Wesson, R. H. (2011). The Impact of Performance Skills on Students' Attitudes towards the Learning Experience in Higher Education. *Issues in Educational Research, 21*(1), 22–22.

Wiberg, M., & Ljungberg, F. (2001). Exploring the Vision of "Anytime, Anywhere" in the context of mobile work. *Knowledge Management and Business Model Innovation*, 157–169.

Wichmand, M. (2017). Urgent Evoke a digital ba? How a game can make space for knowledge creation. In D. Kolbaek (Ed.), *Online Collaboration and Communication in Contemporary Organizations*. Hershey, PA: IGI Global.

Wiesenfeld, B. M., Raghuram, S., & Garud, R. (1999). Communication patterns as determinants of organizational identification in a virtual organization. *Organization Science*, *10*(6), 777–790. doi:10.1287/orsc.10.6.777

Wildman, J. L., & Griffith, R. L. (Eds.). (2015). *Leading Global Teams: Translating Multidisciplinary Science to Practice.* SpringerLink. Retrieved from http://dl.pgu.ac.ir/handle/Hannan/8470

Wilson, J. M., O'Leary, M. B., Metiu, A., & Jett, Q. R. (2008). Perceived proximity in virtual work: Explaining the paradox of far-but-close. *Organization Studies*, *29*(7), 979–1002. doi:10.1177/0170840607083105

Wilson, J., Crisp, C. B., & Mortensen, M. (2013). Extending construal-level theory to distributed groups: Understanding the effects of virtuality. *Organization Science*, *24*(2), 629–644. doi:10.1287/orsc.1120.0750

Winter, R. (2008). Design Science Research in Europe. *European Journal of Information Systems*, *17*(5), 470–475. doi:10.1057/ejis.2008.44

Wood, E., Zivcakova, L., Gentile, P., Archer, K., De Pasquale, D., & Nosko, A. (2012). Examining the impact of off-task multi-tasking with technology on real-time classroom learning. *Computers & Education*, *58*(1), 365–374. doi:10.1016/j.compedu.2011.08.029

Working Virtual. (2014). *VS14 (Report on Working Virtual).* Retrieved September 21, 2017, from http://workingvirtual.dk/uploads/workingvirtual_dk/files/VS14.pdf

Yin, R. K. (2009). *Case study research: Design and methods.* Thousand Oaks, CA: Sage.

Yoo, Y., Boland, R. J. Jr, Lyytinen, K., & Majchrzak, A. (2012). Organizing for innovation in the digitized world. *Organization Science*, *23*(5), 1398–1408. doi:10.1287/orsc.1120.0771

Yuan, Y., & Zheng, W. (2009). Mobile Task Characteristics and the Needs for Mobile Work Support: A Comparison between Mobile Knowledge Workers and Field Workers. In *2009 Eighth International Conference on Mobile Business* (pp. 7–11). IEEE. doi:10.1109/ICMB.2009.9

Yulk, G. (2013). *Leadership in organizations.* Albany, NY: Prentice Hall.

Zammuto, R. F., Griffith, T. L., Majchrzak, A., Dougherty, D. J., & Faraj, S. (2007). Information technology and the changing fabric of organization. *Organization Science*, *18*(5), 749–762. doi:10.1287/orsc.1070.0307

Zax, D. (2013, August 14). *Flickr Cofounders Launch Slack, An Email Killer.* Retrieved from Fast Company: https://www.fastcompany.com/3015730/innovation-agents/flickr-cofounders-launch-slack-an-email-killer

Zheng, L. (2015). A systematic literature review of design-based research from 2004 to 2013. *Journal of Computers in Education*, *2*(4), 399–420. doi:10.1007/s40692-015-0036-z

Zhu, H., Kraut, R., & Kittur, A. (2012) Effectiveness of shared leadership in online communities. In *Proceedings of the ACM 2012 Conference on Computer Supported Cooperative Work, ACM.* (pp. 407- 416). New York, NY: ACM. doi:10.1145/2145204.2145269

Ziek, P., & Smulowitz, S. (2014). The impact of emergent virtual leadership competencies on team effectiveness. *Leadership and Organization Development Journal*, *35*(2), 106–120. doi:10.1108/LODJ-03-2012-0043

Zimmermann, P., Wit, A., & Gill, R. (2008). The relative importance of leadership behaviours in virtual and face-to-face communication settings. *Leadership*, *4*(3), 321–337. doi:10.1177/1742715008092388

About the Contributors

Ditte Kolbaek is an assistant professor at Aalborg University, Denmark, where she conducts research in organizational learning and online leadership and cooperation. Ditte Kolbaek is passionate about learning, especially ICT-mediated learning in the context of work. She calls for making it easy to learn in complex environments, such as the workplace, that often lack formal ways of learning from experience. Prior to enrolling in the doctoral program, Ditte worked at Oracle for 10 years: as Manager of Organizational and Talent Development for three years, and as Manager of Organizational Learning in Oracle (EMEA) for seven years. These positions allowed her to develop and implement processes for learning with peers, including getting new hires up to speed, learning from retiring staff, and introducing Proactive Review, which has been implemented in more than 40 countries in Oracle (EMEA). This working environment required online leadership, online collaboration, and online learning; hence, Ditte has solid experience from practice, and practice is the foundation for her approach to research.

* * *

Bonnie Cheuk is a strategic, hands-on, business & results-driven executive with strong digital transformation, web/digital channel management, information & knowledge management, collaboration, social media and Enterprise 2.0 expertise. Originally from Hong Kong, she has 20+ years of international work experience in HK, Singapore, USA, UK and Europe. Her expertise lies in blending her deep understanding of information/knowledge management, change management, communication/facilitation principles and practices with her knowledge of the latest technologies to deliver user-centric business solutions. She has led the successful roll out of social collaboration and knowledge sharing programmes for Euroclear, Citigroup and ERM. In 2017, she is shaping the digital clients experience and digital workplace vision for a global financial services company. Bonnie's PhD focussed on understanding the information seeking and use behaviour in the workplace context. She has strong interest in, and have brought, information literacy to life in the workplace. Bonnie is the author of "Social Strategies in Action: Driving Business Transformation" (Ark Group, 2013). She publishes journal papers, advises PhD students, delivers keynotes at universities and international conferences. Connect with Bonnie on LinkedIn: https://www.linkedin.com/in/bonniecheuk or twitter: @bonniecheuk

Benjamin Cline lives with his wife, daughter, and pets in the beautiful mountains of Silver City, NM where he is Associate Professor of Speech and Communication at Western New Mexico University. Through teaching and research, Dr. Cline uses humanities-based thinking to philosophically consider how people can and should interact rhetorically in the contemporary, media saturated environment in which we live. Dr. Cline has produced more than 40 publications including book chapters, scholarly articles, conference presentations, and other scholarly productions.

Kimiz Dalkir is an Associate Professor and Director of the School of Information Studies at McGill University. She has a Ph.D. in Educational Technology, an MBA in Management Science and Management Information Systems and a B.Sc. in Human Genetics. Dr. Dalkir wrote Knowledge Management in Theory and Practice, soon to be released as a third edition, which has had an international impact on KM education and on KM practice. She recently published Intelligent Intelligent learner modeling in real-time and co-edited Utilizing Evidence-Based Lessons Learned for Enhanced Organizational Innovation and Change. Dr. Dalkir pursues research on the effectiveness of knowledge processing in both profit and non-profit organizations, learning in peer networks and measurement frameworks for assessing knowledge management success. Prior to joining McGill, Dr. Dalkir worked in the field of knowledge transfer for 17 years. She designed and developed KM systems for clients in Europe, Japan and North America.

Claus Elmholdt, MSc psychology, Ph.D., assistant professor at Aalborg Universitet and CEO at LEAD – enter next level. Claus is a researcher, lecturer and consultant within leadership and organizational psychology. His consultancy expertise and research interests include power, emotions and motivational aspects of leadership, boundary spanning and collective leadership, management teams, organizational learning and creativity. He has published over 90 academic books and articles. His latest's books includes "Effective management teams" (2015), "Power in organizations" (2014), "Leadership psychology" (2013), "Emotions in leadership" (2011).

Yrjö Engeström is Professor Emeritus of Adult Education at University of Helsinki and Professor Emeritus of Communication at the University of California, San Diego. He is Director of the Center for Research on Activity, Development and Learning (CRADLE). In his work Engeström applies and develops cultural-historical activity theory as a framework for the study of transformations in educational settings, work activities and communities. He is known for his theory of expansive learning and for the methodology of formative interventions, including the Change Laboratory method. Engeström's most recent books are From Teams to Knots: Activity-Theoretical Studies of Collaboration and Learning at Work (2008), Learning by Expanding: An Activity-Theoretical Approach to Developmental Research, 2nd Edition (2015), and Studies in Expansive Learning: Learning What Is Not Yet There (2016), all published by Cambridge University Press. Engeström serves currently as Visiting Professor at Rhodes University in South Africa and at University West in Sweden.

Morten Fogsgaard is MSc in Psychology from Aarhus University with further training from The Bayswater Institute, London. He is a Ph.D. fellow at Aalborg University and partner of the consulting firm UKON. He is currently working as a consultant while conducting research in leadership development, the management of interdisciplinary cooperation, the development of leadership teams and the question of power and influence in organizations. Morten has published a series of articles and has recently co-edited a book entitled (Power in Organizations). He teaches in the Master program in Leadership and Organizational Psychology (LOOP) at Aalborg University and in Master of Public Governance at Copenhagen Business School.

Rosemary Francisco is a PhD in Business Administration at Unisinos University with a doctoral internship from the University of Helsinki, Finland. She has been working for more than fifteen years in the area of software processes where she has acquired experience and knowledge in the best practices used throughout the projects' life cycle. Currently, she is working as a Consultant, focused on Business Process Management (BPM) and Knowledge Management and as an Assistant Professor in undergraduate, specialisation and MBA courses at Unisinos University. She is also conducting a research project called mobChangeLab (http://www.mobchangelab.com/index_en.html) with the aim to develop a collaborative approach to identifying improvements and problem-solving in the context of mobile workers. Her research interests include learning organisation, knowledge management, enterprise mobility, business process management, agile project management, and software engineering.

Iben Jensen is Professor, Department of Learning and Philosophy, Aalborg University, Copenhagen, Denmark.

Amarolinda Klein is a Full Professor at Unisinos University (Brazil) and a CNPq (Brazilian National Research Agency) researcher. She has more than 20 years of experience in Management and Information Systems Research. Her research interests are related to social studies of Information Technology, learning processes in organizations and enterprise mobility.

Rikke Lindekilde, PhD, Business Consultant & Owner at Lindekilde & Partners. Masters degree in Psychology, Organizational- and Workpsychology. Rikke's PhD dissertation within the field of global leadership and development of distributed teams has served as an impetus to begin interdisciplinary research in the field of IS, Psychology and Educational theories. As a PhD and a consultant she has global experience from international engineering corporations and from practical training and advising of managers and employees from all parts of the world. She has published several papers and book chapters on the topic including: "The centrality of communication, norm alignment, role clarity, and trust in global project teams"; "Perceived challenges in global collaboration: Implicit theories in virtual meetings and their impact on collaborative behavior."

Jane McKenzie is Professor of Management Knowledge and Learning at Henley Business School. She has spent seventeen years researching ways to enhance organisational learning, improve the contribution of knowledge to business value and the effect of leadership and change programmes on that process. She is particularly interested in the paradoxes and tensions that create complexity and affect change in the knowledgeable enterprise. Half her working life has been spent in industry working in financial and change management roles for various large and medium sized businesses and half in academia. She has published four books and many papers. Connect with Jane at http://www.henley.ac.uk/people/person/professor-jane-mckenzie/ or via https://linkedin.com/in/jane-mckenzie-934423.

Anders Sandborg Nielsen is a free agent, travelling the world with his family. His professional background is in the University College sector in Denmark in the field of ICT based learning.

Anne-Mette Nortvig is an associate professor, Ph.D., at University College Absalon in Denmark. She is a qualified theologian from the University of Copenhagen and has taught religion, philosophy, and citizenship in teacher training between 2001 and 2011. Nortvig has a master's degree in ICT and Learning (Information, Communication & Technology) from Aalborg University and she has worked with e-learning in professional education since 2004. The last seven years, she has also taught subjects relating to ICT, learning, learning design, technology, and education at Aalborg University. Anne-Mette Nortvig's research interests are found in ICT and learning design, and they focus on the influence of technology on people's lives, learning, and education. She has published several papers, articles, and contributions to books in this area.

Jan Pries-Heje is Professor in Information Systems at Roskilde University. Jan Pries-Heje has more than 20 years of experience working as Project and Research manager and doing research in IT and innovation. Jan Pries-Heje holds M.Sc. and Ph.D. degrees from Copenhagen Business School. Jan Pries-Heje's research interests include IT innovation, IT project management, information systems development, and process improvement. His research focuses on designing and building innovative solutions to managerial and organizational IT problems. Previous and current projects explore quality software development @ Internet Speed, innovative capability in projects, the ability for an organization to improve, and how one can design a process for making better sourcing decisions.

Lene Pries-Heje is Associate Professor in Information Systems at The IT University of Copenhagen. She is currently also Head of Studies (Dean of Education) at the IT University. Lene Pries-Heje holds an M.Sc. degree from Copenhagen Business School and a Ph.D. degree from The IT University of Copenhagen. Lene Pries-Heje's research interests include ERP systems implementation, Agile systems development, digital innovation, and process improvement.

Annalisa Sannino is a Professor of Education at University of Tampere. She is a visiting professor at Rhodes University, South Africa and at University West, Sweden. She has worked at the Center for Research on Activity, Development and Learning (CRADLE) and was recently granted a Finnish Academy Research Fellowship. Her research is primarily focused on developing an activity-centered educational and dialectical theory of learners' transformative agency and equity. Her work demonstrates how transformative agency is enacted and how it can be identified, supported and enhanced within communities, educational settings and work activities by means of formative interventions.

Mads Schramm holds a Master Degree in Psychology from Copenhagen University. He has been working as a consultant with virtual teams and virtual leadership in international organisations for the past 15 years. He authored the book Virtual Leadership, which was published in 2012. In his job with online colloboration Mads has also been working with online leadership counselling and online coaching in Netpsykologer.dk and done research in that field documented in the book www.netcommunication.net (2001).

Chrysavgi Sklaveniti is Post-Doctoral Research Fellow in the Institute of Organizational Psychology in the University of St Gallen, Switzerland. Her research interests revolve around the relational aspects of team work (virtual and co-located), technological integration, entrepreneurship, creativity and leadership. Her publications pursue these themes in conceptual and empirical formats.

Mette Wichmand is Assistant Professor at Roskilde University, Denmark at the Department of Communication and Arts. Mette's research is concerned with how to build and sustain communities online with the purpose of generating civic engagement and social change. She is currently researching how social network games can empower players to become post-game social innovators. She holds PhD in Communication but as also spent many years outside academia researching and producing internationally broadcasted documentaries and children's programs on human rights.

Index

www.ingramcontent.com/pod-product-compliance
Lightning Source LLC
LaVergne TN
LVHW062134300926
849161LV00010B/2531